The Organization of
Cell Metabolism

NATO ASI Series

Advanced Science Institutes Series

A series presenting the results of activities sponsored by the NATO Science Committee, which aims at the dissemination of advanced scientific and technological knowledge, with a view to strengthening links between scientific communities.

The series is published by an international board of publishers in conjunction with the NATO Scientific Affairs Division

A	**Life Sciences**	Plenum Publishing Corporation
B	**Physics**	New York and London
C	**Mathematical and Physical Sciences**	D. Reidel Publishing Company Dordrecht, Boston, and Lancaster
D	**Behavioral and Social Sciences**	Martinus Nijhoff Publishers
E	**Engineering and Materials Sciences**	The Hague, Boston, Dordrecht, and Lancaster
F	**Computer and Systems Sciences**	Springer-Verlag
G	**Ecological Sciences**	Berlin, Heidelberg, New York, London,
H	**Cell Biology**	Paris, and Tokyo

Recent Volumes in this Series

Volume 123—The Molecular Basis of B-Cell Differentiation and Function
  edited by M. Ferrarini and B. Pernis

Volume 124—Radiation Carcinogenesis and DNA Alterations
  edited by F. J. Burns, A. C. Upton, and G. Silini

Volume 125—Delivery Systems for Peptide Drugs
  edited by S. S. Davis, L. Illum, and E. Tomlinson

Volume 126—Crystallography in Molecular Biology
  edited by Dino Moras, Jan Drenth, Bror Strandberg,
  Dietrich Suck, and Keith Wilson

Volume 127—The Organization of Cell Metabolism
  edited by G. Rickey Welch and James S. Clegg

Volume 128—Perspectives in Biotechnology
  edited by J. M. Cardoso Duarte, L. J. Archer, A. T. Bull,
  and G. Holt

Volume 129—Cellular and Humoral Components of Cerebrospinal Fluid in
  Multiple Sclerosis
  edited by A. Lowenthal and J. Raus

Series A: Life Sciences

The Organization of Cell Metabolism

Edited by
G. Rickey Welch
University of New Orleans
New Orleans, Louisiana

and
James S. Clegg
University of California, Davis
Davis, California

Plenum Press
New York and London
Published in cooperation with NATO Scientific Affairs Division

Proceedings of a NATO Advanced Research Workshop on
the Organization of Cell Metabolism,
held August 31–September 4, 1985,
in Hanstholm, Denmark

_____

Library of Congress Cataloging in Publication Data

NATO Advanced Research Workshop on the Organization of Cell Metabolism
(1985: Hanstholm, Denmark)
   The organization of cell metabolism.

   (NATO ASI series. Series A, Life sciences; v. 127)
   "Published in cooperation with NATO Scientific Affairs Division."
   Proceedings of a NATO Advanced Research Workshop on the Organization of
Cell Metabolism, held August 31–September 4, 1985, in Hanstholm, Denmark"
—T.p. verso.
   Bibliography: p.
   Includes index.
   1. Cell metabolism—Congresses. I. Welch, G. Rickey. II. Clegg, James S.,
1933-   . III. North Atlantic Treaty Organization. Scientific Affairs Division. IV.
Title. V. Series.
QH634.5.N38   1985   574.87'61   87-2468
ISBN-13: 978-1-4684-5313-3   e-ISBN-13: 978-1-4684-5311-9
DOI: 10.1007/978-1-4684-5311-9
_____

A Division of Plenum Publishing Corporation
233 Spring Street, New York, N.Y. 10013

This volume is dedicated to the memory of the late Dr. Mario di Lullo, Programme Director in the NATO Scientific Affairs Division. Our workshop (like many others in the past) benefited greatly from his advice, generous assistance, and personal care.

FOREWORD

The NATO Advanced Research Workshop on "The Organization of Cell Metabolism" was held 31 August – 4 September, 1985, at the Hotel Hanstholm in Hanstholm, Denmark. From areas of cell biology, bio-chemistry, enzymology, and biophysical chemistry, the Workshop brought together workers whose research focuses on the character of the metabolic infrastructure of the living cell. The organizing committee was composed of the following members:

James S. Clegg

Douglas B. Kell

Paul A. Srere

G. Rickey Welch (Chairman)

There has now arisen an edifice of proof that cell metabolism is highly ordered in space and time. The refinement of cytological extraction methodology and the development of more gentle isolation techniques have led to the demonstration of structural organization for most of the primary metabolic pathways. Moreover, the study of isolated enzyme complexes and membranous arrays has revealed unique functional properties of the organized state. Interest in metabolic organization has heightened very recently, due to the emergence of exciting new cytochemical and electron-microscopic advances in the elucidation of the cytomatrix and associated cell water. The hyaloplasmic space of eukaryotic cells has now been shown to be laced with a dense network of various filamentous structures — one role of which appears to be that of a structural support for the microcompartmentation of metabolic processes. The supramolecular organization of certain processes may exceed the physical confines of intracellular particulates, involving large-scale entrainment of the cytoplasm in space and time.

The Workshop dealt with the following major areas:

1) Organization of the cytomatrix and aqueous compartments. The aqueous milieu in the metabolic microenvironments of the cell may be different from the bulk aqueous condition in vitro. Cell-associated water must play an important role in the integration of metabolic processes. In addition to their roles in scaffolding, organelle motility, and cellular motion, the cytomatrix may provide a structural setting for the organization of metabolic processes.

2) Organization of macromolecular synthesis. Macromolecular (e.g., protein) assemblage is an extremely complex process, involving the coordination of a multitude of enzymatic events. In eukaryotic cells many of these processes are associated with the cytomatrix.

3) <u>Organization of metabolic pathways</u>. Organization of the multienzyme systems in intermediary metabolism may lead to such physiological advantages as "channeling" of intermediates (with the attendant elimination of diffusional transit, segregation of metabolic flow, etc.), unique modes of free-energy transduction, and coordination of metabolic regulation. The organizational state may involve static spatial designs with protein-protein, protein-membrane, and/or protein-cytomatrix interactions. In some cases the organization may not be localized statically, but entail periodic spatio-temporal order extending throughout the bulk cytoplasm.

4) <u>Experimental and theoretical modeling of metabolic organization</u>. Realization of the organized nature of cellular metabolism necessitates the construction of theoretical paradigms which will provide us with a better qualitative and quantitative understanding of the workings of the cell. In addition, theoretical modeling opens up new vistas for experimental analysis of metabolic processes. Moreover, the emergence of immobilized-enzyme technology has fostered the usage of macroscopic model systems for exploring experimentally some of the unique advantages of the organized state.

There were main lectures in each of these four areas. In addition, there were round-table discussions focusing on topics of interest within these headings. The round-tables were steered by invited experts. And, there was a poster session.

These published proceedings include the main lectures,[*] as well as presentations by the invited round-table panelists. As is usual in successful scientific gatherings, the discussions that took place outside the formal structure of the program were at least as important. As evinced from the table of contents herein, this Workshop was rather unique in the scope and breadth of coverage of the subject of "organization of cell metabolism". Our sincere thanks go to the contributing authors, for their time and energy invested in bringing this project to fruition.

G. Rickey Welch, New Orleans

James S. Clegg, Bodega Bay

---

[*]Manuscripts were not submitted for the following two lectures: "Structure and Evolution of Ribosomes and Their Components" (by Dr. H. Wittmann) and "Time Patterns in Cellular Metabolism" (by Dr. B. Hess).

ACKNOWLEDGEMENTS

We would like to express our sincere thanks to NATO for major
financial support. In particular, we are indebted to Dr. Mario Di Lullo,
Programme Director in the Scientific Affairs Division, for his continued
interest and advice. Additional financial support came from the
following sources:

Commission of the European Communities. A special note of thanks is due
to Dr. Paolo Fasella, Director-General for Science, Research, and
Development, for his interest and assistance. He himself has contributed
significantly to the scientific subject of this Workshop. Regrettably,
he was unable to attend due to pressing obligations of his office.

International Union of Biochemistry. We are especially grateful to Dr.
W. J. Whelan of the Executive Committee, for his encouragement and
interest and for his effort in the expeditious handling of our
application to the IUB.

NOVO Industri A/S (Denmark). The kind assistance of Dr. Villy Jensen
(Research and Development Division) is much appreciated.

We are very grateful to Mr. Kjeld Olsen and his staff at the Hotel
Hanstholm, for their important role in the execution of our Workshop. We
cannot say enough about their hospitality and cooperation. The peaceful
seclusion of the North Sea village of Hanstholm was perfect for a small,
intensive scientific meeting.

We would like to thank Mrs. Barbara Kester of the International
Transfer of Science and Technology Office (Belgium), for much logistical
help in the early planning of the Workshop.

Last, but not least, we thank the Department of Biological Sciences
(Chairman: Dr. Michael A. Poirrier), University of New Orleans, for help
with the extensive costs of telephoning, mailing, photocopying, and
printing, as well as for expert typing/clerical assistance. In
particular, the warm smile, care, and skilled typing work of Mrs. Marie
LeBlanc are beyond words.

CONTENTS

ORGANIZATION OF THE CYTOMATRIX
AND AQUEOUS COMPARTMENTS

ORGANIZATION OF MACROMOLECULAR SYNTHESIS

ORGANIZATION OF BIOSYNTHETIC AND
BIODEGRADATIVE PROCESSES

ORGANIZATION OF ENERGY METABOLISM:
ENZYMOLOGICAL APPROACHES

ORGANIZATION OF ENERGY METABOLISM:
IN SITU APPROACHES

EXPERIMENTAL AND THEORETICAL MODELING
OF METABOLIC ORGANIZATION

THE ORGANIZATION OF CELL METABOLISM: A HISTORICAL VIGNETTE

G. Rickey Welch

Department of Biological Sciences
University of New Orleans
New Orleans, LA 70148 U.S.A.

and

James S. Clegg*

Laboratory of Quantitative Biology
University of Miami
Coral Gables, FL 33124 U.S.A.

> A· *reductio ad absurdam* I admit, but I cannot
> conceive how the living cell can behave as
> it does, unless practically all the molecules
> inside it are under some kind of directed
> activity. (Peters, 1930)

Enzymology as a distinct science is just over 100 years old, dating most notably to the work of W. Kühne (Gutfreund, 1976). Originally, enzymes were called "ferments", reflecting the early interest in the nature of alcoholic fermentation. In 1876 Kühne presented to the Heidelberger Naturhistorischen und Medizinischen Verein a paper (reproduced in FEBS Letters, Volume 62, Supplement, 1976), in which he proposed that isolable "ferments" should be called "enzymes". As we learn in elementary biochemistry courses, the word "enzyme" comes from the Greek for "in yeast" or "leavened".

Throughout the history of the conjoined sciences of enzymology and biochemistry, it has been taken as a matter of course that most metabolic activity of the cell results from the superposition of the action of individual enzymes dissolved in an aqueous phase -- with the dynamics being governed by simple mass-action laws and random thermal motion of metabolite molecules in weak-electrolyte solution. Early students of cell metabolism, in the late 1800's, were seemingly justified in adopting this simplistic paradigm. For, physical chemistry had come of age with

---

*Present address: University of California
                   Bodega Marine Laboratory
                   Bodega Bay, CA

the entrenchment of gas-phase kinetics and chemical laws for dilute aqueous solutions. It was only natural to extrapolate to the physicochemical processes in living systems. The isolation of individual enzymes and the observation of "saturation kinetics" in vitro, culminating in the Michaelis-Menten and Briggs-Haldane theoretical treatments, helped to foster this status rerum. This is not to belittle such reductionistic efforts. Indeed, this approach resulted in the gradual recognition and delineation of the major pathways of intermediary metabolism, and also led to our present-day understanding of the mechanism of enzyme action.

The actual origin of the notions of "soluble enzymes" and "soluble pathways" is hard to trace, but Büchner's work on fermentation in cell-free yeast extracts late in the 19th century certainly had a major impact in that regard. A few decades before Büchner carried out his pioneering research, W. Pfeffer produced studies on osmotic behavior in cells. His extensive work provided much of the basis for the subsequent development of thought and experimentation in cell physiology: the cell behaves as if the plasma membrane (the existence of which Pfeffer deduced) separates two ordinary aqueous solutions, one being that inside the cell.

This brief historical note allows us to see how it was that scientists of the early to mid-1900's embraced the idea that the cell is a homogeneous "bag of enzymes" possessing an interior whose aqueous-phase properties are essentially the same as those of an ordinary aqueous solution. The use of electron microscopy and cell fractionation techniques during the 1950's and 1960's modified that picture somewhat: certain metabolic pathways were found to be partly, or completely, associated with structures that could be isolated and identified by electron microscopy. However, the purported existence of soluble metabolic pathways remained, chiefly as a deduction from high-speed centrifugation experiments on cell and organelle extracts; and it is still a widespread and firmly rooted paradigm of cell biology. Thus, "soluble pathways" and "normal aqueous-phase properties" have become conventional wisdom for many contemporary scientists, forming the basis for experimental design and execution, interpretation of data, and construction of metabolic models. But, we must not lose sight of the fact that these beliefs are built more on assumption and extrapolation of 19th-century physical chemistry than on hard evidence, and that the extent to which they are correct remains to be rigorously evaluated.

Let us go back again in time. Although little known, the distinction between isolated vs. organized enzymes actually was of concern from the earliest days of biochemistry and enzymology. From Claude Bernard (1878), the founding father of 20th-century physiology, in his Lecons sur les Phénomènes de la Vie communs aux Animaux et aux Végétaux, we learn something of the historical context underlying the physiology of "organized enzyme systems". By Bernard's time, the term "ferments" (or "fermentation") was being used generally to describe "all organic reactions induced by a material that gained nothing and lost nothing in the process, but which seemed to intervene only by its presence" (Bernard, 1878). Berzelius called this phenomenon "catalytic action", likening it to well-known inorganic surface catalysis; and "ferments" came to be regarded as "organic catalysts". Bernard, being a physiologist rather than a chemist, was bothered by this chemically "purist" definition of enzyme catalysis; for him, it did not relate to the living state; it did not bespeak the relationship of the "ferment" to the "organization, the development, and the multiplication, that is to

say to the life of the cell" (Bernard, 1878). He tells us that,
"Berzelius ... did not know that the ferment yeast was an organized
being; he regarded it as an amorphous principle."

Bernard credits Cagniard de Latour (ca. 1836), and later Pasteur,
with the origination of a "physiological theory of fermentation" -- based
on the concept of the enzyme ("ferment") operating *in vivo*. Bernard
(1878) was led to the following conclusion:

> "Today, two kinds of fermentations are distinguished,
> according to the soluble or insoluble nature of the
> ferment; the one produced by the intervention of an
> organized or *structured* ferment, the other produced
> by nonorganized ferments, liquids, *soluble* products,
> that are elaborated and secreted by the living
> organisms". (italics Bernard's)*

Of course, Bernard had no knowledge of the intricacies of cell infrastructure. Notwithstanding, he relied on what he called "physiological
determinism", to reason that the chemistry of living cells must be
somehow "different" from that *in vitro*, in other words,

> "Chemical phenomena in living organisms can never
> be fully equated with phenomena that take place
> outside them. This means to say, in other words,
> that the chemical phenomena of living beings,
> although they take place according to the general
> laws of chemistry, always have their own special
> apparatus and processes".

The biochemistry/cell biology literature of the past 100 years is
replete with references to the significance of the organization of cell
metabolism -- sometimes a casual mention, sometimes a commentary,
sometimes a serious ratiocination (see Welch, 1977, for additional
historical aspects). Although, it has been only in the past 20 years or
so that the empirical evidence has mounted. As evinced from the
proceedings of this Workshop, as well as from a perusal of the
literature, there is now evidence for organization (structure) in most of
the major metabolic pathways (see Srere, 1984, 1987, for an up-to-date
tabulation of evidence.) Analogous organizational principles have been
extended to the processes of bioenergetic transduction (Kell, 1979; Kell
and Westerhoff, 1985). A "newness" has come upon the classic study of
intermediary metabolism and its union with intracellular architecture.

The origination of the term "multienzyme complex" dates to 1947 with
David Green (see Green, 1957). He used the appellation "cyclophorase
complex" for the aggregated system of enzymes carrying out the Krebs
cycle. Later, this "aggregate" was seen to be the very mitochondrion
itself. Green visualized an enzyme complex as "an organized mosaic of
enzymes in which each of the large number of component enzymes is
uniquely located to permit efficient implementation of consecutive
reaction sequences." Notably, the presentations by Dr. Srere and by Drs.
Weitzman and Barnes in these proceedings report the existence of a Krebs-
cycle complex of the kind originally envisioned by Green. The search for

---

*Interestingly, the aforementioned paper by Kühne (1876) is aptly
entitled, "Ueber das Verhalten verschiedener organisirter und sogenannt
ungeformter Fermente."

complete "pathway particles" has been a major motivational factor in the study of organized enzyme systems (Srere, 1985).

Interest in the roles of the cytomatrix and cell-associated water in the integration of metabolism has followed a similar course in its history, within the same timespan as above. A true appreciation of the elemental form of the cytoskeleton and of its many functions in cell architecture, motility, and metabolism has come only in recent decades. However, again, one finds a plethora of references, dating to the last century, as to the existence and role of a proteinaceous filamentous matrix in cell structure and function. (For historical perspectives see Clegg, 1984a,b, and Porter, 1984.) The prescient writings of R. Peters (1930) deserve particular mention. It is pleasing, holistically speaking, to witness the rapid convergence of cell biology and biochemistry/enzymology occurring during the last 10 years or so, as it becomes more and more evident that subcellular particulate surfaces are the location for most (perhaps all) of intermediary metabolism. It seems appropriate to resurrect the old cell biology term, ergastoplasm (originally applied to regions of the endoplasmic reticulum), to designate the proximate zones of metabolic activity in the vicinity of particulate structures in general. Terms such as "cytoplasm" and "cytosol" are too banal, conjuring a sense of amorphism and homogeneity.

Contemporary thinking about metabolic organization has been heavily influenced by images of cells obtained by transmission electron microscopy. It is well to remember that the aqueous compartments of cells, comprising about 75% of their mass and volume, are missing from those images. Thus, our impressions of cellular infrastructure are influenced profoundly by this sudden cut-off, at a completely artificial and misleading level of resolution. It is inescapable, that a continuum must exist in the real biological situation, and that the sharp boundaries implied by the static images of microscopy are as deceptive as those implied from results based on the violent destruction of cells by the usual extraction techniques. We must somehow avoid the use of these destructive methods, as a first step in our experimental analysis, if we are ever to comprehend the organization of cell metabolism as a whole, rather than to attempt a reconstruction from what we know about the bits and pieces. That opinion is not to be construed as anti-reductionistic. We recognize that certain questions about cell metbolism are best answered by that approach. We do suggest, however, that the more subtle and delicate aspects of metabolic organization will only be manifested by the intact cell and cannot be reconstructed, let alone identified, through the convenient but inadequate application of the "grind-and-find" approach.

Of course, identifying and characterizing anatomically the character of metabolic organization is only the first stage. Next, we must develop the necessary theoretical constructs to describe the thermodynamic-kinetic quality of biochemical processes extant therein. Most surely, we will eventually have to modify such familiar tools as " ΔG " and "Michaelis-Menten kinetics" (Welch, 1984; Welch and Kell, 1986).

The authors contributing to these proceedings represent different opinions on the many issues at stake, and it is clear that no unified conception has resulted from our efforts. Nevertheless, one theme does appear to be emerging: the cytomatrix, the intracellular disposition of enzymes, and the properties of the aqueous phase within the cell all appear to be intimately related; and these organizational interrelationships are vital to normal metabolic function. We have only begun to appreciate the remarkable subtleties of cell metabolism.

REFERENCES

Bernard, C., 1878, "Lecons sur les Phénomènes de la Vie communs aux Animaux et aux Végétaux", translated from the French by H.E. Hoff, R. Guillemin, and L. Guillemin (Charles C. Thomas Publisher, Springfield, Illinois, 1974).

Clegg, J.S., 1984a, Properties and metabolism of the aqueous cytoplasm and its boundary, *Amer. J. Physiol.*, 246: R133.

Clegg, J.S., 1984b, Intracellular water and the cytomatrix: some methods of study and current views, *J. Cell Biol.*, 99: 167s.

Green, D.E., 1957, *Harvey Lectures*, Ser. 52: 177.

Gutfreund, H., 1976, Wilhelm Friedrich Kühne: an appreciation, *FEBS Lett.*, 62: E1 (Supplement).

Kell, D.B., 1979, On the functional proton current pathway of electron transport phosphorylation: an electrodic view, *Biochim. Biophys. Acta* (Reviews on Bioenergetics), 549: 55.

Kell, D.B. and Westerhoff, H.V., 1985, Catalytic facilitation and membrane bioenergetics, *in*: "Organized Multienzyme Systems", G.R. Welch, ed., Academic, New York.

Kühne, W., 1876, see Gutfreund (1976), *op. cit.*

Peters, R.A., 1930, Surface structure in the integration of cell activity, *Trans. Faraday Soc.*, 26: 797.

Porter, K.R., 1984, The cytomatrix: a short history of its study, *J. Cell Biol.*, 99:3s.

Srere, P.A., 1984, Why are enzymes so big?, *Trends Biochem. Sci.*, 9: 387.

Srere, P.A., 1985, The metabolon, *Trends Biochem. Sci.*, 10: 109.

Srere, P.A., 1987, Complexes of sequential metabolic enzymes, *Ann. Rev. Biochem.*, Vol. 56, to appear.

Welch, G.R., 1977, On the role of organized multienzyme systems in cellular metabolism: a general synthesis, *Prog. Biophys. Mol. Biol.*, 32: 103.

Welch, G.R., 1984, Biochemical dynamics in organized states: a holistic approach, *in*: "Dynamics of Biochemical Systems", J. Ricard and A. Cornish-Bowden, eds., Plenum, New York.

Welch, G.R. and Kell, D.B., 1986, Not just catalysts — molecular machines in bioenergetics, *in*: "The Fluctuating Enzyme", G.R. Welch, ed., Wiley, New York.

ORGANIZATION OF THE CYTOMATRIX
AND AQUEOUS COMPARTMENTS

STRUCTURAL ORGANIZATION OF THE CYTOMATRIX

Keith R. Porter

Department of Biological Sciences
University of Maryland
Catonsville, Maryland 21228

INTRODUCTION

During the period when centrifugal fractionation was the popular
approach to the investigation of the cellular contents and their
function, some misconceptions regarding the cytomatrix became part of an
established dogma. It said that mitochondria, microsomes, and secretory
granules were floating around in a cytosol. Subsequently those
procedures were adapted for the isolation of ribosomes, the Golgi
complex, the plasma membrane, and certain components of the cytoskeleton.
The results were impressive and contributed to a large literature which
has perpetuated the notion that the cell is simply an assemblage of
isolatable parts. Questions of how they interact or whether they are
tied together for communication are seldom, if ever, asked. The "dogma
of fractions" has dominated the thinking, in spite of long recognized
facts about cells and their organization. It is well known, for example,
that the cellular organelles are not randomly distributed in the
cytoplasm, that cells possess anisometric forms, that they have a
polarity, and that animal cells show in most instances a center
(cytocentrum) around which other components (dense bodies or MTOC's) are
arranged, sometimes in a radial fashion (Porter et al., 1983). It has
since been demonstrated that the nucleus (the nucleoplast) deprived of
the bulk of this organized cytoplasm (the cytoplast) survives for only 48
hrs, even though it has ribosomes, mitochondria, as well as vesicles of
the ER, some cytomatrix, and an intact plasma membrane around it all.
One might imagine that, with all that, the nucleoplast might regenerate
the rest. But not so; without the cell center and its associated mass of
organized cytoplasm, the nucleoplast rounds up and is nonviable. The
cytoplast, on the other hand, adopts the form of the parent cell; and,
even though viable for only 48 hrs, it moves about independently for part
of the time. An E.M. study shows it to possess centrioles and
microtubules (Shay et al., 1974).

These comments will suffice as an excuse for initiating, as we did a
few years ago, a search for a structural ground substance in what had
been identified as the domain of the cytosol (Buckley and Porter, 1975;
Wolosewick and Porter, 1979). We took advantage of high voltage electron
microscopy, because the beam of the microscope involved can penetrate
thick specimens (whole cultured cells or thick sections) and can provide
pairs of stereo images which when viewed together give a three-dimensional

depth image of the specimen. In the case of whole cultured cells, the procedures of specimen preparation preserved the inherent structure of a living unit without the use of an embedding matrix (Figs. 1, 2, and 3). This meant that we were viewing the electron scattering properties of everything natively present. Only subsequently was it recognized that the resins used in the section-generating procedures were hiding some of the structure by the simple property of scattering electrons to the same extent as the structures embedded in it (Figs. 4 and 5). Over the years of early microscopy then, the investigator was looking at the cytomatrix and recognizing that some "wispy" component was present but concluding from its nature that it was simply a product of protein precipitation. He little realized that more was there than appeared in the image.

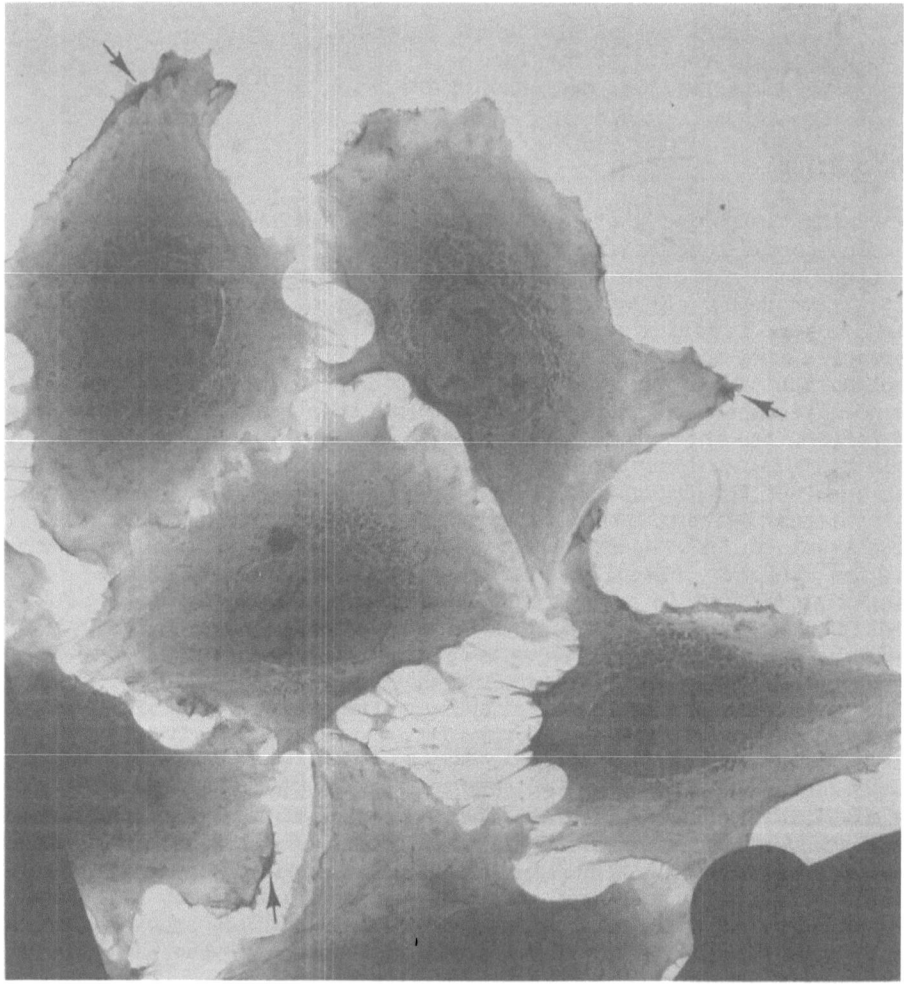

Fig. 1.   A population of NRK cells grown on a gold "finder" grid that was coated with formvar and carbon. A 1000 kV microscope was used to produce the images, which accounts for the fact that structure is evident even in the thickest parts of the cell and that the resolution is excellent despite the thickness. At regions indicated by arrows, the cell margins show ruffling (see Fig. 2 for a higher power image). Magnification 1200X.

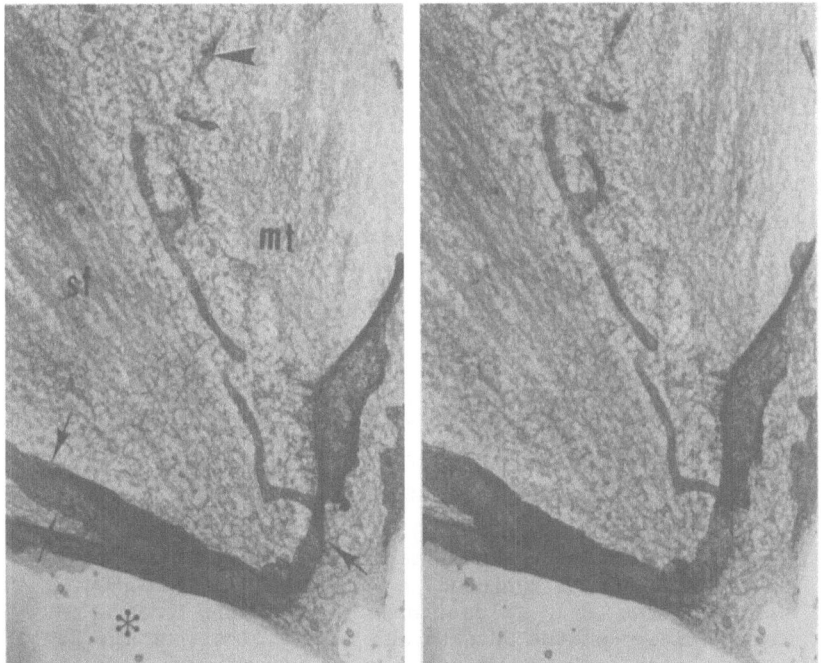

Fig. 2.   These are stereo images of the cytoplasmic matrix near the
           margin of an NRK cell. The extracellular space is marked by
           an asterisk.   Major ruffles are imaged along what was the
           advancing edge of the cell (arrows). Stress fibers
           (actin-rich) are present in the cytomatrix at sf; areas rich in
           microtubules are evident in parallel array at mt.   These and
           other elements of the matrix are apparently contained in the
           microtrabecular lattice (mtl) that pervades all parts of the
           cytoplasm.   Minor ruffles are present at arrow head.
           Magnification 18,750X. (The reader is advised to
           examine these stereo pairs with viewers available from Abrams
           Instrument Corporation in Lansing, Michigan. Such viewers
           magnify, as well as assist in the fusion of the two images.)

Furthermore allowance was seldom made for the obvious fact that thin
sections include only fragments of any continuous structure that might be
present (Guatelli et al., 1982).

     These considerations, which developed retrospectively from several
studies of whole cells and thick sections, have kept alive an interest in
the structure of the cytomatrix and an increasing faith in believing what
we see (Porter and Anderson, 1980, 1982; Porter et al., 1982, 1983;
Porter and McNiven, 1982; Porter and Tucker, 1981).

     What I propose to do in occupying the rest of the assigned space, is
to review what we and others have observed and to speculate on its
significance for the preservation of cell form and function (Porter et
al., 1983; Porter and McNiven, 1982; Porter and Tucker, 1981). Our
thesis is that a structural matrix, with unity and with an extraordinary
capacity to maintain its design and form, is required to contain and
control the disposition of these otherwise unconnected components of the
cytoplasm. Though not as yet shown by any in situ studies, it seems

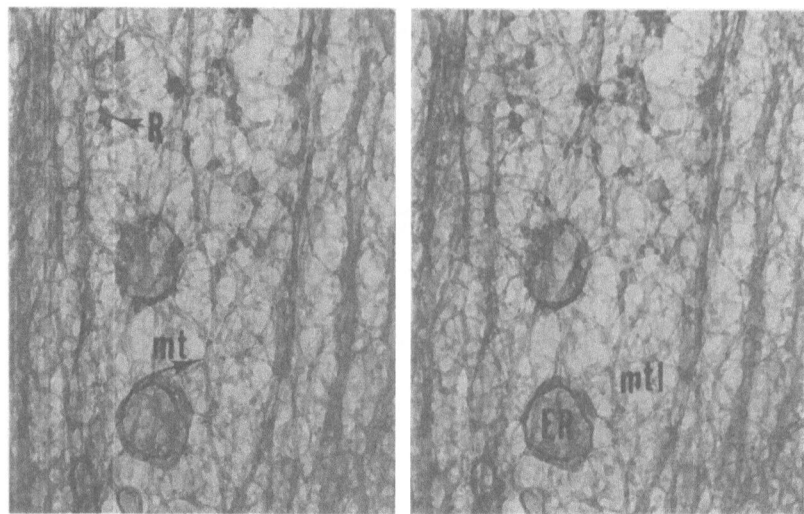

Fig. 3.   High magnification stereo images of a small area of
           the cytomatrix as depicted in Fig. 2. It is designed to show
           microtubules (mt), filamentous units in the stress fibers (sf),
           two vesicles of the ER detached from the rest of the ER system
           and polysomes (R). All of these components are part of
           (contained by) the protein-rich continuous phase called the
           microtrabecular lattice (mtl). This consists of numerous
           pleomorphic bridges or links which appear to tie things
           together. The intertrabecular spaces are structure-free and
           probably represent a water-rich phase which together with the
           mtl give the cytoplasm its gel-like viscosity. It could
           influence diffusion rates and provide, as a unit structure, a
           structural basis for the non-random distribution of the ER and
           other systems and organelles of the cytoplasm. Magnification
           70,000X.

probable from what is known that enzymes with a single purpose are
localized in domains and distributed according to a plan (Clegg, 1982;
Kempner and Miller, 1968; Masters, 1984).

THE MICROTRABECULAR LATTICE (MTL)

    This, the major component of the cytoplasmic matrix, is a complex of
slender strands which vary from 5-10 nm in diameter and 50 to 100 nm in
length. They interconnect one another and the surfaces of the various
organelles and membrane limited systems (ER and Golgi). Thus, one can
identify filaments of the MTL running from microtubules to the ER and
from microtubules to the inside surface of the cytoplasmic cortex. They
comprise a space-filling lattice that extends to all parts of the
cytoplast and even into marginal ruffles (Figs. 1-3, 6 and 7). We refer to
it as the protein-rich phase, to set it apart from the water-rich phase
which occupies the space between the trabeculae. The lattice fills about
20% of the space and the water-rich phase the rest (Gershon et al., 1985).

    There is a concept of the MTL that some of our contemporaries find
difficult to understand and accept. It is that the ribosomes, ER

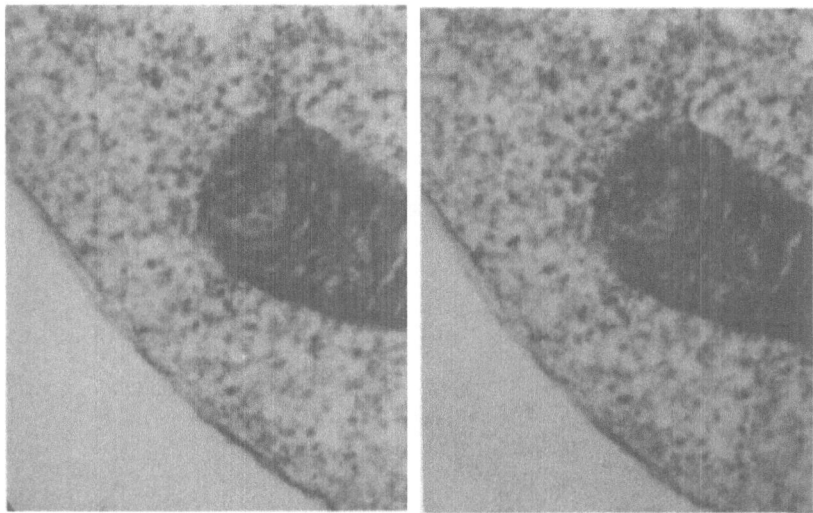

Fig. 4.   This stereo image is of the cytoplasmic matrix of a rat
           lymphocyte sectioned from cells embedded in epoxy resin. The
           micrograph illustrates the influence of the epoxy resins on
           image quality, which derives in part from the fact that it
           scatters electrons to the same degree as the objects embedded
           in it. This image coincides in its major features with that
           shown frequently in published images of sections.
           Magnification 60,000X.

cisternae, microtubules and pigment granules are contained in (are a part
of) the lattice - i.e., they reside not between the trabeculae in the
water-rich phase but in the substance of the protein-rich phase (Gershon
et al., 1985; Porter and McNiven, 1982; Stearns and Ochs, 1982).
Isolated from cells with microtubules, the associated trabeculae can be
equated with latter-day microtubule associated proteins, or MAPS. If a
microtubule is caused to disassemble by low temperature or nocodazole (or
both), the monomeric or dimeric tubulin remains an integral part of the
associated lattice and appears to be kept localized for rapid reassembly
into microtubules (Clegg, 1982; Ellisman and Porter, 1980) (Fig. 3).

     There is another important feature of observations on cultured cells:
the lattice varies in degree of compactness from cell periphery to cell
center. Thus at the outside it is relatively open, whereas in and
around the cell center (the centrioles) it is more finely divided and
compact. The limits of the centrioles grade-off into, and are continuous
with, the MTL and the same is true of the dense bodies or satellites
from which microtubules are initiated in their assembly. As their
structure would suggest, these are the more highly gelled regions of the
cytoplast.

     The relationship between lattice and microtubules is apparently
special. We think of the microtubules, the straight and relatively rigid
elements of the complex, as acting like struts and giving the whole its
anisometric form. In other words, when the microtubules assemble in a
pattern prescribed by dense bodies and MTL, they tend in their association with the matrix to distort it (i.e., the total lattice); the two act
synergistically to produce the requisite cell shape, the MTL being the
prominent anisometric element.

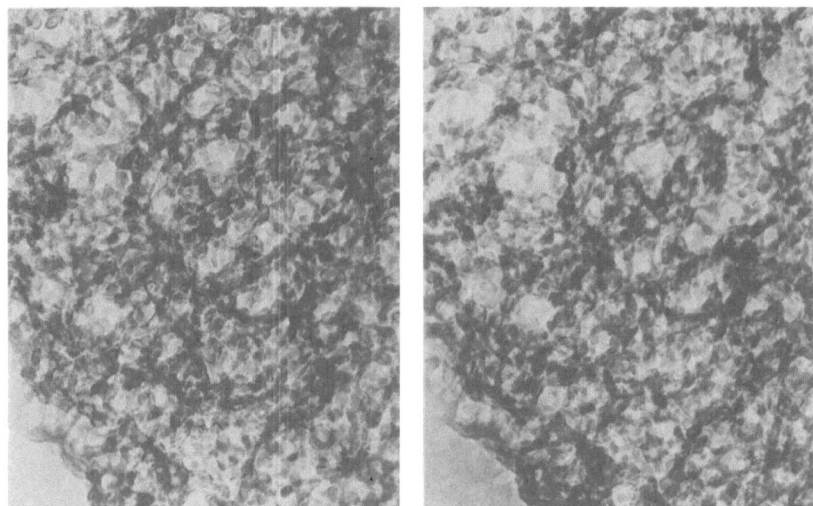

Fig. 5.   This stereo image is of the cytoplasmic matrix of a rat
           lymphocyte sectioned from cells embedded in polyethylene
           glycol, which was subsequently dissolved out. The specimen had
           no associated embedding matrix when examined. As a
           consequence, the contrast is better and the structures
           represent the 3-dimensional lattice of the cytoplasmic matrix.
           It was dried by the critical point method. What is included in
           the image represents what was in the specimen and nothing
           additional. It has basically the same structure as the
           cytoplasmic matrix of the thinly spread cultured cells (Figs. 2
           and 3). Magnification 87,000X.

IS THE MTL AN ARTIFACT OF FIXATION OR OTHER TECHNICAL PROCEDURE?

The older literature (1930's) contains an observation by Kopac and
Chambers that is pertinent here (Chambers, 1940; Kempner and Miller,
1968). They found that when small oil droplets (globules) were injected
into living cytoplasm no membrane (of adsorbed protein) formed on their
surface, and none appeared when part of the oil was withdrawn. If,
however, they injected the same oil droplets into the cytoplasm of a cell
whose integrity had been destroyed, a membrane formed which crinkled when
oil was removed (the Deveaux effect). Chambers concluded in a subsequent
paper (Chambers, 1940), that the proteins in a living (intact) cell are
somehow not available to accumulate on "experimentally introduced
surfaces." He suggests boldly that "the cytoplasmic proteins of living
cells may be bound together to form some kind of continuous phase."

There are, in the preparation of cells for electron microscopy,
procedures that could (and probably do) induce artifacts (Buckley and
Porter, 1975; Kopac, 1938; Stearns and Ochs, 1982). These are
unavoidable in electron microscopy. H. Ris has made much of the
necessity to use anhydrous alcohol and CO_2 in critical point drying.
This requirement was well known to the author before Dr. Ris recognized
it. In Boulder, Colorado, the dry air guaranteed our working under dry

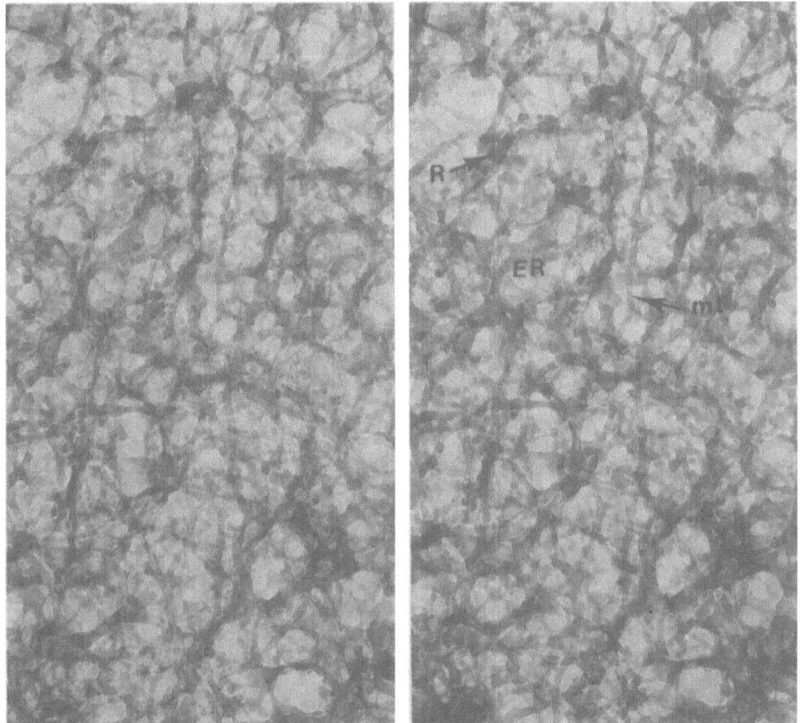

Fig. 6. This stereo pair shows the structure preserved in NRK cells (cultured) that were frozen instantly by plunging the e.m. grid and cells into propane cooled to -185°C. Thereafter the specimen was kept at -95°C during drying. Sublimation of the amorphous ice required 48 hours. Regular components of the cytoplasm are readily identified. Microtubule at mt, ER at ER, polysomes at R identified only by size and greater density. The rest of the space is occupied by the mtl and the intertrabecular spaces. The dimensions of the microtrabeculae, especially diameters, run larger than in cells fixed with glutaraldehyde and dried by the critical point method. Thus the majority of the trabeculae have measure 20 to 40 nm. The microtubules are uniform in diameter at 25 nm. This is probably as close as we have ever gotten to the preservation of the structured matrix. Compare with Fig. 7. Magnification 70,000X.

conditions, whereas on the shores of Lake Mendota in Madison, Wisconsin, humidity does become a problem. To avoid CPD altogether we prepared specimens by freezing the cultured cells at liquid N_2 temperatures and then subliming away the amorphous ice while maintaining the frozen state at -95°C. The drying took about 48-72 hrs., but the results in terms of morphology were essentially the same as those in cells first fixed with glutaraldehyde and OsO_4 and then frozen and dried (Figs. 6 and 7). The major difference is in the tapering of some of the microtrabeculae – on which H. Ris focuses his attention. Considering the relatively slow diffusion of glutaraldehyde into the cell, it is no wonder that some of the trabeculae contract or stretch to give a tapered form. This is a minor variation on a theme that remains intact; the cytomatrix is

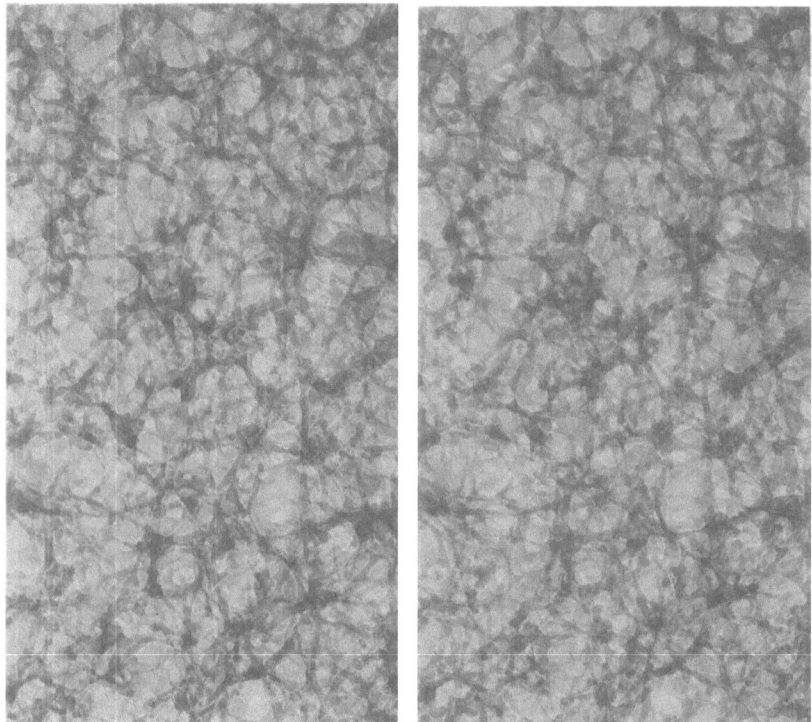

Fig. 7.   The NRK cytomatrix depicted here was fixed in glutaraldehyde
and OsO_4 before freezing at $-185^{\circ}C$ and then treated the same as
the cell (cytomatrix) depicted in Fig. 6. It was never exposed
to alcohol or critical-point drying. Distortions of the
structure shown in Fig. 6 are fairly obvious and result
presumably from the slowness with which glutaraldehyde
penetrates and fixes. This gives the matrix time to react by
contracting locally into clumps of lattice elements and thus
generating larger spaces. Microtubules are less linear and
more difficult to follow. Magnification 70,000X.

structured, as described above. Even our severist critic admits that the
matrix is structured.

THE RESPONSE OF THE LATTICE TO ENVIRONMENTAL FACTORS INCLUDING TONICITY

     One of the more interesting features of the MTL (and one that
supports its reality) is its response (morphologically) to altered
conditions of the environment. For example, in cells that are subjected
to low temperatures $(2-4^{\circ}C)$ over periods of two to three hours, the MTL
essentially collapses. It loses its integrity and associates more
closely with the inner surface of the plasma membrane (the cytoplasmic
cortex) (Porter and Anderson, 1982). Relatively large lakes appear from
a confluency of the water-rich phase. The cell withdraws its margins (if
free and not attached to other cells) and tends to become spherical.
Brownian motion of cytoplasmic granules is the rule. When, now, the cell
(still viable) is returned to the incubator at $37^{\circ}C$ the MTL recovers its
earlier 3-D form, and Brownian motion is again inhibited. In other words,

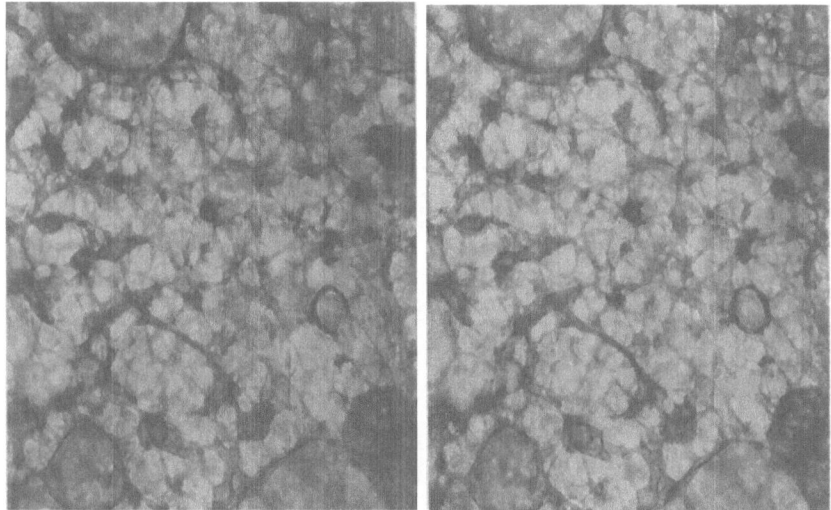

Fig. 8.   When cultured cells of the NRK type are exposed to an
           environment minus Mg^{++}, they immediately round up. The
           microtubules disappear and the cytoplasmic matrix, represented
           by the mtl acquires a coarseness in its structure that is shown
           in this stereo pair. The material of the microtrabeculae is
           apparently pulled into slender strands; other parts of the
           lattice gather into clumps; the intertrabecular spaces become
           enlarged. The characteristic form of the lattice is greatly
           altered (compare with Fig. 9). Magnification 50,000X.

the cytomatrix reacquires its former gel-like viscosity. Within 15
minutes the cell has again spread out over the substrate, and its MTL has
returned to characteristic form.

Departures from this characteristic form are equally striking when
the (NRK) cultured cells are deprived of Mg^{++} (Porter and Anderson,
1982). The lattice in this response becomes more coarsely divided with
prominent coalescences. Returned to 1 mM Mg^{++} the cell stretches out so
thinly that fenestrae appear in its thinner margins, and the MTL
reassumes its usual morphology (Figs. 8 and 9).

Cytochalasin and its capacity to bind actin has a distinctive effect
that also is difficult to interpret. Again the MTL is distorted and only
returns to normal when the cytochalasin is replaced by normal culture
medium (Porter and Anderson, 1982).

These phenomenal behaviors of the lattice are under further study.
At this stage of our understanding, little more than a descriptive
treatment can be expected.

All of these responses, accompanied by structural changes in the MTL,
remove it from the category of gross artifacts and give it the status of
a dynamic component of the cytomatrix.

As mentioned earlier, it would find more ready acceptance if it
displayed some repeating pattern or some order in its structure.
Without that, it lacks some credibility. A structure in the same size

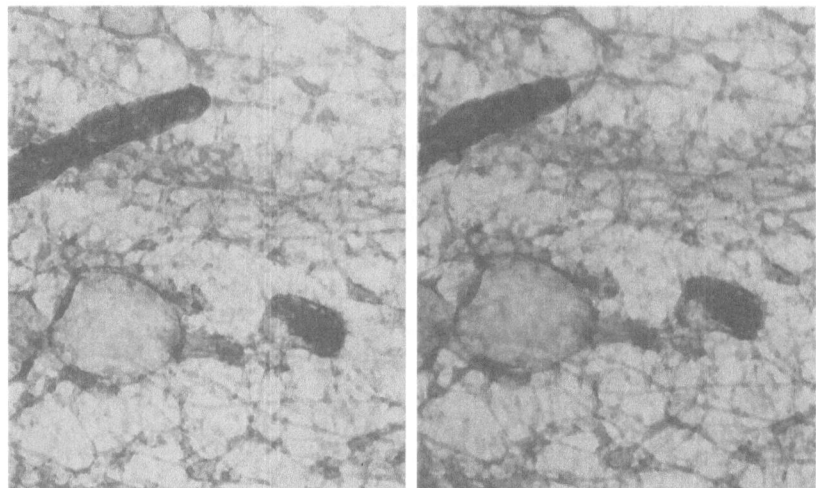

Fig. 9.   When Mg^{++} is returned to the medium in which the cells are
           bathed in concentrations of 1 mM, the cells quickly spread out,
           the microtubules (mt) return, and the lattice adopts a more
           nearly normal form. This stereo pair depicts the change. It
           appears that the Mg^{++}, perhaps through a dependent ATPase,
           provides the energy required for the cell to adopt a more
           anisometric form. Spreading is so extreme that fenestrations
           appear in the cell. Magnification 52,000X.

range is found in cilia in the form of slender spokes, which connect the
peripheral doublets with the axial cylinder surrounding the central pair.
Their existence has never been questioned, because (a) there are always
nine, and (b) they are disposed along the axis of the cilium (or
flagellum) in groups of three in an axial repeat every 86 Å. Nothing of
the sort has yet been observed in cultured cells, perhaps because they
spread out thinly and are usually growing and proliferating.

This may suggest that the whole MTL is special in cultured cells.
For this reason, among others, it has been important to examine sections
(thicker than usual) of fixed (0.25 - 0.5 cm thick) cells, embedded in
PEG. This is special because with water the PEG can be dissolved from
the section, which is thereafter dried in anhydrous alcohol and carried
through CO_2 by the critical point method. The resulting image is
outstanding for a couple of reasons. First, it is evident that the MTL
is there and in a form that repeats the basic features of its morphology
in cultured cells. Second, in resin-free sections it is evident that in
comparison with the structure in resin-embedded sections, the
microtrabeculae are much more prominent and easily interpreted as a
continuous three-dimensional lattice. The individual trabeculae are
totally convincing rather than "wispy" (see Figures 4 and 5).

THE MTL IN CHROMATOPHORES

Chromatophores of some of the invertebrates and lower vertebrates
possess the ability to move their pigment from a dispersed to an
aggregated condition in a matter of seconds. When dispersed, the pigment

granules are more or less uniformly distributed throughout the cytoplasm and lend their color to the whole organism. When aggregated, the pigment comprises a tiny mass 2 μm in diameter which has essentially no influence on the color of the animal. The distances the pigment is moved vary with the size of the chromatophore, but 10 μm per second is not unusual. How it is moved, how the cell generates and applies the motive force required to move the pigment, has been a mystery.

Again, high voltage microscopy has given some clues. It is possible to bring whole chromatophores into _in vitro_ culture and into the path of the electron beam. Microtubules are shown to be very numerous and are arranged radially around the cell center. They parallel the direction of pigment motion during aggregation and dispersion (Fig. 10). It has been shown, moreover, that microtubules are essential to the controlled motion of the pigment. Like other cells so far examined, fish (_Holocentrus_) chromatophores possess a microtrabecular lattice; and here it is possible to identify trabeculae that connect pigment granules to microtubules as

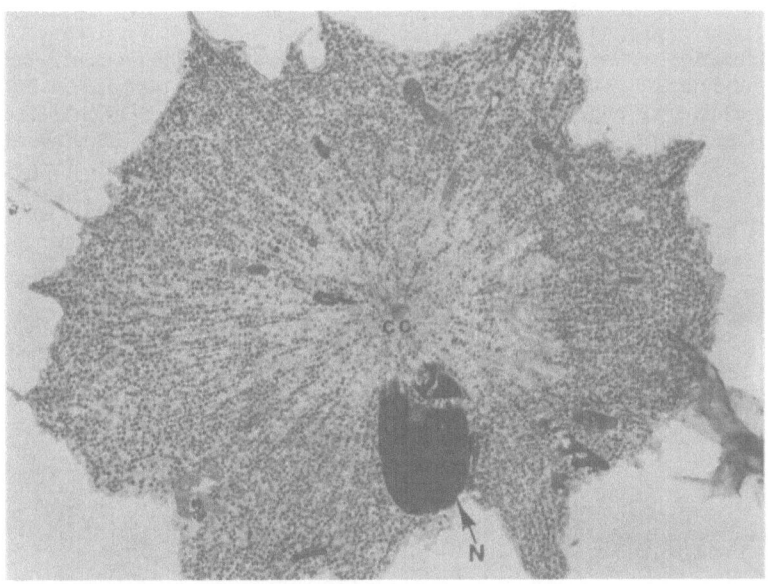

Fig. 10. This figure is a micrograph of a thick section through a fish chromatophore (erythrophore) cut parallel to the basal surface. The pigment is dispersed and distributed radially around the cell center (cc). The dark object is the cell nucleus. These cells are able to aggregate all their pigment into a central mass within 2-3 seconds after exposure to epinephrin. The cell thus becomes an object of exceptional interest, especially since it can subsequently disperse its pigment in about 5-6 seconds. It appears that two somewhat different mechanisms are involved in moving the pigment. When these cells are examined by high voltage microscopy, it is found that the pigment is suspended in a form of the MTL. During aggregation of the pigment the individual microtrabeculae adopt a beaded form as an accompaniment of shortening. During dispersion of the pigment the beads disappear. For students of the MTL, these changes are important because they show that the morphology changes with a function in which they are apparently involved. Magnification 3,000X.

well as to one another. In other words, the pigment is contained in a lattice of trabeculae which moves with the pigment during aggregation and dispersion. The trabeculae that attach to the microtubules give the motion its radial direction. During aggregation the trabeculae shorten and become beaded with small (200 Å) varicosities (Fig. 11). These disappear during dispersion. Toward the center of aggregation the lattice takes on an added configuration; the trabeculae gather into "clots" (Shay et al., 1974) or agglutinations which contribute, it would seem, to the completion of aggregation.

So, in these phenomena, one recognizes a morphological change associated with a highly dynamic function, and one gains faith in the reality of the structure involved. Furthermore, these changes in form are best preserved by the techniques of freeze substitution, where the rapid motion is instantaneously stopped by freezing at -185°C (Porter et al., 1982).

THE MTL AND THE WATER-RICH PHASE

Measurements made from cultured PTK cells that had been frozen (at -185°C) and dried, while kept at -95°C, indicate that roughly 80% of the cell's volume is represented by the so-called water-rich phase. From such preparations, also, it is evident that the spaces of the water-rich phase are free of residue after drying. It occurred to us as probable,

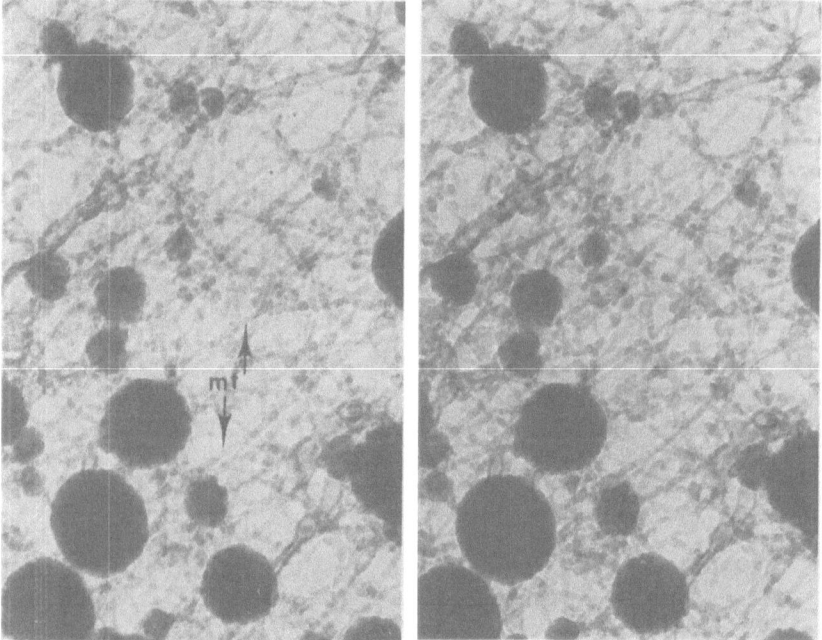

Fig. 11. Stereo pair showing high voltage micrographs of a small area of erythrophore cytoplasm preserved while the cell was aggregating its pigment. The dark spheres are pigment granules, the microtubules (mt) are the straight filaments of uniform diameter, and the residual filaments (which belong to the MTL) are beaded (diam. 200 Å). Magnification 78,000X.

if our interpretation of the micrographs is correct, that water might be added to or subtracted from this space by exposing the cells to hypotonic or hypertonic solutions.

When this was done, using in different series of experiments, cultured PTK, NRK, or chick embryo fibroblasts, we noted as expected that the water-rich phase diminished in volume at osmolarities above 300 mO_sM and increased in volume at O_sM's below 300. The resulting changes in dimensions were striking and easily shown to be significant when measured and treated statistically. One-minute exposure to the osmotically active environment was used uniformly.

Recovery from various mO_sM's was also examined and found to take place in three minutes. In each instance the structure of the cytomatrix returned to that of the normal control kept at 300 mO_sM. The basic lattice structure of the cytomatrix survives all these changes (Figs. 12, 13, 14). These seem also not to be changes associated with cell death, for the cells exposed to the trypan blue viability test showed dye exclusion. Furthermore, cells in cultures taken through 150 mO_sM exposure, as well as 600 mO_sM, continued to grow and multiply when returned to 300 mO_sM.

AN INTERPRETIVE DRAWING OF THE CYTOMATRIX

The drawing in Figure 15 was made several years ago (Porter and

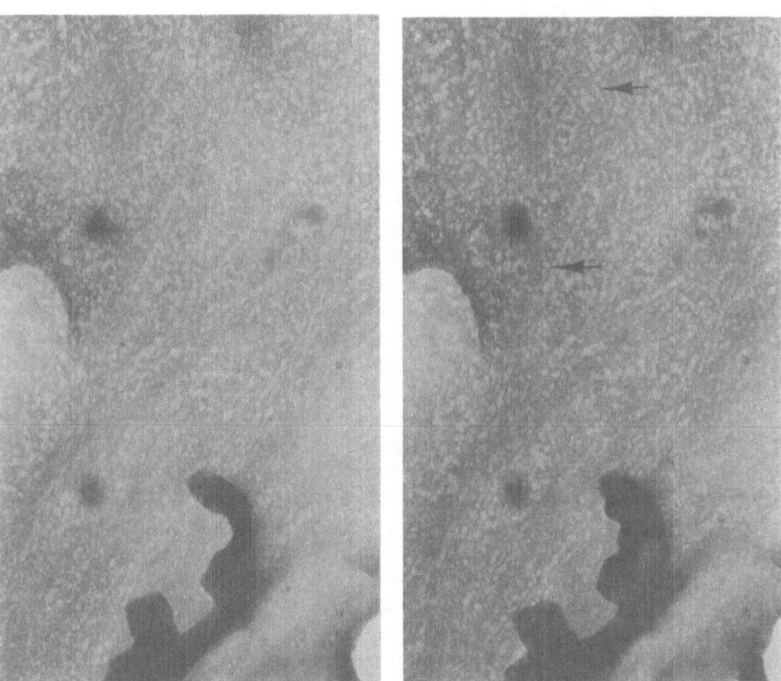

Fig. 12.   This stereo image depicts changes in the form (dimensions) of the MTL and the intertrabecular spaces, which occur in response to changes in the osmolarity of the extracellular environment. The cytomatrix shown is in a cell (NRK) exposed to 600 mO_sM's for one minute before fixation.

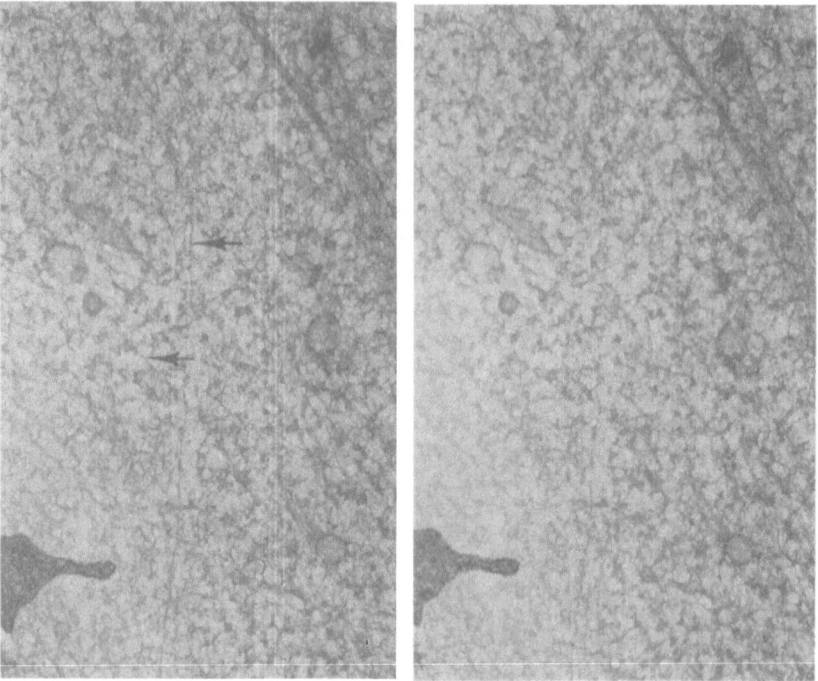

Fig. 13. Same as in Fig. 12, except that the cell was maintained in an
isotonic 300 mO$_s$M medium.

Tucker, 1981), and we have found no good reason to change it. In a few
respects the proportions are in error, e.g., the microtrabeculae are too
slender and the intervening spaces are too large. The identification of
the structures is possible without labels, but a few will help the
uninitiated.

The drawing represents a small cube of cytoplasm about 2 μm's on a
side. The upper plasma membrane is peeled back at one corner, chiefly to
emphasize that it is underlain by a thin cortex (C) which fixes without
evidence of structure. It can be seen in stereo images of whole cells
and varies in thickness in such images. It is shown as well at the
ventral surface of the drawing and next to the substrate. Usually there
is evidence of actin filaments in this lower cortex. They occur in so-
called "stress fibers" (SF). The microtrabeculae (Mt) of the matrix are
represented as slender strands connecting with other elements of the
cytoplasm. Some run between microtubules and the inner surface of the
cortex, while others run from microtubules to vesicles of the ER.
Ribosomes are frequently found at the cross-points of the trabeculae. It
is evident that the trabeculae constitute a space-filling, 3-D lattice in
which other components of the cytoplasm are suspended. Whether this is
true of mitochondria (m) has not been determined with certainty.

We have become accustomed to calling the lattice the protein-rich
phase and the intertrabecular spaces the water-rich phase. The image
indicates that microtubules, actin filaments, ribosomes and ER are
clothed in the protein-rich phase. The bulk of the protein in the
cytoplasm is part of this phase. When a microtubule disassembles, the

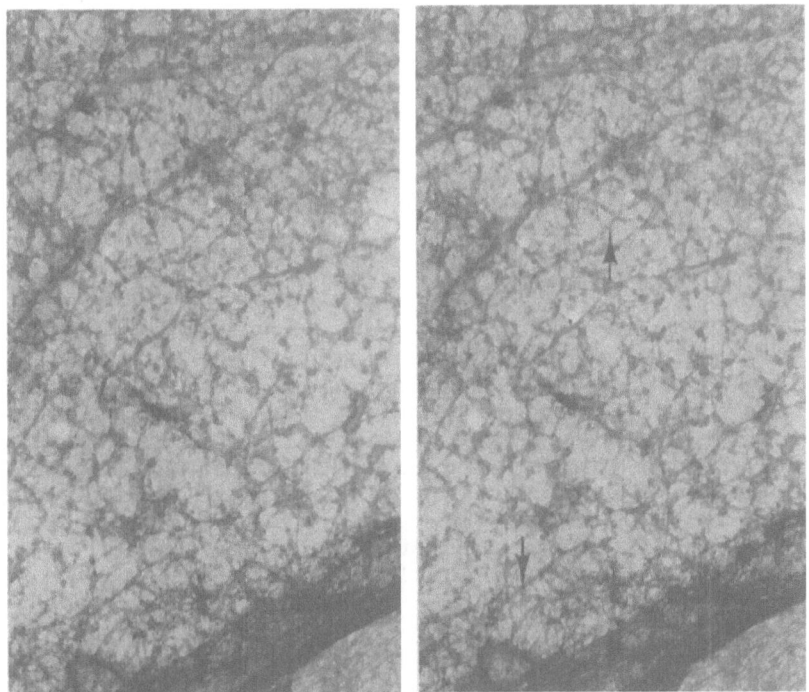

Fig. 14. Same as in Fig. 12, except that the cell was fixed one minute
after exposure to a 150 mO_sM medium. - It is obvious that in
response to hypertonic conditions, the intetrabecular space
loses volume whereas exposure to hypotonicity increases its
volume.

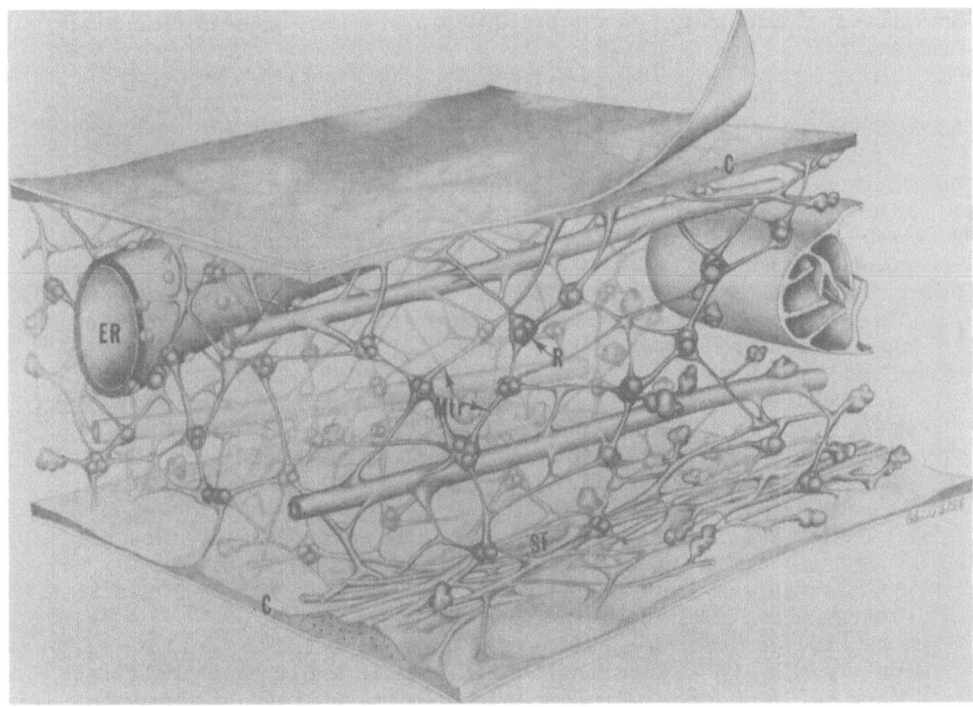

Fig. 15. A realistic model of the cytomatrix. (See text for comments.)

monomer or dimeric tubulin becomes part of the lattice and may be kept in place (though not polymerized) by the lattice, in anticipation of the next assembly. Time for diffusion and availability of tubulin greatly in excess of what is needed would be required, were not some provision made for having the tubulin where it is needed (DeMey et al., 1980). A combination of ribosomes and mRNA should be available in the lattice to synthesize the protein where it is needed. How the RNA gets there and by what device it and ribosomes may be positioned defies our imagination. This requirement aside, it does make sense to have a system of this nature present in the cytoplasm to control the non-random positioning of the products of synthesis.

The water-rich phase, which occupies about 80% of the cytoplasmic volume, is only that as near as one can tell. Its volume can be manipulated, as noted above, by varying the osmolarity of the surrounding medium. The phase is continuous throughout the cytomatrix and probably provides for the diffusion of small metabolites. Its existence in this form contributes homeostasis to the internal milieu. Such metabolites as glucose, amino acids, and inositol immediately come to mind. The diffusion of large molecules in this phase is retarded by some property of the matrix, and it has been suggested that the huge surfaces of the lattice may adsorb such molecules at least momentarily (Guatelli et al., 1982; Porter and Anderson, 1980).

ACKNOWLEDGMENTS

The author is pleased to acknowledge that the micrographs in several instances came from collections taken by Karen Anderson and Mark McNiven. Preparation of the manuscript was expedited by Audrey Ellis and Marcia Ames.

This work was supported by Grant No. GM 34422 from the National Institutes of Health and Grant No. CD-191A from the American Cancer Society.

REFERENCES

Baccetti, B., Porter, K.R., and Ulrich, M., 1985, High voltage electron microscopy of sperm axoneme, J. Submicrosc. Cytol., 17:171.
Buckley, L.K., and Porter, K.R., 1975, Electron microscopy of critical-point dried whole cultured cells, J. Micros., 104:107.
Byers, H.R., and Porter, K.R., 1977, Transformations in the structure of the cytoplasmic ground substance in erythrophores during pigment aggregation and dispersion, J. Cell Biol., 75:541.
Byers, H.R., Fujiwara, K., and Porter, K.R., 1980, Visualization of microtubules of cells in situ by indirect immunofluorescence, Proc. Natl. Acad. Sci. USA, 77:6657.
Chambers, R., 1940, The micromanipulation of living cells, in "The Cell and Protoplasm," F.R. Moulton, ed., A.A.A.S. Publication No. 14. Science Press, Lancaster, PA.
Clegg, J.S., 1982, Interrelationships between water and cell metabolism in Artemia cysts. IX. Evidence for organization of soluble cytoplasmic enzymes, in: Cold Spring Harbor Symposium on Quantitative Biology, Vol. XLVI, Cold Spring Harbor Laboratory.
DeMey, J. Wolosweick, J.J., DeBrabender, M., Geuens, G., Joniau, M., and Porter, K.R., 1980, Tubulin localization in whole, glutaraldehyde fixed cells, viewed with stereo high-voltage electron microscopy, in:

"Cell Movement and Neoplasia," J. DeBrabender, ed., Pergamon Press, Oxford.

Ellisman, M.H., and Porter, K.R., 1980, Microtrabecular structure of the axoplasmic matrix: Visualization of cross-linking structures and their distribution, J. Cell Biol., 87:464.

Gershon, N.D., Porter, K.R., and Trus, B.L., 1985, The cytoplasmic matrix: Its volume and surface area and the diffusion of molecules through it, Proc. Natl. Acad. Sci. USA, 82:5030.

Guatelli, J.C., Porter, K.R., Anderson, K.L., and Boggs, D.P., 1982, Ultrastructure of the cytoplasmic and nuclear matrices of human lymphocytes observed using high voltage electron microscopy of embedment-free sections, Biol. Cell, 43:69.

Kempner, E.S., and Miller, J.H., 1968, The molecular biology of Euglena gracilis. V. Enzyme localization, Exp. Cell Res., 51:150.

Kopac, M.J., 1938, The Deveaux effect at oil-protoplasm interfaces, Biol. Bull., 75:372.

Luby, K.J., and Porter, K.R., 1982, The control of pigment migration in isolated erythrophores of Holocentrus ascensionis (Osbeck). II. The role of calcium, Cell, 21:13.

Luby-Phelps, K., and Porter, K.R., 1982, The control of pigment migration in isolated erythrophores of Holocentrus ascensionis (Osbeck). II. The role of calcium, Cell, 29:441.

Masters, C., 1984, Interactions between glycolytic enzymes and components of the cytomatrix, J. Cell Biol., 99:1:Pt.2:2225.

Mastro, A.M., and Keith, A.D., 1981, Spin label viscosity studies of mammalian cell cytoplasm, in "The Transformed Cell," I. Cameron and T. Pool, eds., Academic Press, New York.

Porter, K.R., and Anderson, K.L., 1980, Structural responses of the microtrabecular lattice (mtl) to changes in temperature, concentration of divalent cations and to cytochalasin, Proceedings of 2nd ICCB, Berlin, Eur. J. Cell Biol., 22:351.

Porter, K.R. and Anderson, K.L., 1982, The structure of the cytoplasmic matrix preserved by freeze-drying and freeze-substitution, Eur. J. Cell Biol., 29:83.

Porter, K.R., Beckerle, M.C., and McNiven, M.A., 1983, The cytoplasmic matrix, in: "Spatial Organization of Eukaryotic Cells," J.D. McIntosh, ed., Alan R. Liss, New York.

Porter, K.R., Boggs, D.P., and Anderson, K.R., 1982, The distribution of water in the cytoplasm, Proc. 40th Ann EMSA Meeting, Washington, D.C.

Porter, K.R. and McNiven, M.A., 1982, The cytoplast: A unit structure in chromatophores, Cell, 29:23.

Porter, K.R., and Tucker, J.B., 1981, The ground substance of the living cell, Sci. Am., 244(3):57.

Ris, H., 1985, The cytoplasmic filament system in critical point dried whole mounts and plastic-embedded sections, J. Cell Biol., 100:1474.

Schliwa, M., and Euteneuer, U., 1978, A microtubule-independent component may be involved in granule transport in pigment cells, Nature, 273:556.

Shay, J.W., Porter, K.R., and Prescott, D.M., 1974, The surface morphology and fine structure of CHO (Chinese Hamster Ovary) cells following enucleation, Proc. Natl. Acad. Sci. USA, 71:3059.

Stearns, R.E., and Ochs, R.L., 1982, A functional in vitro model for studies of intracellular motility in digitonin permeabilized erythrophores, J. Cell Biol., 94(3):727.

Wolosewick, J.J., and Porter, K.R., 1979, The microtrabecular lattice of the cytoplasmic ground substance: Artifact or reality, J. Cell Biol., 82:114.

THE CYTOSKELETON

Mary Osborn and Klaus Weber

Max Planck Institute for Biophysical Chemistry
D-34 Goettingen, F.R.G.

INTRODUCTION

The cell is a macromolecular organization where the whole seems greater than the sum of the parts. Cells contain a cytoskeleton, made from microfilaments, microtubules, and intermediate filaments, which together with the nucleus remains behind if cells are treated with certain detergents. Cells commit 10-40% of their proteins to these structures, and currently some 50-100 proteins which either form or bind to these structures have been characterized. The elements of the cytoskeleton interconnect and form a three-dimensional network. Cytoplasmic organelles and enzymes, such as ribosomes, motochondria, glycolytic enzymes, and calmodulin, may be bound at least transiently to particular cytoskeletal fibers. Receptors and transmembrane proteins may also be in intimate contact with the cytoskeleton. The cytoskeleton is often in a dynamic and continually changing state, although microscopy gives only static snapshots of its three-dimensional dynamic components.

This picture of the cytoplasmic organization — which we have referred to elsewhere as "the new anatomy" — evolved in large part in the 1970s, as a result of the application of different microscopical techniques to animal cells in culture. Transmission electron microscopy, quick-freeze deep-etch microscopy (Heuser and Kirschner, 1980), high-voltage electron microscopy (Wolosewick and Porter, 1976), and immunofluorescence microscopy (Lazarides and Weber, 1974) result in static pictures; and it is important to realize that the overview obtained will depend in part on the method used. For example, is one studying only a part of the cell at high resolution, as in transmission electron microscopy or quick-freeze deep-etch microscopy, or is one using a whole-mount technique at low resolution, as in high voltage or immunofluorescence microscopy, where the whole cell can be viewed at once? Whereas in the electron microscopical techniques all structures which survive the fixation and embedding process and have sufficient contrast to stand out against the embedding material will be visualized, in immunofluorescence microscopy it is possible by the use of an appropriate antibody to view the arrangement of a particular cytoskeletal element or protein selectively. This fact, as well as the possibility of constructing a biochemical anatomy of a particular cytoskeletal structure by identifying specific proteins which belong to it, or are associated with it, account for the dramatic increase in the use of this method since Lazarides and Weber (1974) described the use of actin antibody to study stress fiber distribution in animal cells. Microinjection of fluorescently labeled proteins or cytoskeletal constituents into living cells (Taylor and Wang, 1978; Wehland and Weber, 1980), where their fate is studied as a function of time by high-intensity video microscopy, has extended this approach to the living cell and increased the confidence with which results obtained from immunofluorescence studies performed on fixed cells can be interpreted. The reader is urged to consult the references, to regard the original micrographs on which the view of the cytoskeleton developed in this manuscript is based.

CELL COMPARTMENTALIZATION

While the most obvious division of the cell is between nucleus and cytoplasm, the functional pore size of the nuclear envelope is about 45 Å, allowing many molecules to cross this boundary (Paine et al., 1975). Artificial manipulation of the nuclear-cytoplasmic boundary can be achieved, for instance by treating the cell with dimethylsulfoxide. This leads to a disappearance of stress fibers in the cytoplasm and actin accumulation in the cell nucleus, in the form of paracrystals (Osborn and Weber, 1980; Fukui and Katsumura, 1979). In the cytoplasm of normal cells transmission electron microscopy sometimes shows zones from which ribosomes, glycogen, or cell organelles are excluded. Cells on a substratum can demonstrate a variety of different surface motions, such as producing and moving filopodia, blebbing, or ruffling. The same cell often shows different surface motions on different parts of its surface. In a moving cell such as a fibroblast, ruffling is usually associated with the leading edge. If the same cell is observed over a longer period, the cell can change direction; and another part of the cell margin will then develop ruffles. If, however, as shown by Albrecht-Buhler (1980), cells are treated with cytochalasin, disrupted into small pieces by forcing a jet of liquid at them, and if then the pieces are trypsinized and allowed to reattach, pieces with as little as 2% of the cell's volume (i.e. 100-200 μm^3 cytoplasm) will display a single characteristic motion, such as blebbing or ruffling, which will then be repeated time and time again for periods as long as 6 hours. These observations seem best interpreted in terms of local differences in cytoplasmic constituents. Unfortunately, however, since currently there is no way to purify fragments of a particular type, one cannot determine the necessary molecules for blebbing or ruffling by direct analysis of the constituents of each fragment type.

MICROFILAMENTS, MICROTUBULES, INTERMEDIATE FILAMENTS

The three fibrous elements of the cytoskeleton also serve to define different regions of the cytoplasm. Electron microscopic examination of fibroblasts or glial cells, after a brief extraction with a detergent, has emphasized the highly ordered actin network of single 6 nm filaments that fill the edge of the cell (Small et al., 1979, 1982; Hoglund et al., 1980). These 6 nm diameter polymers of actin, referred to as F-actin or microfilaments, are enriched in the cortical net underlying the plasma membrane. In some cell types they may be further aggregated laterally into thick actin cables or stress fibers (Lazarides and Weber, 1974), or the highly ordered bundles underlying the microvilli of intestinal epithelial cells (Weber and Glenney, 1982).

Biochemical studies and immunofluorescence microscopy have identified at least thirty actin-associated proteins (for review see

Weeds, 1982). These actin-associated proteins show different
distributions when different actin-containing structures within the cell
are examined. Thus, for instance, more stable actin-containing
structures such as stress fibers have myosin and tropomyosin, while
ruffles usually have reduced levels of these two actin-associated
proteins (Heggeness et al., 1977). Microvilli of intestinal epithelium
totally lack these proteins and instead have a totally different
structural principle (Weber and Glenney, 1982). Actin-associated
proteins are usually subclassified into groups which exert different
effects on actin molecules. Thus, gelation factors form flexible but
tight links between criss-crossed filaments. Bundling factors laterally
aggregate microfilaments into highly ordered bundles, while severing and
stabilizing factors bind to single filaments where they govern their
length and stability. Calcium ion fluxes are involved in the control of
at least some of the actin-binding proteins, and in particular of the
severing and of certain capping and spacing proteins (e.g., see Bretscher
and Weber, 1980). Myosin occupies a special place among the actin-
associated proteins, in that it together with F-actin converts the
chemical energy of ATP into contractile movement. Direct *in vitro* proof
of the involvement of myosin kinase in this process has recently been
obtained (Holzapfel et al., 1983). Other *in vitro* experiments have shown
that it is possible to reconstruct a system in which myosin-coated beads
can move along ordered arrays of F-actin filaments when ATP is present
(Spudich et al., 1985). These experiments leave little doubt that actin
and its multitude of actin-associated proteins are involved in locomotion
and remodelling of cell morphology, in agreement with the finding that
drugs such as phalloidin, which bind directly to F-actin (Wieland, 1977),
inhibit translocation of the cell on a substratum (Wehland et al., 1977).

The two other fibrous systems — microtubules and intermediate
filaments — fill the internal cytoplasm but can be very different in
their arrangements within the cytoplasm. Like F-actin, microtubules are
polar structures; and like F-actin, microtubules are built from a single
major protein, tubulin, which is formed from two closely related (but not
identical) subunits, alpha- and beta-tubulin. Like F-actin, microtubules
require a triphosphate for assembly. Within a typical interphase tissue-
culture cell, microtubules running from the centriole, or cytocenter, to
close to the plasma membrane are often seen (Weber and Osborn, 1979). As
shown by both immunofluorescence and some high-voltage micrographs
(Osborn et al., 1978; Schliwa and Van Blerkom, 1981), single microtubules
appear continuous between these two points in cells; and this, together
with the polarity they display in regrowth experiments (Osborn and Weber,
1976), suggests that they may be used by the cell as tracks along which
organelles, vesicles, or smaller particles are transported. Direct
confirmation of this idea has been obtained by experiments in which
microinjection of tubulin antibodies has been shown to stop saltatory
motion (Wehland and Willingham, 1983), and also by recent elegant
experiments in which it has been possible to demonstrate microtubule-
determined organelle movement in *in vitro* systems derived from axoplasm
(Allen et al., 1985; Vale et al., 1985; Schnapp et al., 1985). All
organelles, regardless of size, moved in these experiments with a speed
of approximately 2.2 μ/sec. Interestingly, in these experiments two
particles appear able to move in different directions along the same
microtubule without collision. In mitosis the interphase microtubules
are depolymerized, and the tubulin is reutilized in the microtubules of
the mitotic spindle. A variety of tubulin-associated proteins have been
described, including the high molecular-weight MAP proteins and the tau
proteins. At least the MAPs have been shown to project from the
microtubule wall for a distance approximately equal to the microtubular

diameter, and it may be that these proteins help to define a special zone or compartment around the microtubule.

Intermediate filaments (IFs) are the third fibrous system. While microfilaments and microtubules are ubiquitous, there appear to be a few cell types which lack IFs. IFs are cytoplasmic structures which, depending on the cell type, can be laterally aggregated. Most striking is their cell-type specificity (Franke et al., 1982; Holtzer et al., 1982; Osborn et al., 1982). Thus epithelial cells have keratin IFs, neuronal cells neurofilaments, glial cells glial fibrillary acidic protein, most cells of muscle origin have desmin IFs, while many cells of mesenchymal origin have only vimentin IFs. This strict cell type-specific expression of IFs has resulted in their use as markers of histogenetic origin, both in cell biology and also in routine pathology where they can distinguish the major different human tumor types (for review see Osborn and Weber, 1983). All IF proteins so far examined display a common structural motif, i.e. a rod-like central alpha-helical portion of constant length, responsible for the common structural appearance of many of these filaments when viewed in electron microscopy, flanked by non-helical head and tail regions of variable lengths (Geisler and Weber, 1982; Hanukoglu and Fuchs, 1983; Steinert et al., 1983). For instance, neurofilaments in mammals are built from three neurofilament proteins with apparent molecular weights on SDS gels of 68 kd, 160 kd and 200 kd. All three proteins display the constant rod-like region but have variable-length tails which project from the filament, and which in the case of the 200 kd subunit tail can be identified with cross-bridges visualized in electron micrographs of neurofilament bundles (Geisler et al., 1983; Hirokawa et al., 1984). Microinjection of antibodies to particular IF-proteins or to certain IF-associated proteins results in no inhibition of cell movement or of cell division (Gawlitta et al., 1981; Lin and Feramisco, 19 ; Klymkowsky, 1981). Thus, the question of IF function in the interphase cell is left open, although undoubtedly they do provide tensile strength in tissues such as epithelium where they abut onto the desmosomes which link neighboring epithelial cells.

THREE-DIMENSIONAL CYTOSKELETAL NETWORKS

While in general a coherent picture of the overall arrangement and composition of the main cytoskeletal elements has emerged (i.e. the 6 nm microfilaments, the 20-22 nm microtubules, and the 7-11 nm intermediate filaments), there is still among those interested in the cytoskeleton considerable discussion as to the physical appearance and the composition of structures which serve to link the main cytoplasmic cytoskeletal elements. To draw attention to these interconnecting structures and to their possible functions within the cytoplasm, Porter and his colleagues have coined the terms "microtrabeculae", "ground substance", and "cytomatrix" (Wolosewick, 1979; Wolosewick and Porter, 1979; Schliwa et al., 1981). Microtrabeculae are defined as fibers or filaments interconnecting microfilaments, microtubules, intermediate filaments and various organelles, including ribosomes and internal membranes. They form a space-filling lattice and have diameters that range between 2 and more than 10 nm. Individual fibers are usually thinner at the center and thicker where they abut other cytoskeletal elements. Much of the evidence concerning microtrabeculae has been collected with the high-voltage electron microscope, and the discrepancies between Porter's group using the Colorado instrument and Ris' group using the Wisconsin instrument (Ris, 1985) are well documented in the literature. Ris argues that it is crucial that water is not present in the sample and describes microtrabeculae as "a distorted image of the cytoplasmic filament network produced during critical-point drying by traces of water or ethanol in the CO_2". To support this idea he shows micrographs of cytoplasm and of F-actin processed for electron microscopy by different fixation techniques (Ris, 1985). Further support for this point of view can be derived from the varying images obtained of two structures where F-actin is very abundant, i.e. the differing profiles of the vaccinia-induced macrovilli obtained using high-voltage microscopy (Stokes, 1976), or by using a combination of immunofluorescence and electron microscopic techniques (Hiller et al., 1979). A similar situation is apparent if one compares micrographs of the leading edge of the cell taken by high voltage (Wolosewick and Porter, 1979; Schliwa et al., 1981; Ris, 1985) with electron micrographs obtained in Small and Lindberg's laboratories (Small et al., 1982; Hoglund et al., 1980; Hoglund, 1985) on cells fixed by exposing them to a detergent fixative mixture followed by negative staining with a tungstate salt. Whereas by high voltage microtrabeculae are visualized, the other method reveals ordered arrays of F-actin filaments. Furthermore, examination of two-dimensional gels of the protein components from a variety of cell lines (Schliwa et al., 1981; other references not cited) provides no support for the idea that there is a single novel protein present in sufficient amount to be the building block of microtrabeculae. An alternative possibility is of course that microtrabeculae are formed not from a single protein but from several proteins, which may themselves either project from the three major cytoskeletal fibrous systems or be rod-like themselves. Some obvious candidates, such as the MAP proteins, or IF protein extensions such as the neurofilament tailpieces, have already been discussed; others that might be considered in this context include other IF protein tailpieces, alpha-actinin, plectins, and even proteins which are not an integral part of any cytoskeletal element but which are transiently bound and are trapped at the moment of fixation. Thus a final evaluation of the morphology and the protein composition of elements connecting the three main filament systems within the cytoplasm requires more experimentation.

Is there a continuous three-dimensional cytoskeletal framework running from the nucleus to underneath the plasma membrane in the interphase tissue-culture cell? Tantalizing hints suggest that microtubules sometimes associate with centrioles and radiate outwards from these structures, whereas at least some IF types associate with the nuclear envelope (Small et al., 1982; Lehto et al., 1978) or even share some structural features with nuclear lamins. In one very specialized cell type — the mammalian red blood cell, which lacks a nucleus — biochemical evidence (Lehto et al., 1978; Branton et al., 1981) suggesting the existence of a submembraneous skeleton has been recently supplemented by striking electron micrographs (Byers and Branton, 1985), directly demonstrating the previously postulated networks of spectrin and ankyrin underlying the red blood cell membrane in a geodesic dome-like fashion. Very short filaments of F-actin appear associated with this network at the intersection points, and the whole network is probably held in place by the transmembrane protein Band III. In the same system the IF protein vimentin has been shown to bind to inverted membrane vesicles from human erythrocytes through ankyrin (Georgatos et al., 1985). If these observations can be extended to nucleated cells, such as fibroblasts which are known to contain both spectrin analogs as well as IFs, a major part of the puzzle will have been solved.

THE CYTOSKELETON AS A SCAFFOLD FOR CELL METABOLISM

The arguments for wanting to associate cell metabolism with the cytoskeleton are mostly summarized elsewhere in this volume. Thus Masters (1978, 1981) demonstrated interactions _in vitro_ among F-actin, actin-associated proteins, and certain glycolytic enzymes. Masters also raises the interesting possibility that the activities of these enzymes could be modulated by interactions with the cell's structural components. There is also evidence that the glycolytic enzymes may be compartmentalized within the cell. Thus, certain membrane fractions from erythrocytes and yeast cells are capable of catalyzing the reactions of the glycolytic pathway with specific activities greater than those displayed by total cell homogenates (Green et al., 1965). Glyceraldehyde phosphate dehydrogenase binds to the cytoplasmic tail of Band III protein of the red blood cell, and the binding is sufficiently strong so that this enzyme is one of the major proteins identified in red blood cell ghosts. Interestingly, in axonal transport (Hoffmann and Lasek, 1975) some glycolytic enzymes move at the same speed as actin and certain actin-associated proteins (for review see Weiss, 1982) and therefore appear in the same compartment or window, again suggesting a potential for interaction of the glycolytic enzymes with these cytoskeletal constituents.

Perhaps, therefore, one of the very important points about the cytoskeletal network is not only its three-dimensionality but also that filaments or fibrous structures generate large surfaces suitable for at least transiently binding many so called "soluble" proteins. Thus, the cytoskeletal network could be thought of as providing the scaffold on which much of cell metabolism could occur. Two points seem of particular importance in considering this hypothesis. First, the binding of enzymes and soluble proteins needs not to be very tight to allow for a transient binding, which could in turn result in a modulating effect. Second, intracellular pH, salt, and ionic conditions may have an effect different than the buffers usually used _in vitro_. Third, the intracellular protein concentrations can be much higher than those used in standard enzyme tests. Thus, there is still some way to go before the pictures of the cytoskeleton obtained from the different microscopic techniques and the pictures emerging from _in vitro_ biochemical studies can be brought together in a common framework.

REFERENCES

Albrecht-Buhler, G., 1980, Autonomous movements of cytoplasmic fragments, _Proc. Nat. Acad. Sci. USA_, 77:6639.

Allen, R.D., Weiss, D.G., Hayden, J.H., Brown, D.T., Fujiwake, H., and Simpson, M., 1985, Gliding movement of and bidirectional transport along single native microtubules from squid axoplasm: Evidence for an active role of microtubules in cytoplasmic transport, _J. Cell Biol._, 100:1736.

Branton, D., Cohen, C.M., and Tyler, J., 1981, Interaction of cytoskeletal proteins on the human erythrocyte membrane, _Cell_, 24:24.

Bretscher, A., and Weber, K., 1980, Villin is a major protein of the microvillus cytoskeleton which binds both G- and F-actin in a calcium-dependent manner, _Cell_, 20:839.

Byers, T.J., and Branton, D., 1985, Visualization of the protein associations in the erythrocyte membrane skeleton, _Proc. Nat. Acad. Sci. USA_, 82:6153.

Franke, W.W., Schmid, E., Schiller, D.L., Winter, S., Jarasch,
E.D., Moll, K., Denk, H., Jackson, B.W., and Illmensee, K, 1982,
Differentiation related patterns of expression of proteins of
intermediate filaments in tissues and cultured cells, Cold Spring
Harbor Symp. Quant. Biol., 46:431.

Fukui, Y., and Katsumura, H., 1979, Nuclear actin bundles in Amoeba,
Dictyostelium and human HeLa cells induced by dimethylsulfoxide.
Exp. Cell Res., 120:451.

Gawlitta, W., Osborn, M., and Weber, K., 1981, Coiling of intermediate
filaments induced by microinjection of a vimentin-specific antibody
does not interfere with locomotion and mitosis, Eur. J. Cell Biol.,
26:83.

Geisler, N., and Weber, K., 1982, The amino acid sequence of chicken
muscle desmin provides a common structural model for intermediate
filament proteins, EMBO J., 1:1649.

Geisler, N., Kaufmann, E., Fischer, S., Plessmann, U., and Weber, K.,
1983, Neurofilament architecture combines structural principles of
intermediate filaments with carboxy-terminal extensions increasing in
size between triplet proteins, EMBO J., 2:1295.

Georgatos, S.D., Weaver, D.C., and Marchesi, V.T., 1985, Site specificity
in vimentin-membrane interactions: Intermediate filament subunits
associate with the plasma membrane via their head domains, J. Cell
Biol., 100:1962.

Green, D.E., Murer, E., Hultin, H.O., Richardson, S.H., Salmon, B.,
Brierly, G.P., and Baum, H., 1965, Arch. Biochem. Biophys., 112:635.

Hanukoglu, I., and Fuchs, E., 1983, The DNA sequence of a type II
cytoskeleton keratin reveals constant and variable domains among
keratins, Cell, 33:915.

Heggeness, M.H., Wang, K., and Singer, S.J., 1977, Intracellular
distribution of mechanochemical proteins in cultured fibroblasts.
Proc. Nat. Acad. Sci. USA, 74:3883.

Heuser, J.E., and Kirschner, M.W., 1980, Filament organization revealed in
platinum replicas of freeze-dried cytoskeletons, J. Cell Biol., 86:
212.

Hiller, G., Weber, K., Schneider, L., Parajsz, C., and Jungwirth, C.,
1979, Interaction of assembled progency pox viruses with the cellular
cytoskeleton, Virology, 98:142.

Hirokawa, N., Glicksman, M.A., and Willard, M.R., 1984, Organization of
mammalian neurofilament polypeptides within the neuronal
cytoskeletons, J. Cell Biol., 1523.

Hoffmann, P.N., and Lasek, R.V., 1975, The slow component of axonal
transport. Identification of major structural polypeptides of the
axon and their generality among mammalian neurones, J. Cell Biol.,
66:351.

Hoglund, A.S., 1985, The arrangement of microfilaments and
microtubules in the periphery of spreading fibroblasts and glial
cells, Tissue & Cell, 17:649.

Hoglund, A.-S., Karlsson, R., Arro, E., Fredriksson, B.-A., and Lindberg,
U., 1980, Visualization of the peripheral weave of microfilaments
in glia cells, J. Musc. Res. Cell Motility, 1:127.

Holtzer, H., Bennett, G.S., Tapscott, S.J., Croop, J.M., and Toyama, Y.,
1982, Intermediate sized filaments: change in synthesis and
distribution in cells of the myogenic and neurogenic lineages, Cold
Spring Harbor Symp. Quant. Biol., 46:317.

Holzapfel, G., Wehland, J., and Weber, K., 1983, Calcium control of actin-
myosin based contraction in Triton models of mouse 3T3 fibroblasts is
mediated by the myosin light chain kinase (MLCK)-calmodulin complex,
Exp. Cell Res., 148:117.

Klymkowsky, M., 1981, Intermediate filaments in 3T3 cells collapse after intracellular injection of a monoclonal antibody, _Nature_, 291: 249.

Lazarides, E., and Weber, K., 1974, Actin antibody: The specific visualization of actin filaments in non-muscle cells, _Proc. Nat. Acad. Sci. USA_, 71:2268.

Lehto, V.P., Virtanen, L., and Kurki, P., 1978, Intermediate filaments anchor the nucleus in nuclear monolayers of cultured human fibroblasts, _Nature_, 272:175.

Lin, J.-C, and Feramisco, J.R., 19 , Disruption of the _in vivo_ distribution of the intermediate filaments in fibroblasts through the microinjection of a specific monoclonal antibody, _Cell_, 24:185.

Masters, C.J., 1978, Interactions between soluble enzymes and sub-cellular structure, _TIBS_, 206.

Masters, C.J., 1981, Interactions between soluble enzymes and sub-cellular structure, _CRC Critical Review in Biochemistry_, 105.

Osborn, M., and Weber, K., 1976, Cytoplasmic microtubules in tissue culture cells appear to grow from an organizing structure toward the plasma membrane, _Proc. Nat. Acad. Sci. USA_, 73:867.

Osborn, M., and Weber, K., 1980, Dimethylsulfoxide and the ionophore A23187 affect the arrangement of actin and induce nuclear paracrystals in PtK2 cells, _Exp. Cell Res._, 129:103.

Osborn, M., and Weber, K., 1983, Biology of disease. Tumor diagnosis by intermediate filament typing: A novel tool for surgical pathology, _Lab. Invest._, 48:372.

Osborn, M., Geisler, N., Shaw, G., Sharp, G., and Weber, K., 1982, Intermediate filaments, _Cold Spring Harbor Symp. Quant. Biol._, 46: 413.

Osborn, M., Webster, R.E., and Weber, K., 1978, Individual microtubules viewed by immunofluorescence and electron microscopy in the same PtK_2 cell, _J. Cell Biol._, 77:R27.

Paine, P.L., Moore, L.C., and Horowitz, S.B., 1975, Nuclear envelope permeability, _Nature_, 254:109.

Ris, H., 1985, The cytoplasmic filament system in critical point-dried whole mounts and plastic-embedded sections, _J. Cell Biol._, 100: 1474.

Schliwa, M., and Van Blerkom, J., 1981, Structural interaction of cyto-skeletal components, _J. Cell Biol._, 90:222.

Schliwa, M., Van Blerkom, J., and Porter, K.R., 1981, Stabilization of the cytoplasmic ground substance in detergent opened cells and a structural and biochemical analysis of its composition, _Proc. Nat. Acad. Sci. USA_, 78:4329.

Schnapp, B.J., Vale, R.D., Sheetz, M.P., and Reese, T.S., 1985, Single microtubules from squid axoplasm support bidirectional movement of organelles, _Cell_, 40:455.

Small, J.V., Celis, J.E., and Isenberg, G., 1979, Aspects of cell architecture and locomotion. _in_: "Transfer of Cell Constituents into Eucaryotic Cells." Plenum Press, New York.

Small, J.V., Rinnerthaler, G., and Hinssen, H., 1982, Organization of actin meshworks in cultured cells: the leading edge, _Cold Spring Harbor Symp. Quant. Biol._, 46:599.

Spudich, J.A., Kron, S.J., and Sheetz, M.P., 1985, Movement of myosin coated beads on oriented filaments reconstituted from purified actin. _Nature_, 315:584.

Steinert, P.M., Rice, R.T., Roop, D.R., Trus, B.L., and Steven, A.C., 1983, Complete amino acid sequence of a mouse epidermal keratin subunit and implications for the structure of intermediate filaments, _Nature_, 302:794.

Stokes, G., 1976, High-voltage electron microscope study of the release of vaccinia virus from whole cells, _J. Virol._, 18:636.

Taylor, D.L., and Wang, Y.L., 1978, Molecular cytochemistry: Incorporation of fluorescently labeled actin into living cells, Proc. Nat. Acad. Sci. USA, 75:857.

Vale, R.D., Schnapp, B.J., Reese, T.S., and Sheetz, M.P., 1985, Movement of organelles along filaments dissociated from the axoplasm of the squid giant axon, Cell, 40:449.

Weber, K., and Glenney Jr., J.R., 1982, Calcium-modulated multifunctional proteins regulating F-actin organization, Cold Spring Harbor Symp. Quant. Biol., 46:541.

Weber, K., and Osborn, M., 1979, The intracellular display of microtubular structures revealed by indirect immunofluorescence microscopy, in: "Microtubules", (K. Roberts, J.S. Hyams, eds., Academic Press, New York.

Weeds, A., 1982, Actin binding proteins - regulators of cell architecture and motility, Nature, 296:811.

Wehland, J., and Weber, K., 1980, Distribution of fluorescently labeled actin and tropomyosin after microinjection in living tissue culture cells as observed with TV image intensifaction, Exp. Cell Res., 127: 397.

Wehland, J., and Willingham, M.C., 1983, A rat monoclonal antibody reacting specifically with the tyrosylated form of α-tubulin. II. Effects on cell movement, organization of microtubules, and intermediate filaments, and arrangement of Golgi elements, J. Cell Biol., 97:1476.

Wehland, J., Osborn, M., and Weber, K., 1977, Phalloidin-induced actin polymerization in the cytoplasm of cultured cells interferes with cell locomotion and growth, Proc. Nat. Acad. Sci. USA, 74:5613.

Weiss, D.G., 1982, General properties of axoplasmic transport, in: "Axoplasmic Transport in Physiology and Pathology," D.G. Weiss and O. Gorio, eds. Springer-Verlag, New York.

Wieland, T., 1977, Modification of actins by phallotoxins, Naturwissenschaften, 64:303.

Wolosewick, J.S., 1979, Microtrabecular lattice of the cytoplasmic ground substance, J. Cell Biol., 82:114.

Wolosewick, J.S., and Porter, K.R., 1976, Stereo high-voltage electron microscopy of whole cells of the human diploid line, WI-38, Am. J. Anat., 147:303.

Wolosewick, J.S., and Porter, K.R., 1979, Microtrabeculae lattice of the cytoplasmic ground substance: artifact or reality?, J. Cell Biol., 82:114.

INTERACTIONS WITHIN THE

CYTOSKELETON

Alice B. Fulton

Department of Biochemistry
University of Iowa
Iowa City, Iowa 52242, U.S.A.

INTRODUCTION

The complex meshwork of the cytoskeleton displays three forms of interaction: structural, biochemical, and functional. As our knowledge of each form grows, it becomes increasingly clear that the cytoskeleton often behaves as a single, integrated organelle.

Structural interactions are readily visible in the electron microscope, both in intact cells and in cytoskeletons. Porter and coworkers have reported extensively on the anastomosing nature of the cytoplasmic matrix, and emphasized the numerous contacts made between elements of the matrix (Porter, 1984). Schliwa has observed cytoskeletons closely under conditions in which specific cytoskeletal elements can be identified. He has observed contacts between microfilaments and microtubules, microfilaments and intermediate filaments, and between microtubules and intermediate filaments (Schliwa and Van Blerkom, 1982). The exact frequency of each of these classes of contacts depends upon the cells observed. Osborn and Weber have also observed contacts between different cytoskeletal elements, thus extending these observations to additional cell types (Webster et al., 1978).

Observations of so many structural contacts between different cytoskeletal filaments suggest that it should be possible to detect biochemical interaction as well. Several groups have studied the bifunctional nature of cross-linking proteins (Pollard et al., 1984). It is now known that the high molecular-weight "microtubule associated protein," MAP 2, can aggregate actin filaments and can lead to contacts between microtubules and microfilaments. MAP 2 can also interact with intermediate filaments so that it can cross-link microtubules with these filaments. Ankyrin is also bifunctional; it binds to spectrin, an actin binding protein, but can also bind to microtubules. Ankyrin has homology to MAP 1. It is unlikely however that the full range of bifunctional proteins has been exhausted.

The presence of structural and biochemical interactions between different members of the cytoskeleton suggests that it should also be possible to observe behavioral interaction between different filament classes. In fact, several such interactions are now known to exist. I will describe in some detail the interactive aspects of our own system, and then review other reports that indicate that ours is not an isolated phenomenon.

We observe sarcomere development in synchronized muscle cultures. In these cultures greater than 60% of the cells acquire their first sarcomeres during a 24 h window, allowing closely spaced developmental stages to be distinguished. Using these synchronized cells and an antibody specific for muscle myosin, we have observed that myosin is initially diffusely distributed in the cell and then recruited into cables of actin and myosin. These cables are the locus for the formation of the first sarcomeres in the cell, which we call sarcomere initiation.

Because of the synchrony of the system, it is possible to perturb the system for brief periods with anti-cytoskeletal drugs and observe their effects on the development of sarcomeres. We have examined both myosin recruitment and sarcomere initiation for sensitivity to cytochalasin D, an anti-microfilament drug, nocodazole, a microtubule-depolymerizing drug, and taxol, a microtubule-stabilizing drug. Because of previous reports of interactions beween these classes, we also examined cytochalasin D's effects in the presence of either nocodazole or taxol.

We observed that cytochalasin rapidly disassembled actin-myosin cables and also rapidly disassembled young sarcomeres (Denning, Lilleg, and Fulton, submitted). Nocodazole prevented recruitment of myosin into cables but did not disrupt existing cables. Taxol stimulated the formation of cables. Most striking, however, was that both nocodazole and taxol prevented cytochalasin from disrupting actin-myosin cables, although these two microtubule drugs act in opposite directions upon microtubules. When we examined sarcomere initiation, it was clear that cytochalasin was quite disruptive to young and moderately developed sarcomeres. Nocodazole caused young sarcomeres to lose their periodicity and revert to cables. Taxol slightly stimulated sarcomere initiation. As in the case of cables, both nocodazole and taxol antagonized the disruptive effects of cytochalasin.

These striking drug interactions place severe constraints on potential models for sarcomere initiation. We have developed one that emphasizes a contribution of both microtubules and microfilaments to myosin recruitment and sarcomere initiation. We propose that microtubules deliver myosin to the actin-myosin cables. Once in the cable, myosin is held there by interactions with microfilaments that terminate outside the cable. These latter microfilaments occasionally terminate on microtubules. Thus, myosin has both direct and indirect contact with microtubules. We believe this explains the effects of the drugs as follows. Nocodazole prevents recruitment of myosin by blocking the transport of myosin to the cable. Cytochalasin leads to the loss of cables by unbalancing the tension relationships within the cytoskeleton, pulling myosin out of the cables. Nocodazole antagonizes disruption by reducing the tension that can be generated by the microfilament networks. Taxol antagonizes disruption directly, by providing stable contacts to anchor the myosin in place. We believe similar interactions can account for sarcomere initiation, although we do not know whether the contacts between microtubules at sarcomere initiation are the same or different than those needed for transport to the cable.

One striking feature of the model is that it invokes only one new form of contact in the cytoskeleton: that between myosin and microtubules involved in the transport to the cable. It is known that myosin and tubulin can interact directly _in vitro_, so it is possible that there is no intermediating molecule. Aside from that, the model is conservative of new entities and is of interest for that reason. It also predicts that certain classes of molecular interactions should be observable; we are pursuing these with cross-linking experiments in which cytoskeletons are cross-linked and then immunoprecipitated to select out myosin and those molecules cross-linked to myosin. Most striking of all perhaps is that the model proposes that the sarcomere is not assembled solely as a consequence of the information content of its component proteins. Rather, that it is strongly dependent upon the balanced and appropriate interactions of other cytoskeletal elements; these other cytoskeletal interactions remain important to the sarcomere for the first several days of life. We do observe that sarcomeres become drug resistant after several days.

It would be important to know whether such interactions are significant in other cells; in fact, numerous other examples have been reported. It is reasonable to suspect functional cytoskeletal interactions whenever the effects of theses drugs show marked synergy, that is, drug effects in combination that are either greater or less than the simple additive sums of the effects of the drugs. Macrophages, for instance, are less inhibited by the combination of colchicine and cytochalasin than they are by either alone (Cheung et al., 1978). Neuroblastoma cells require a functional microfilament system to retract neurites in the absence of microtubules (Solomon and Magendantz, 1981). A delicate interaction between microfilaments and microtubules is required to maintain the circle of nuclei in multi-nucleated giant cells (Rigby and Papadimitriou, 1984). Microfilaments and microtubules interact extensively in conferring bipolarity and the elongated cell shape when fibroblasts are cultured in hydrated collagen gels (Tomasek and Hay, 1984). An intact microfilament system is required for the nuclear extrusions seen when suspended cells are treated with microtubule poisons (Mori et al., 1984).

For a few of these systems, an explicit model has been outlined in the detail which we have proposed for the muscle system. However, for all of them, it seems likely that something comparable must be invoked to explain the behavior of the cells. Finally, such interactions are not simply a consequence of obligatory interactions within the cytoskeleton; there are numerous cell behaviors which are effected by only one and not both classes of anti-cytoskeletal drugs, and there are examples known in which the drugs together do not have synergistic effects. Perhaps the last class of cytoskeletal interactions to be alluded to are the well known ones between microtubules and intermediate filaments. It has long been known that most agents which perturb microtubules will lead to a collapse of the vimentin intermediate filaments in cultured cells, suggesting extensive connections between microtubules and intermediate filaments.

All of these cases together indicate that many aspects of the cytoskeleton are interactive and depend upon the precise connections between different elements of the cytoskeleton and the precise balance of tensions between these elements. It becomes increasingly difficult to model the cytoskeleton used by cells as a homogeneous isotropic gel under static conditions. Novel methods of analysis may well prove essential for undertstanding these aspects of cytoskeletal function.

REFERENCES

Cheung, H. T., Cantarow, W. D., and Sundharadas, G., 1978, _Exp. Cell Res._, 111:95.

Mori, Y., Akedo, H., Matsuhisa, T., Tanigaki, Y., and Okada, M., 1984, _Exp. Cell Res._, 153:574.

Pollard, T. D., Selden, S. C., and Maupin, P., 1984, _J. Cell Biol._, 99:33s.

Porter, K. R., 1984, _J. Cell Biol._, 99:3s.

Rigby, P. J. and Papadimitriou, J. M., 1984, _J. Pathologys_, 143:17.

Schliwa, M. and Van Blerkom, J., 1982, _J. Cell Biol._, 90:222.

Solomon, F. and Magendantz, M., 1981, _J. Cell Biol._, 89:157.

Tomasek, J. J. and Hay, E. D., 1984, _J. Cell Biol._, 99:536.

Webster, R. E., Osborn, M., and Weber, K., 1978, _Exp. Cell Res._, 117:47.

ON THE PHYSICAL PROPERTIES AND POTENTIAL ROLES OF

INTRACELLULAR WATER

James S. Clegg*

Laboratory for Quantitative Biology
University of Miami
Coral Gables, FL 33124 USA

INTRODUCTION

Compared to our current understanding of the properties and roles of cellular macromolecules, we know very little indeed about water-the most abundant and pervasive molecule in cells. It is inescapable that water must be involved, directly or indirectly, in all forms of metabolic organization. That is so because it is an indispensable component of virtually all macromolecules, a substrate or product of many metabolic reactions, and obviously provides a dielectric continuum for diffusion and other translocation processes. Moreover, we can expect the aqueous phases of cells to operate importantly in the hydrophobic interactions involved with assembly processes, such as enzyme-enzyme and enzyme-cytomatrix associations.

This paper will be concerned chiefly with the "bulk aqueous phases" of animal cells, the goal being to summarize what we know about them and how they might be involved in metabolic organization. I will not deal with the very important matter of macromolecular hydration and the catalytic participation of water at the immediate interface between enzyme and the surrounding bulk medium. These considerations have been taken up by others, whose work has been reviewed by Welch et al. (1982). I will attempt to evaluate the properties and roles of intracellular water more distance (> 6 Å) from the surfaces of macromolecules and other ultrastructural components; it is this I refer to by the phrase "bulk aqueous phase". Obviously, an important consideration here will be to inquire into the solute composition of such phases: are they very concentrated solutions containing a wide variety of dissolved, freely diffusing macromolecules, metabolites, inorganic ions, and so forth, as many believe? I cannot expect to provide a definitive listing of such solutes, but it will be worthwhile at least to evaluate what evidence exists on the subject.

---

*Present address: University of California, Bodega Marine
                   Laboratory, Bodega Bay, California 94923.

A BRIEF HISTORICAL INTERLUDE

It seems appropriate to inquire into the development of thought concerning the aqueous intracellular environment. Historical accounts of science usually attempt to provide a specific date or dates, involving a specific researcher(s), that mark the birth of an idea or concept. For instance, in 1897 Büchner showed that alcoholic fermentation occurred in cell-free yeast extracts, providing an important and new experimental approach. (That discovery also gave birth to the idea of "soluble enzymes", a topic that pervades this volume in one form or another). How about cell water? When did its presence become documented, and by whom? The answer seems lost in antiquity, long before the dawn of science. It has "always" been recognized that water and life are inextricably linked. This familiarity seems to have bred a form of contempt: "right, water is vital for life...now let's go on to more important things". Do I protest too much? I think not; as I write, the latest issue of Scientific American, fresh from the press (October, 1985), is devoted to "The Molecules of Life". There are 11 articles, none of them dealing with cellular water. Albert Szent-Györgyi (1971) was apparently correct: "Biology has forgotten water, or never discovered it."

I believe most biologists assume that almost all intracellular water is simply a solvent, being no different than water in aqueous solutions. The assumption is commonly not even stated, let alone recognized. This position seems to have been arrived at through tradition, and not by experiment: there is absolutely no justification for that assumption, and quite good reason to conclude that it is not only incorrect, but misleading.

But not all have slipped into such a traditional viewpoint. Gilbert Ling is perhaps the most active contemporary opponent of the traditional view of cell water, and he has compiled an excellent historical account of the subject (Ling, 1985). His book summarizes his research over the last 30 years, and describes a detailed model of his conception of cell structure and function. While one may disagree with all the details of his "association-induction hypothesis", I believe he makes a solid case for the critical and varied importance of cell water. Others have also provided evidence that intracellular water exhibits extraordinary properties compared to the pure liquid, and these views have been described in several books (Keith, 1979; Drost-Hansen and Clegg, 1979; Franks and Mathias, 1982). While I find the collective weight of this evidence to be compelling, the fact remains that the scientific community, in general, remains unconvinced or, perhaps more likely, unaware. It is fair to accept the admonishment that the burden of proof rests upon those who would counter the prevailing wisdom.

In the next section I will summarize a biological system which provides several advantages to those who would explore the nature of cell water.

ARTEMIA CYSTS: A USEFUL MODEL SYSTEM

Artemia is a primitive crustacean found in natural brine pools and solar salt operations world wide. Its biology is well known (Bagshaw and Warner, 1979; Persoone et al. 1980). Under certain conditions the adults produce encysted embryos (cysts) which enter a period of dormancy and are commonly blown on shore where they are harvested, in some cases in ton amounts. Figure 1 is a diagrammatic representation of a cyst from the San Francisco Bay population. Figure 2 illustrates the ultrastructure of two adjacent cells in a cyst, revealed by transmission electron microscopy. The structure is typical of yolky embryonic cells: a large nuclear-to-cytoplasmic volume ratio, abundant glycogen deposits, and relatively scanty cytoplasmic membrane systems (such as endoplasmic reticulum). Yolk platelets are abundant but are not seen in this particular section. There is nothing extraordinary about the morphology of these cells at this level of resolution, a point to which I will return. Each cyst contains about 4000 such cells surrounded by a complex shell, most of which can be chemically removed. Further, there is virtually no extracellular space. Hence, results from experiments that probe cell water in this system can be directly interpreted in terms of intracellular water. These cells have the natural ability to undergo virtually complete dehydration, in a fully reversible way, allowing for experimental situations that normally are not possible. Finally, this system is reasonably well described with respect to ultrastructure, biochemistry and experimental manipulation (see Bagshaw and Warner, 1979; the three volumes edited by Persoone et al. 1980).

SUMMARY OF THE PHYSICAL PROPERTIES OF WATER IN ARTEMIA CYSTS

Studies on cell water are made complicated by the rather elaborate methodology that must commonly be employed. Furthermore, each technique requires that scientists skilled in that area be involved not only with the experiments, but also with the interpretation of the data obtained. Therefore, interdisciplinary cooperative research programs are called

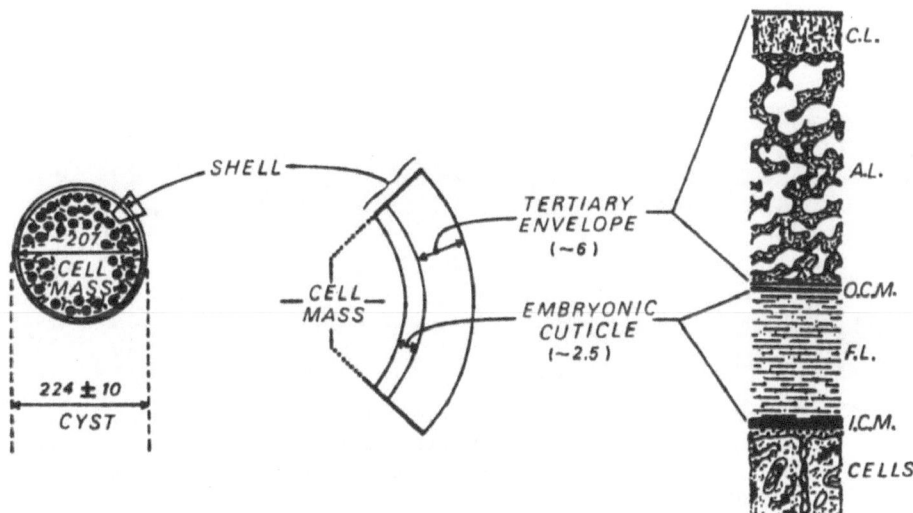

Fig. 1. Diagrammatic representation of an Artemia cyst (left) and its surrounding shell (right). All numbers are in micrometers and refer to cysts from the San Francisco Bay Population at maximum water content. Abbreviations: C.L. (cuticular layer); A.L. (alveolar layer); O.C.M. (outer cuticular membrane); F.L. (fibrous layer); I.C.M. (inner cuticular membrane).

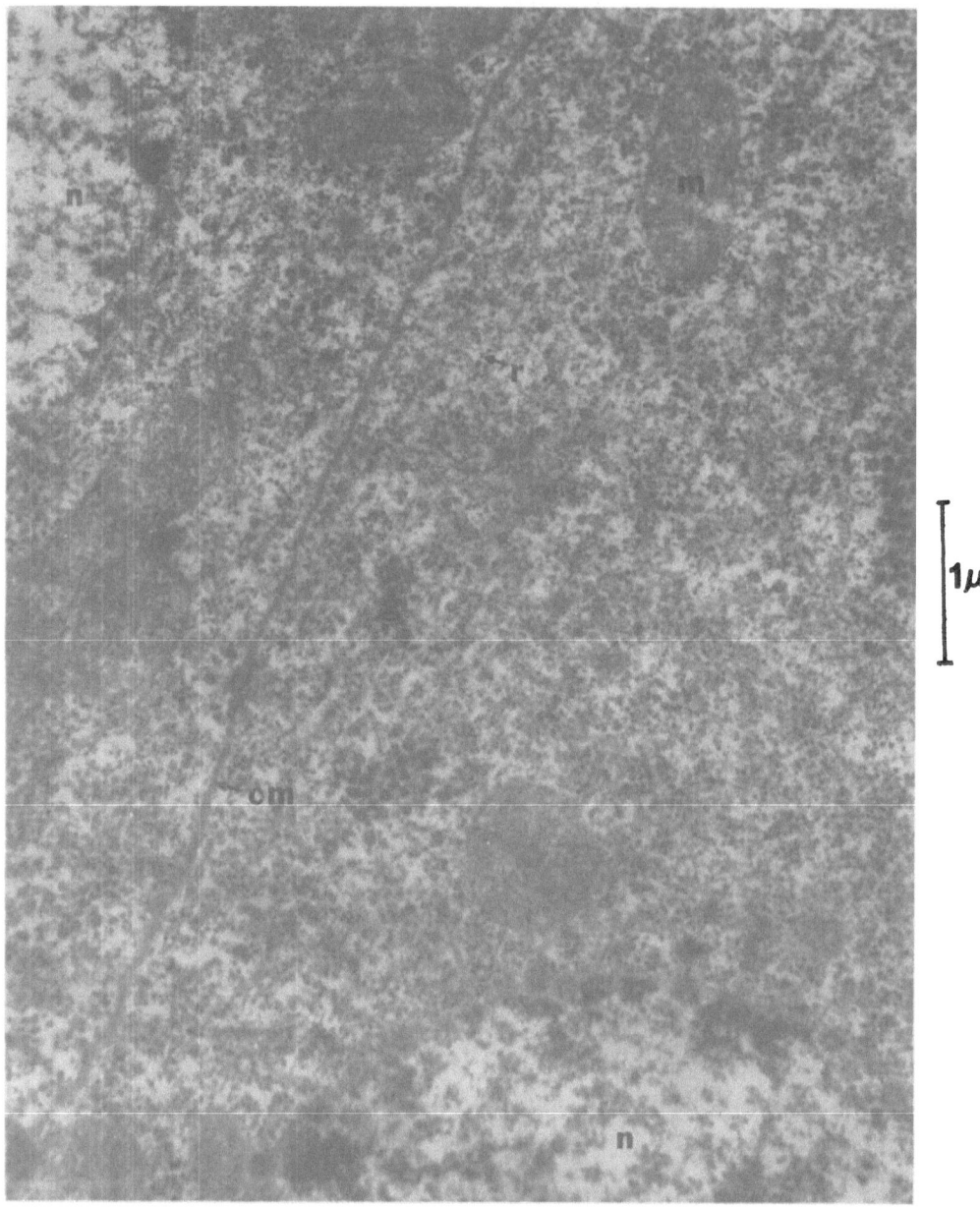

Fig. 2. Transmission electron photomicrograph of two adjacent cells in
an __Artemia__ cyst at maximum water content. (n = nucleus; cm =
cell membrane; m = mitochondrion; g = glycogen granule; r =
ribosome).

for, and I have been fortunate in enlisting the help of a number of laboratories in this regard. Before describing some of our results, I will list the techniques used and my colleagues:

a. Nuclear Magnetic Resonance (NMR) Spectroscopy: Drs. C. F. Hazlewood (Baylor College of Medicine), H. E. Rorschach (Rice University), and their students and associates.

b. Quasi-elastic Neutron Scattering (QNS): as above, and Drs. R. M. Nicklow and N. Wakabayashi (Oak Ridge National Laboratory).

c. Microwave Dielectrics (MD): Drs. E. H. Grant, R. J. Sheppard, S. Szwarnowski, V. E. R. McClean, and their associates (Queen Elizabeth College and King's College, University of London).

d. Differential Scanning Calorimetry (DSC): Dr. C. A. J. Hoeve (Texas A&M University).

e. Density measurements: Dr. W. Drost-Hansen (University of Miami).

Table 1 summarizes results on _Artemia_ cysts at their maximum water content (about 1.4g H_2O/g dry cysts, at which the _cells_ contain about 1.65g H_2O/g dry weight, roughly the water content of mammalian liver cells). Brief comment will be made on each result, and then an integrated interpretation will be advanced.

NMR. This technique can be used to probe the motion of water protons by a "pulse" method, the details of which have frequently been described (see Hazlewood, 1979; Beall & Hazlewood, 1983). It has been applied to dozens of cells and tissues during the last 15 years without much resolution. That is chiefly because it is necessary to interpret the data within the confines of a model, and several have been constructed. Three parameters can be obtained from pulse NMR experiments that provide information about the motion of cell water: two "relaxation times", T_1 and T_2, and the self-diffusion coefficient, D. Simply put, the extent to which these parameters deviate from pure water will reflect altered motion of cell water. Measurements of T_1 and T_2 for pure water are in the vicinity of 3000 msec, but for water in a wide variety of cells and tissues the T_1s are roughly 150-1000 msec and T_2s are 20-250 msec. _Artemia_ is no exception (Table 1). These reductions have been interpreted to mean that all of the cell water exhibits reduced motion or, that all cell water (~95%) has the same motion as pure water. In the latter models reductions in T_1 and T_2 are commonly explained by the rapid exchange of "bulk" water molecules with a small "tightly bound" fraction (about 5% of the total), whose existence greatly and disproportionately influences the relaxation times measured. Variations on this theme have also been proposed.

These various models carry different predictions about the self-diffusion coefficient which, unlike relaxation times, is a simple average of the total cell water. For example, models that interpret T_1 and T_2 reductions as the result of "fast-exchange" also must predict that D for cell water should be nearly the same as pure water. Measurements of D in cells reveal 2-7 fold reductions by NMR, as we observe for _Artemia_ cells (Table 1). Fast-exchange models explain these reductions by obstruction and compartmentation effects, which are indeed plausible because NMR diffusion coefficients are ordinarily measured over distances of the order of, or greater than, the cell diameter. Therefore, the critical test

Table 1. Results on the Physical Properties of Water in _Artemia_ Cysts, and of Pure Water, Using a Variety of Techniques.

Parameter[+]	A. Artemia Water[*]	B. Pure Water	B/A
NMR:			
T_1 (msec)	275	3000	11
T_2 (msec)	53	1800	34
D ($10^{-5}cm^2$/sec)	0.4	2.41	6
QNS:			
D ($10^{-5}cm^2$/sec)	0.8	2.40	3
τ (ps)	4	1	0.25
MD:			
ε'(2GHz)	40	78	1.95
τ (ps)	10-25	8	0.8-0.3
α	0.46	0.02	0.04
M-V:			
ρ (g/cm^3)	0.97	1.00	1.03
DSC:			
C_p (cal/g/deg)	0.874	1.00	1.14

[*]All data are for cysts at maximum water content (~1.4g H_2O/g dry mass)

[+]NMR (nuclear magnetic resonance); QNS (quasi-elastic neutron scattering); MD (microwave dielectrics); M-V (mass-volume studies); DSC (differential scanning calorimetry). Units for τ are picoseconds (10^{-9} second), and α is a dimensionless measure of deviation from simple Debye relaxation ($\alpha = 0$). Values of the dielectric parameters τ and α were obtained from analysis of the dispersion between 0.8 and 70 GHz; most of the cyst water relaxes at much lower frequencies. These data are taken from Seitz et al. 1980, 1981 (NMR); Trantham et al. 1984 (QNS); Clegg et al., 1982, 1984a,b (MD); and Clegg, 1984a,b (M-V).

of these models is to measure D over very short distances (and times), thereby not allowing the water molecules to encounter barriers. However, that is very difficult to do using current NMR technology. Thus, NMR studies alone would not appear to provide a compelling answer.

QNS. In principle, this method should distinguish between the models advanced to explain NMR results. That is because it gives information on the diffusive motion of water over periods of about 10^{-12} second, and distances of an angstrom or two. Interpretation is not free from difficulty, but the limitations are less than those of NMR. A major reason why this technique has not been used to resolve the controversy is that the sample must remain close-packed and sealed in the measuring cell for several days (and usually about a week or more) to obtain sufficient data for analysis. Most living systems cannot tolerate such treatments, but _Artemia_ encounters no difficulty whatsoever; fully hydrated cysts under QNS conditions for 13 days exhibit no significant decrease in viability, an extraordinary capability. Inspection of Table 1 shows that the self-diffusion coefficient of water (D) obtained from QNS measurements on _Artemia_ cysts is 3 times slower than pure water. We advanced these conclusions: first, as fast-exchange models have predicted, some of the reduction in the diffusive motion of cyst water evaluated by NMR does indeed appear to be due to obstruction and similar effects, being roughly one-half of the 7-fold reduction; second, even over distances of a few angstroms there is still a 3-fold reduction in D. Because the latter cannot be due to anything but the motions of the water molecules themselves, these data provide good evidence that the diffusion of at least a large fraction of the water in these cells, possibly all of it, differs markedly from that of water in dilute aqueous solutions. That being the case, T_1 and T_2 reductions cannot easily be accounted for by fast-exchange models which are, therefore, rejected as inapplicable for this system.

MD. Microwave dielectric relaxation studies also provide a means of evaluating the motion of water in cells. However, like the NMR work, interpretation is difficult. Also, to obtain unambiguous data it is necessary to make measurements over the frequency range of water relaxation (roughly 1 to 100 GHz), which poses some serious technical problems. Almost all of the published data have been obtained at frequencies below about 10 GHz, requiring considerable extrapolation for interpretation. Thus, dielectric studies do not, in and of themselves, provide a direct and unambiguous description of the behavior of cell water.

Nevertheless, the NMR and QNS results on _Artemia_ cysts predict that the dielectric relaxation of their water should differ from that of water in dilute aqueous solutions. That result was obtained from studies over the frequency range of 0.8-70 GHz (Table 1). The dielectric relaxation time (τ) of cyst water is longer (most of it being much longer), and the average permittivity (ε' , dielectric constant) appears to be considerably lower than that of pure water, roughly by one-half. Although the caveat must be made that the use of highly over-simplified mixture equations weakens the certainty of our interpretations, these dielectric results remain consistent with the NMR and QNS observations: most of the water in the cells of _Artemia_ cysts exhibits rotational and translational motions that differ appreciably from those of the pure liquid. It is quite possible that all of it behaves that way.

Density. Very few studies have been made on the density of cell water in any system (see Pocsik, 1967; Hansson-Mild and Lovtrup, 1981). The problem here is to sort out the density of the water from the complex nonaqueous components of cells. That cannot be done with any confidence at present; however, we have carried out the appropriate measurements on _Artemia_ cysts (Clegg, 1984a) and summarize here our "best guess" as to what we think they mean. Using high-precision sedimentation-velocity measurements to obtain cyst density, as a function of cyst hydration, the method of slopes and intercepts was used to extract the contributions of cyst water (the "partial density" of cyst water). The result is that the average density of water in these cells is about 3% less than pure water (Table 1), the experimental error being only about 1%. Such a change in density must reflect a marked change in water structure. We interpret this to mean that cell water, on average, has more extensive hydrogen bonding than the pure liquid.

DSC. The result shown in Table 1 for <u>Artemia</u> cysts is somewhat preliminary. Nevertheless, it does appear that the partial heat capacity (C_p) of water in this system (at maximum hydration) is significantly less than that of pure water at 25°C. As always, interpretation is fraught with difficulty. I think it is possible that such a result could reflect fewer degrees of freedom for cyst water, perhaps due to enhanced hydrogen bonding as indicated by the density measurements. But there are several alternative explanations.

A few words in summary: the water in <u>Artemia</u> cysts definitely exhibits a marked reduction in both its translational and rotational motion compared to pure water, and these reductions appear to be the result of increased hydrogen bonding; the possible reduction in the permittivity suggests that at least a large fraction of cyst water might exhibit solvent properties that differ from those of ordinary dilute aqueous solutions. We find very little, if any, evidence that cyst water behaves like the latter.

An important matter should be dealt with next: do results on <u>Artemia</u> cysts have general application? This system is, indeed, "unusual" in many ways, notably in being able to dehydrate reversibly. That question is not easily answered. I have argued elsewhere that such data obtained from the <u>fully hydrated</u> cysts will, more likely than not, apply to animal cells in general (Clegg, 1979, 1982, 1984a,b). In this regard I believe the most significant feature of work on <u>Artemia</u> concerns a comparison of NMR and QNS results. Many dozens of different tissues and cells, from a variety of taxa, have greatly reduced T_1 and T_2 values, with more modest reductions in D as measured by NMR. The cysts do not deviate from this range, and they do not exhibit bizarre or "unusual" behavior when viewed by NMR. Thus, it seems reasonable to suppose that if these dozens of "usual" cells and tissues could be studied by QNS, that they would, like the cysts, reveal a comparable outcome.

Having some reason to believe that intracellular water is not mimicked by that in dilute solutions, we can inquire into the reasons for that behavior: what is the water experiencing in cells that alters its properties?

CELL WATER AND THE CYTOMATRIX

Addressing the question just posed requires that we know something about the intracellular environment, the nonaqueous architecture, and dissolved solutes which the water "sees" and with which it may interact. As is the case for the properties of cell water, there is by no means uniform agreement on this matter. My perception is that most view the cell as a complex system of various organelles and cytoskeletal elements (having relatively few structural interconnections between them), and that in between all this architecture there exists a crowded aqueous solution highly concentrated with respect to macromolecules and the location of the "soluble metabolic pathways". I believe there is ample experimental evidence to reject that conception of the eucaryotic cell and have gone to some effort to compile that evidence (Clegg, 1984b). I adopt here the image presented by Porter and his associates as being a much better approximation of what cells are like than the "alphabet soup" description just mentioned and considered by many to be the prevailing wisdom (see Porter, et al. 1983; and the article by Porter, in this volume). That representation of the eucaryotic cell considers virtually all of the architecture to be interconnected (the "microtrabecular lattice"), with the intervening aqueous phase(s) being extremely dilute with respect to dissolved macromolecules. I hasten to point out that I do not envision a sharp, rigid discontinuity between the "architecture" and the surrounding "aqueous phase". On the contrary, we can suppose that the interactions between these components must be dynamic and substantial. Electron photomicrographs, at best, represent a snapshot of those relationships. Nevertheless, to me, the MTL represents the most plausible account we have at present and, importantly, it tells us that cells contain a formidable amount of surface which the water must experience in its rapid motion throughout the cell. Just how extensive this interface actually is has been estimated by Gershon et al. (1982): for a cell 16 μm in diameter, having a 10 μm nuclear diameter, the cytoplasmic cytomatrix alone represents about 50,000-100,000 μm^2, truly an astonishing figure. Elsewhere (Clegg, 1984b), I have calculated that a monolayer of water on this surface would consume about 2-4% of the total cytoplasmic water. We now arrive at a crucial consideration: how far from such surfaces can we expect the water to be influenced? If we know that figure, we can make some useful approximations; we will have an ultrastructural basis upon which to interpret the experimental data on the properties of cell water.

SURFACES AND THE PROPERTIES OF WATER

Classical colloid and surface chemistry indicates that all surface influence on the surrounding water structure decays to near zero at about 6 Å or so from the surface (i.e., two molecular layers of water). However, that view has been seriously challenged by the elegant work of Israelachvilli, Ninham, Pashley and their asociates (see Israelachvilli and Pashley, 1982) and Parsegian and colleagues (see Lis et al. 1982). Most recently, Pashley et al. (1985) demonstrated that, for planar, non-charged hydrophobic surfaces, the influence extends much farther, in excess of 100 Å from the surface with an exponential decay length of about 14 Å. The situation is less dramatic for "mixed" surfaces, or strictly polar ones. Nevertheless, these findings provide us with a basis for estimating the influence of intracellular surfaces upon the surrounding aqueous phase. Thus, if we accept an effective distance of 60 Å over which water can be perturbed by intracellular surfaces, then well over one-half of cytoplasmic water will be so effected (see Clegg, 1984b). Even though we cannot expect these estimates to be precise, they do suggest a basis for the results obtained with Artemia cysts.

There are no observations for Artemia cells that are comparable to those of Gershon et al. (1982). However, the results of water-sorption isotherms provide one alternative approach (Clegg, 1974). Figure 3 shows a B.E.T. plot (Brunauer et al. 1938; Brunauer, 1945) for water sorption by severely dried cysts. From such data one can calculate "monolayer coverage" (a_1, Fig. 2), which for Artemia cysts is 0.057 g H_2O/g dry mass. It is easy to calculate the area that would be covered by this amount of water, the "specific surface area" (S):

$$S = N \Theta a_1 / v_o$$

where N is Avogadro's number, Θ the area covered by a water molecule, a_1 is "monolayer coverage", and v_o the molar volume of water vapor (STP). S for the cells in 1g dry wt. of cyst is 128 m^2. Knowing the number of cysts per gram dry weight (2.8×10^5) and the number of cells per cyst (4000), we calculate the number of cells/g dry cysts: 1.1×10^9. Using S, it can now be calculated that the specific surface area per cell is about

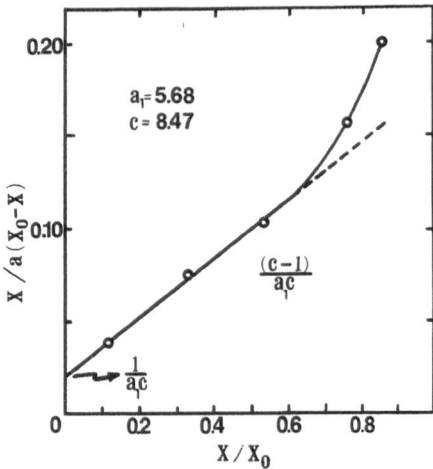

Fig. 3. B.E.T. plot using cyst sorption isotherm data (25°C). X is the vapor pressure of a saturated salt solution over which cysts are equilibrated to some hydration value, a (g H_2O/100g dried cysts) and X_0 is the vapor pressure of pure water. The intercept and slope are used, as indicated, to extract the B.E.T. constants, c and a_1 (monolayer coverage).

110,000 μm^2. Although it is clear that the surface area estimated in this fashion is by no means comparable to that estimated by Gershon et al. (1982) using computer analysis of electron photomicrographs, the value is large enough to convince us that the cells of Artemia cysts also contain an impressive surface area. We deduce that it is these surfaces that produce the alterations in water structure and dynamics that we have measured (Table 1).

What difference does it make if cell water has properties unlike those of an ordinary solution? Several answers come to mind:

1. Much of what is known about macromolecular function in cells is based on data obtained in vitro, almost always in highly dilute aqueous solutions. If intracellular water differs from that in test tubes, as some of us believe, then information obtained in vitro may not allow us to construct (or better "reconstruct") an accurate description of these molecules and their activities within cells.

2. Direct interactions between macromolecules and their surrounding water appear to be even more important than has commonly been believed. Welch et al. (1982) have recently reviewed the abundant evidence for this, and their analysis makes it very likely that water plays subtle but important roles in metabolism: to understand those we must know the details of the aqueous microenvironment in which most of this activity occurs.

3. Available evidence suggests that the solvent properties of at least a large fraction of the total cell water, notably in cytoplasm, differ from those of ordinary aqueous solutions (see Ling, 1985; Horowitz, this volume). At least some contribution to the uneven distribution of certain solutes across the plasma membrane, as well across membranes within cells (organelles), could arise from such

"solvent" differences. Thus, small metabolites might "partition" between various intracellular aqueous phases, a possibility generated by the work of Garlid (1979) on mitochondria. Even protein distribution within cells may be influenced in this fashion. A speculative "model" on the organization of enzymes in the aqueous cytoplasm includes the possibility that a loose association of enzymes with the cytomatrix may be driven by water interactions involving their respective surfaces (Clegg, 1979, 1982a,b), similar to those involved in association by hydrophobic interactions.

4. Assembly-disassembly processes are influenced by the properties of the aqueous phase within which they occur. Such mechanisms should be important to enzyme-enzyme associations and the dynamic turnover of the cytomatrix and possibly other cell structures.

5. A great many of the molecular interactions in cells involve ionic bonds which are sensitive to the dielectric properties of the surroundings. Thus, the possibility that the permittivity of cell water is reduced, relative to dilute solutions, may not be trivial.

6. A reasonably good correlation exists between modifications in the cytomatrix and changes in the amount and properties of cell water, both of which commonly, although not always, accompany cell transformation by viruses or carcinogens. While that may be fortuitous, it is notable that the usual observation is a reduction in cytomatrix surface area (see Brinkley, 1982) and an increase in the amount of cell water that has "bulk-like" properties (see Hazlewood, 1979; Beall et al. 1982; Beall and Hazlewood, 1983). That is consistent with the proposed relationship between the cytomatrix and the properties of its surrounding aqueous environment. It has not escaped our attention that many of the metabolic changes accompanying the transformation process are associated with "soluble" enzymes which, in the view of some of us, are not really "soluble" at all but are instead part of the water-cytomatrix system.

SOME OLD EXPERIMENTS: THE "DEVAUX EFFECT"

I conclude this article by summarizing some work published by R. Chambers and M. J. Kopac, performed at the Marine Biological Laboratory, Woods Hole, during the late 1930s. These elegantly simple observations tell us a great deal about cytoplasmic organization and, as a result, indicate bounds for the construction of models on metabolic organization.

Almost 50 years ago Kopac and Chambers injected droplets of oil into various echinoderm eggs and observed the result under a variety of experimental conditions. Figure 4 is useful in summarizing their findings, which were reviewed by Chambers (1940) who cites the specific papers to be considered here. If the cytoplasm of injected eggs was not injured the oil drops remained spherical; however, if the cytoplasm was intentionally damaged the droplet would then undergo a surface crinkling, known as the "Devaux effect", within seconds. It was pointed out that this crinkling occurs when proteins are adsorbed at an oil-water interface at, and above, monolayer concentrations (Fig. 4). Kopac and Chambers proposed that the absence of the Devaux-effect in undamaged cells reflected the absence of significant concentrations of diffusible proteins of the size that would adsorb onto the droplets. Moreover, Kopac devised a way to evaluate protein adsorption on the droplets in amounts well below the criticial monlayer value. That was accomplished by the "drop retraction" method depicted diagrammatically in Figure 4.

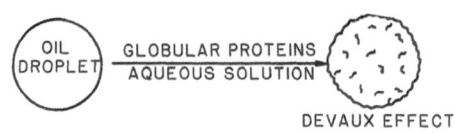

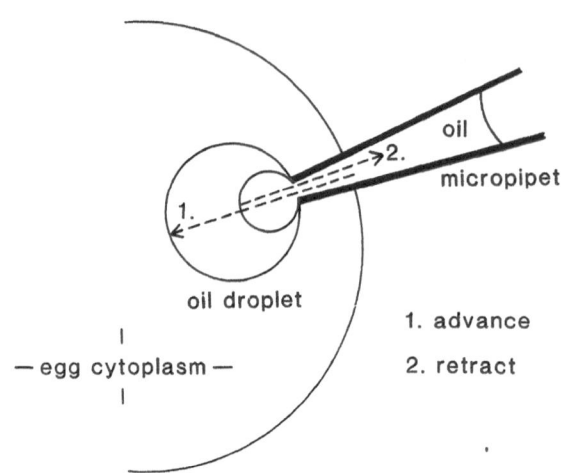

Fig. 4   Summary description of the oil droplet injection experiments of
          Kopac (1938) and Chambers (1940). See text for discussion.

An oil droplet was injected to a certain size (arrow 1 in Fig. 4) and
then, after suitable time, the droplet volume was gradually diminished by
withdrawing oil back into the pipet (arrow 2 in Fig. 4). If some protein
had been adsorbed, but less than a monolayer, then the Devaux effect
would in principle suddenly occur when the volume of the droplet was
reduced to the critical level (a decrease in surface area over which
monolayer adsorption could occur). Kopac calculated that an oil droplet
10 μm in diameter, about 3×10^{-6} cm^2 surface area, would require only
about 3×10^{-6} μg of protein for monolayer coverage. However, the Devaux
effect was not detected in normal cells by the oil retraction method down
to very small droplet sizes. No mention of the total protein of these
eggs is made, but we can suppose it to be roughly 0.1 μg/cell. Thus,
assuming Kopac could _not_ have detected one-tenth of a monolayer by the
drop-retraction method, we can estimate that only about 3×10^{-4}% of the
total egg protein was available for adsorption (i.e., diffusing in three
dimmensions). The calculation takes the egg to be 110 μm in diameter, a
water content of 75%, and the protein to be 20% of the wet weight, and of
a density of 1.3 g/cc.

    It is difficult to understand why the oil droplet would not quickly
adsorb proteins, if these were present in solution in the cytoplasm of
healthy cells at concentrations of 10-20%, as the prevailing wisdom
advocates. The significance of their findings did not escape Kopac and
Chambers. When asking what such experiments tell us about protoplasmic
structure, Kopac (1938) wrote:

        "The first and most obvious one is that proteins do
    not accumulate on experimentally introduced oil-water
    interfaces while protoplasm is intact. Apparently, proteins

are bound in the living cell and are released only upon cytolysis."

Chambers (1940) extended this conclusion by noting:

"This suggests that proteins are not freely diffusible and adsorbable in protoplasm, and that, therefore, these proteins may be bound together to form some kind of continuous phase."

Chambers' prophetic statement anticipated the image of the eucaryotic cell that Porter and his colleagues have energetically advanced — the microtrabecular lattice — and the one we have adopted for the interpretation of our studies on intracellular water and the properties of the aqueous phases of cells. These early studies obviously bear on the topic of this Workshop and its published proceedings. Although they do not tell us that metabolic organization exists, they do predict that models of metabolic organization and regulation based on simple solution chemistry are bound to fail.

ACKNOWLEDGEMENTS

The skilled assitance of Vivian Roe in the preparation of this manuscript is appreciated. Supported by a grant from the National Science Foundation (PCM 821733).

REFERENCES

Bagshaw, J.C. and Warner, A.H., eds., 1979, "Biochemistry of Artemia Development," University Microfilms International, Ann Arbor, Michigan.

Beall, P.T., Hazlewood, C.F. and Rutzkey, L.P., 1982, NMR relaxation times of water protons in human colon cancer cell lines and clones, Cancer Biochem. Biophys., 6:7.

Beall, P.T. and Hazlewood, C.F., 1983, Distinction of normal preneoplastic and neoplastic states by water proton NMR relaxation times, in "Nuclear Magnetic Resonance Imaging," C.L. Partain, A.E. James, F.D. Rollo, and R.R. Price, eds., W.B. Saunders Co., London.

Brinkley, B.R., 1982, The cytoskeleton: an intermediate in the expression of the transformed phenotype in malignant cells, in "Chemical Carcinogenesis," C. Nicolini, ed., Plenum, N.Y.

Brunauer, S., 1945, "Physics of Adsorption of Gases and Vapors," Oxford University Press.

Brunauer, S., Emmett, P.H., and Teller, E., 1938, Adsorption of gases in multimolecular layers, J. Amer. Chem. Soc. 60:309.

Chambers, R., 1940, The micromanipulation of living cells, in: "The Cell and Protoplasm," F.R. Moulton, ed., Publ. 14, AAAS, Washington, D.C.

Clegg, J.S., 1974, Interrelationships between water and cellular metabolism in Artemia cysts. I. Hydration - Dehydration from liquid and vapor phase, J. Exp. Biol., 61:291.

Clegg, J.S., 1979, Metabolism and the intracellular environment: the vicinal-water network model. in: "Cell Associated Water," W. Drost-Hansen and J.S. Clegg, eds., Academic Press, New York.

Clegg, J.S., 1982a, Intrerrelationships between water and cellular metabolism in Artemia cysts: IX. Evidence for the organization of soluble cytoplasmic enzymes, Cold Spring Harbor Symp. Quant. Biol., 46:23.

Clegg, J.S., 1982b, Alternative views on the role of water in cell function. in "Biophysics of Water," F. Franks and S. Mathias, eds., John Wiley & Sons, New York.

Clegg, J.S., 1984a, Interrelationships between water and cellular metabolism in Artemia cysts. XI. Density measurements, Cell Biophys., 6:153.

Clegg, J.S., 1984b, Properties and metabolism of the aqueous cytoplasm and its boundaries, Amer. J. Physiol., 246:R133.

Clegg, J.S., Szwarnowski, S., McClean, V.E.R., Sheppard, R.J., and Grant, E.H., 1982, Interrelationships between water and cell metabolism in Artemia cysts. X. Microwave dielectrics, Biochem. Biophys. Acta, 721:458.

Clegg, J.S., McClean, V.E.R., Szwarnowski, S., and Sheppard, R.J., 1984, Microwave dielectric measurements on Artemia cysts at variable water content, Phys. Med. Biol., 29:1409.

Drost-Hansen, W. and Clegg, J.S., eds., 1979, "Cell-Associated Water," Academic Press, New York.

Franks, F. and Mathias, S. eds., 1982., "Biophysics of Water," Wiley, New York.

Garlid, K.D., 1979. Aqueous phase structure in cells and organelles, in: Cell-Associated Water," W. Drost-Hansen and J.S. Clegg, eds., Academic Press, New York.

Gershon, N., Porter, K.R. and Trus, B., 1982, The microtrabecular lattice and the cytoskeleton: their volume, surface area and the diffusion of molecules through it, J. Cell Biol., 95:406a.

Hansson-Mild, K. and Lovtrup, S., 1981, The density of water in amphibian eggs, J. Exp. Biol., 91:361.

Hazlewood, C.F., 1979, A view of the significance and understanding of the physical properties of cell water, in: "Cell-Associated Water," W. Drost-Hansen and J.S. Clegg, eds., Academic Press, New York.

Israelachvilli, J.N., and Pashley, R.M., 1982, Double layer, van der Waals and hydration forces between surfaces in electrolyte solutions, in: "Biophysics of Water," F. Franks and S. Mathias, eds., John Wiley & Sons, N.Y.

Keith, A.D., ed., 1979, "The Aqueous Cytoplasm," Marcel Dekker, New York.

Kopac, M.J., 1938, The Devaux effect at oil-protoplasmic surfaces, Biol. Bull., 75:351.

Ling, G.N., 1985, "In Search of the Physical Basis of Life," Plenum Press, New York.

Lis, L.J., McAlister, M., Fuller, N., Rand, R.P. and Parsegian, V.A., 1982, Interactions between neutral phospholipid bilayer membranes, Biophys. J., 37:657.

Pashley, R.M., McGuiggan, P.M., Ninham, B.W., and Evans, D.F., 1985, Attractive forces between uncharged hydrophobic surfaces: direct measurements in aqueous solution, Science, 229:1088.

Persoone, G., Sorgeloos, P., Roels, O., and Jaspers, E., eds., 1980, "The Brine Shrimp, Artemia," Vols. 1-3, Universa Press, Wettern, Belgium.

Pocsik, S., 1967, Density of water in skeletal muscle, Acta Biochim. Biophys. Acad. Sci. Hung., 2:149.

Porter, K.R., Beckerle, M., and McNiven, M., 1983, The cytoplasmic matrix, Mod. Cell Biol., 2:259.

Seitz, P.K., Hazlewood, C.F., and Clegg, J.S., 1980, Proton magnetic resonance studies on the physical state of water in Artemia cysts, in "The Brine Shrimp, Artemia," Vol. 2, G. Persoone, ed., Universa Press, Wettern, Belgium.

Seitz, P.K., Chang, D.C., Hazlewood, C.F., Rorschach, H.E., and Clegg, J.S., 1981, The self-diffusion of water in Artemia cysts, Arch. Biochem. Biophys., 210:517.

Trantham, E.C., Rorschach, H.E., Clegg, J.S., Hazlewood, C.F., Nicklow, R.M. and Wakabayashi, N., 1984, The diffusive properties of water in _Artemia_ cysts as determined from quasi-elastic neutron scattering spectra, _Biophys. J._, 45:927.

Welch, G.R., Somogyi, B., and Damjanovich, S., 1982, The role of protein fluctuations in enzyme action: a review, _Prog. Biophys. Molec. Biol._, 39:109.

DIFFUSION OF A SMALL MOLECULE IN THE AQUEOUS

COMPARTMENT OF MAMMALIAN CELLS

Andrea M. Mastro and David J. Hurley

Department of Molecular and Cell Biology
Pennsylvania State University
University Park, PA   16802

INTRODUCTION

Cell biologists, biochemists, biophysicists, and other life
scientists who study the mammalian cell are in general concerned with the
basic questions, "What is the cell made of?" "How does it work?" In
order to answer these questions, scientists are continuing to develop new
techniques and approaches which are revealing a variety of information.
It is interesting that some of this information remains buried, only to
be rediscovered a decade or so later as new or even old models of cell
structure and function are proposed (see Conklin, 1940, for an historical
review).

One thing is certain: recent models of cell structure must
accommodate many more components than were previously believed to exist.
Advances in electron microscopy and in immunofluorescence microscopic
techniques have made it clear that what was once visualized as an
amorphous ground substance containing a mixture of many molecules is in
fact very structured (Figure 1). Apart from the nucleus, the rest of the
cell—the "gelatinous material," the "ground substance"--has been found
to contain many organelles. In fact, as the resolving power of the
microscope has increased, the discovery of new particles has also
increased. The most recent finding that the cytoplasm has, in addition
to organelles and vesicles, an organized cytoskeleton has lead to the
consideration of cytoplasmic crowdedness (Fulton, 1982). Not only does
the cytoplasm contain this sort of material, but it also appears to be
highly organized. There are microfilaments, intermediate filaments, and
microtubules—all polymers which associate and dissociate into filaments
depending on the state of the cell. In addition, it has been proposed
that a complex network of proteins, termed the microtrabecular lattice
(MTL), exists as an organizing framework for proteins and smaller
organelles in the cell (Wolosewick and Porter, 1979; Tucker and Porter,
1981).

On the other hand, equally impressive discoveries have been made
using a biochemical approach. Enzymes, ions, and metabolites have been
extracted, purified, measured, and analyzed. From a metabolic point of

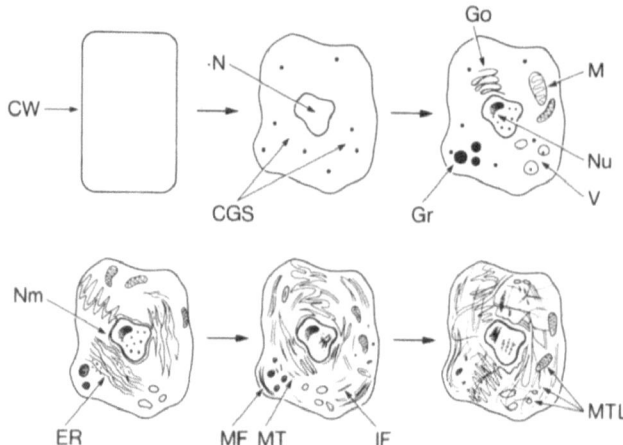

Fig. 1.   Schematic representation of the discovery of mammalian cell organelles following the original discovery of the cell wall of plant cells. CW, cell wall; N, nucleus; CGS, cytoplasmic ground substance; Go, Golgi apparatus; G, granules; M, mitochondria; Nu, nucleolus; V, vesicles; Nm, nuclear membrane; ER, endoplasmic reticulum; MF, microfilaments; MT, microtubules; IF, intermediate filaments; MTL, microtrabecular lattice.

view, the cell appears as complex as it does from a structural one. Thus, there is an immensely structured biochemistry within an immensely structured cytoplasm. This complexity brings up the question of how structure and function relate. How does dynamic movement occur in a very organized system?

We have been interested in answering one specific question which relates to the larger question of structure and function. In spite of the long list of other items, the major component of the cell is water. Therefore one important question is "How do small, aqueous, soluble molecules (e.g., the size of glucose) move in the cytoplasm?"

Many people have, over the years, asked how fast things move in the cell. The general approach has been to inject (or ask the cell to ingest) material and to follow its movement over time. For example, movements of iron filings have been followed in cultured chicken cells (Crick and Hughes, 1950). Diffusion of dyes or radioactive probes has been followed in larger cells (e.g., Hodgkin and Keynes, 1956; Horowitz, 1972; Horowitz and Moore, 1974; Caille and Hinke, 1974).

In order to answer the question of how a small molecule moves in the cytoplasm, we have used electron spin resonance (ESR) spectroscopy. We have applied small spin-labeled molecules (Figure 2) to mammalian cells. These water soluble molecules rapidly enter the cells by diffusion. In the magnetic field of an ESR spectrometer, they act as reporter molecules and emit a spectrum characteristic of their environment (Figure 3). From the spectrum one can gain information about the movement of the molecules both over a very close range (e.g., a few Å) and over a relatively longer range (e.g., 50 to 100 Å), depending on the concentration of spin label. The former can be calculated from the rotational correlation time, τ_c, of the molecule. The latter can be calculated from quenching of the spin label signal with increasing spin label concentration (see Keith et al.,

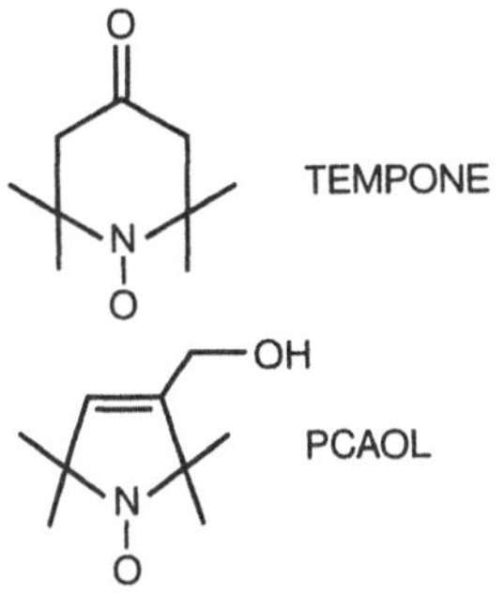

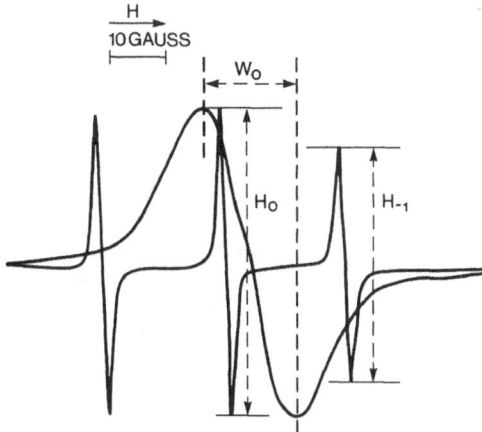

Fig. 2.   Structural formulae of two spin labels, tempone, 2,2,6,6,-
tetramethylpiperidine-N-oxyl; and PCAOL, 2,2,5,5,-tetramethyl-
3-methanol pyrroline-N-oxyl.

Fig. 3.   The signal of a partially immobilized spin label. A full scan
of all three spectral lines is shown, complete with gauss
marker and H field direction. The full scan is a 50-gauss scan,
and the amplification of the midfield line is a 5-gauss scan.
Measurement methods for midfield line height, H_o, low-field
line height, H_{-1}, and midfield line width, W_o, are given. The
sweep time for the full scan is 10 minutes and for the
amplified scan 2.5 minutes. The amplitude modulation is
0.5 gauss.

1977; Mastro and Keith, 1981; Mastro and Keith, 1984; Mastro et al.,
1984; Mastro and Keith, 1986). From calculated diffusion constants, D,
one can obtain the microviscosity.

We took advantage of the fact that both τ_c and D can be calculated
from the same spectrum. Theoretically, one ought to be able not only to
determine the diffusion constant and viscosity, but also to distinguish
between motion dictated by local viscosity and that caused by cytoplasmic
barriers. Spin label-spin label interactions, as well as τ_c, will be
affected by an increase in the solvent viscosity, for example the
viscosity of sucrose or glycerol solutions (Figure 4). However, the
placement of barriers 100 Å or so of the spin label may slow down
translational diffusion with essentially no effect on rotational

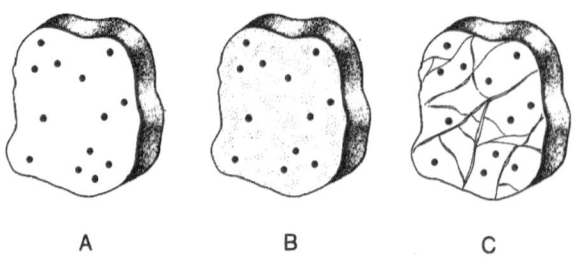

A   B   C

Fig. 4.   Representation of spin-label molecules randomly distributed in
           cytoplasm composed of (A) a low viscosity solution (e.g.
           water), (B) a higher viscosity sucrose solution, or (C)
           microbarriers in a low viscosity solution. The conditions in
           both B and C result in lower apparent translational diffusion
           than seen in A. However, τ_c in C is similar to A while reduced
           in B.

diffusion. The actual distance over which barriers can have this effect
is based on the cubic lattice spacing of the molecules, which in turn is
determined by their concentration and the volume of the aqueous
compartment (Keith et al., 1977). With these ideas in mind, we asked
first, if one could use ESR to determine diffusion and viscosity in model
systems; and second, if these same approaches could be used in mammalian
cells.

METHODS

Cells

     Fibroblast cell lines, mostly Swiss 3T3 cells and their SV40
transformed counterparts, were used for the experiments. Other clones of
3T3, as well as of BHK fibroblasts, were also used. Tempone or PCAOL,
two spin label probes, were used (Figure 2). $NiCl_2$ was used as the
quenching agent for extracellular spin label. The details of these
methods have been published (Mastro and Keith, 1981).

     For some experiments primary cultures of lymphocytes were tested.
These cells were isolated from bovine lymph node as previously described
(Mastro and Pepin, 1980). The freshly isolated cells were washed,
resuspended in phosphate buffered saline (PBS), and centrifuged. The
cell pellet, approximately 10^8 cells per point, was resuspended in 1.0 ml
of labeling solution consisting of trioxalochromate (TOC) (50 mM),
deuterated tempone (1 to 40 mM), and PBS. The final concentration of the
components was calculated after measuring the absorbance of the TOC at
567 nm in solution after the spin label measurements were taken. This
method takes into account the dilution of the labeling solution by cell-
associated buffer. The cell suspension was transferred to a glass
capillary pipet which was sealed on one end. Measurements were made of
the cells in suspension in the labeling solution, the extracellular spin
label being quenched by the TOC. The cells were greater than 90% viable
under the conditions of labeling.

Spin Labeling Procedures

Two aqueous spin labels, PCAOL and tempone (or deuterated tempone), were used. These were synthesized in our laboratory (Hammerstadt et al., 1979). The spin label was used over a range of concentrations. The change in the midfield line width (ΔH) of the ESR signal was plotted as a function of the spin label concentration. The diffusion constant (D) for the spin label was calculated from the slope of this line using the relation, $D = (K \cdot \Delta H)/M$. ΔH is the line width component contributed by the spin label concentration and is calculated from H_m minus H_{min}, where H_m is the line width at the given molar concentration of the spin label and H_{min} the minimum line width of a very dilute spin label concentration; M is the molarity of the spin label; K is a constant of proportionality relating spin-label collision frequency with molar concentration (Mastro and Keith, 1986). τ_c was calculated from the same spectra, but at the lowest spin label concentration—0.1 to 1.0 mM in water and 1.0 to 4 mM in cells, for PCAOL and deuterated tempone, respectively.

RESULTS

Differentiation Between Constraints on Rotational Motion and

Translational Motion in Model Systems

Model systems were used to ascertain, that the effects of viscosity could be distinguished from the effects of barriers using a spin labeling technique (Figure 4). In order to simulate the effect of high viscosity, the spin label, PCAOL, was added to solutions of 10%, 40%, or 70% sucrose in water. The change in midfield line broadening, ΔH, was measured over a range of spin label concentrations for each sucrose solution (Figure 3). Both τ_c and D were calculated from each spectrum. As predicted, both rotational and translational motions were inhibited with increasing concentration of sucrose (Figure 5).

In order to simulate the presence of microbarriers, spin label dissolved in water was trapped in polyacrylamide beads of various pore size (Figure 6). The ESR spectra were measured over a range of spin label concentrations and the τ_c and ΔH calculated. Under thse conditions, as the pores decreased in size, the translational diffusion was inhibited significantly more than the rotational diffusion. In contrast, τ_c changed little compared to that in water until the smallest pore-sized bead was used.

When the changes in rotational and translational diffusion were compared in both of these systems, it was apparent that under conditions of high viscosity (e.g., sucrose solutions) both values were changed by about the same amount. However, the imposition of barriers inhibited translational movement to a much greater degree than it affected the local microviscosity (Figure 7).

Measurement of Rotational and Translational Motion in Mammalian Cells

Fibroblast lines were used for most of the experiments. These cells have been well studied with regard to cytoskeletal components as well as to their growth control. In addition, numerous clones of normal and transformed cells are available for comparative purposes. These cells, which are substrate-dependent for growth, were removed from the plate

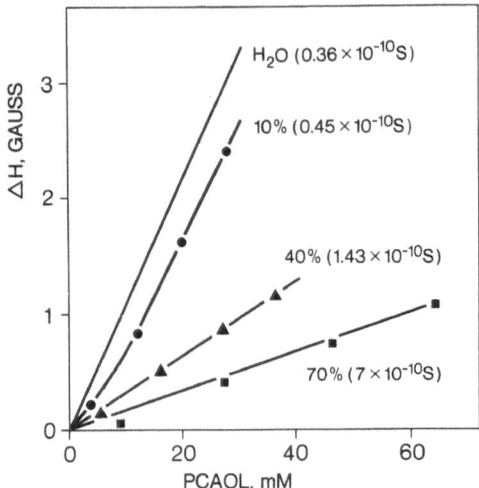

Fig. 5.   The effect of increasing concentrations of sucrose on spin-
           label motion. Line broadening of spin label is shown in water
           (—) and in aqueous solutions of sucrose (wt/vol) at 10% (●),
           40% (▲) and 70% (■). ΔH (gauss) is directly proportional
           to translation diffusion, D, at any given spin-label
           concentration. Numbers in (   ) are values for τ_c measured at
           the lowest spin label concentration for each solution (adapted
           from Mastro and Keith, 1981).

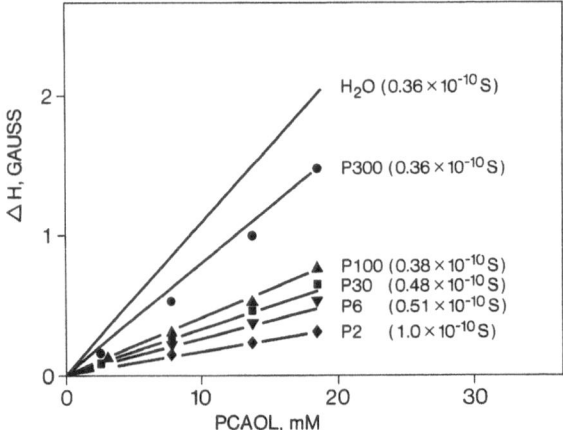

Fig. 6.   The effect of confining space on the motion of spin labels.
           Aqueous solutions of spin label were added to dehydrated
           polyacrylamide beads (Biorad), and placed in capillary tubes
           which were then sealed at one end and centrifuged to remove the
           interbead aqueous volume. $\underline{P}$ notation refers to the equivalent
           spherical molecular weight exclusion limit. Numbers in (   )
           are τ_c . The spin label motion in water is shown for reference
           (adapted from Mastro and Keith, 1981).

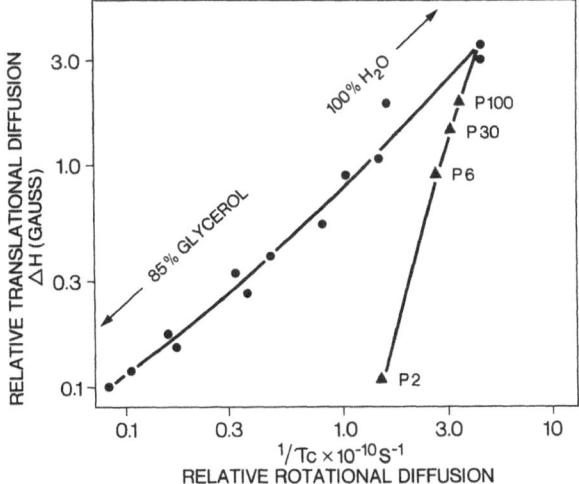

Fig. 7.   Comparison of the effects of barriers vs solvent viscosity on
the translational and rotational diffusion of a spin label.
The spin label was dissolved in water and glycerol (vol/vol)
(●) or in water and trapped in polyacrylamide beads (■)
(see legend to Figure 6). In the glycerol-water solutions the
translational and diffusional motions were slowed down about
equally as the solution viscosity increased. In contrast, in
the bead system, when translational diffusion was inhibited as
much as 30-fold in P100 beads compared to diffusion in water,
the rotational correlation time changed less than three-fold
(adapted from Hammerstedt et al., 1979).

before labeling. In some experiments, quiescent cells (G_o) grown to
confluence were compared with duplicate plates which had been stimulated
by addition of serum 12-16 hr previously. Spin label was added to the
cells; extracellular spin label was quenched with $NiCl_2$. Cells remained
viable under these conditions (Mastro and Keith, 1981). ESR signals were
recorded at several spin label concentrations, from approximately 4 to 40
mM.

In regard to the translational rotation, it was found that the line
broadening was less, at a given spin label concentration, in cells than
in water (Figure 8). That is, the motion of spin label was inhibited
more in cells than in water. An average diffusion constant of 3.4 x 10^{-6}
cm^2/sec was calculated for PCAOL in quiescent 3T3 cells. In serum-
stimulated cells in the same experiment, this value increased to 3.9 x
10^{-6} cm^2/sec. These values are about two-fold lower than that found for
spin label in water in the same experiments. A similar effect was noted
for τ_c where values were about 2-to 2.5-fold greater in the fibroblasts
than in water.

This same experiment was repeated with Swiss 3T3 cells, as well as
with other cell lines. Quiescent, serum-stimulated, and exponentially-
growing cultures were tested, as were various lines of SV40 virus-
transformed 3T3 cells (Table 1). There was a consistent difference of
about 15 to 20% in both τ_c and D between the quiescent and stimulated
cells (Mastro and Keith, 1984; Mastro et al., 1984). However, when the
values for D for numerous experiments were averaged, regardless of the

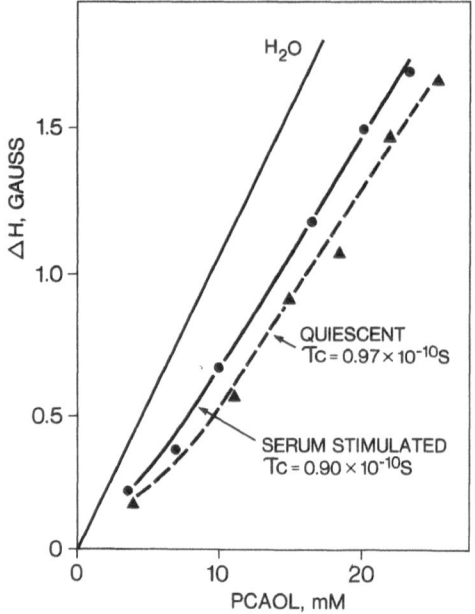

Fig. 8.　Spin-label motion in Swiss 3T3 cells, quiescent and 12 hours
after stimulation with serum. PCAOL was added to cells at the
concentration shown. $NiCl_2$ was used as the external quenching
agent. The averaged data of four separate experiments are
shown. (Adapted from Mastro and Keith, 1981).

Table 1.　Translational Diffusion of Spin Labels in Mammalian Cells[a]

Cell	Growth State	N[b]	Diffusion Constant ($\times 10^{-6} cm^2$/sec)	Viscosity (centipoise)[c]
Swiss 3T3	Quiescent	13	3.4 ± 0.4	2.0
Swiss 3T3	Serum-stimulated	5	3.4 ± 0.5	2.0
Swiss SV40 3T3	Growing	7	3.2 ± 0.4	2.1
Swiss 3T3 (beads)	Growing	4	3.8 ± 1.9	1.8
BALB/c 3T3[d]	Quiescent	3	3.4 ± 0.2	2.0
BALB/c SV40 3T3[d]	Growing	3	3.2 ± 0.4	2.1
BALB/c MCA[d]	Growing	1	3.5 ± 0.3	1.9
BALB/c 3T3[e]	Quiescent	1	3.3 ± 0.3	2.1
BALB/c SV40 3T3[e]	Growing	1	2.2 ± 1.3	3.1
BHK	Growing	5	3.6 ± 0.3	1.8
Lymphocytes	Quiescent	3	3.3 ± 1.5	2.2

[a]Adapted in part from Mastro and Keith, 1984; Mastro et al., 1984.
[b]Number of experiments.
[c]Calculated from Stokes-Einstein relation: $D = kT/(6 \pi r \eta \cdot f/f_o)$.
[d],[e]Indicates different clones.

growth state, the value was 3.3 x 10^{-6} cm^2/sec (Table 1). The value of D for several clones of transformed 3T3 cells was sometimes greater and sometimes less than the value of D in the matched, non-transformed 3T3 cells, depending on the clones. On average, the diffusion parameters for the transformed cells were similar to those of the non-transformed cells (Table 1). The behavior of spin label in 3T3 cells grown on polyacrylamide beads (Biosilon, NUNC) and labeled in situ was approximately the same as for cells which had first been removed from the culture plate and labeled in the test tube. BALB/c 3T3 cells in various states of growth, transformed or normal gave similar values to the Swiss 3T3 cells. BHK cells, another fibroblast line, also gave similar results (Table 1).

Several experiments were done to determine whether a change in the diffusion of spin label could be varied by treating the cells with compounds known to change cytoplasmic structure. Treatment with trypsin (0.25%), colcemid (1 x 10^{-6}M) or vinblastin (1 x 10^{-6}M) caused no significant change in D or τ_c (data not shown). However, when cells were treated with cytochalasin B (which can affect microfilaments), there was a consistent increase of about 20% in both rotational and translational diffusion (Figure 9). This effect was seen with SV40 3T3 cells and with BHK cells treated the same way (data not shown).

In another series of experiments primary cultures of lymphocytes were used. These cells can be maintained in suspension and need not be removed from a substrate to be inserted in the ESR machine cavity. These cells were labeled in suspension with deuterated tempone and with trioxalochromate (TOC) as the external quenching agent. Cell viability was greater than 95% under the labeling conditions. Although lymphocytes are considerably smaller than 3T3 cells, the deuterated tempone is a somewhat more sensitive probe than PCAOL. Thus, a strong signal was determined under these conditions. For lymphocytes, the calculated value of D was found to be approximately two to three times that in water. These values were similar to those seen with 3T3. In the same way, the values for τ_c were reduced by about two-fold, indicating a proportional change in the relative rotational motion.

Apparent Cytoplasmic Viscosity Calculated from τ_c and D

From either τ_c or D, the apparent viscosity of the solvent can be calculated using the Stokes-Einstein relation (Keith and Snipes, 1974). From the translational motion, the apparent viscosity of the cytoplasm was calculated to be 2-3 centipoise (Table 2). The viscosity calculated from τ_c was about the same (Table 2). These results correspond best to those seen in the model system when the spin label was dissolved in sucrose and not when it was trapped in polyacrylamide beads, where the apparent viscosities determined by the two methods were not differentially affected (Table 2).

There was a condition, however, under which the D value could be changed by a relatively large amount, compared to little or no change in τ_c. This differential effect was seen when cells were placed under hypertonic conditions. Under these conditions as the cell volume shrunk, the change in line broadening was greater than the change in τ_c (Figure 10). In BHK cells, when the volume decreased by about two-fold, the translational motion decreased by four-fold. The fact, that D changed more than the volume, is related to the fact that the aqueous volume can change under these conditions, but that the solid matter remains constant. In contrast to D, τ_c changed by about 20% compared to that measured in cells under isotonic conditions.

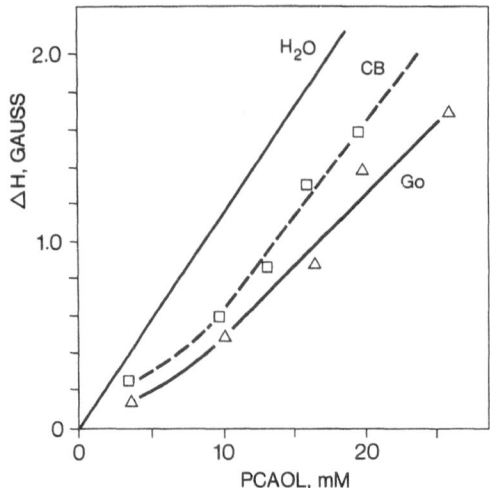

Fig. 9.   The effect of cytochalasin B on spin label motion in 3T3 cells.
Cytochalasin B (5 µg/ml, 0.4% DMSO) was added to cultures of
quiescent 3T3 cells 5 hrs before preparation for labeling.
DMSO alone had no effect on the spin label diffusion. (Adapted
from Mastro et al., 1984; Mastro and Keith, 1984).

Table 2.   Apparent Viscosities of Solutions or Cytoplasm as Determined
from ESR Measurements[a]

	Solute	Viscosity (centipoise)		$\eta D / \eta \tau_c$
		$\eta(\tau_c)$[b]	$\eta(D)$[c]	
A.	Water	1.1	1.1	1.0
	10% sucrose	1.3	1.5	1.2
	40% sucrose	4.1	3.9	1.0
	Water in P300 beads	1.1	2.7	2.5
	Cytoplasm 3T3 (isosmotic)	2.6	2.2	0.8
B.	Cytoplasm, BHK, 300 mosmol	3.4	2.9	0.8
	Cytoplasm, BHK, 550 mosmol	4.0	10.0	2.5

[a]Adapted from Mastro and Keith, 1984; Mastro et al., 1984. For
determination in A, PCAOL was used; for B, deuterated tempone was
used.

[b]Viscosity calculated from rotational correlation time, τ_c .

[c]Viscosity calculated from translational diffusion coefficient, D.

A similar effect was seen using lymphocytes. In a salt solution of
410 mosmol the cell volume decreased about 25%, compared with that measured

in an isotonic solution. In contrast, under these hypertonic conditions, the viscosity calculated from D was 2.3-fold greater than under isotonic conditions, while the viscosity calculated from τ_c changed by only 30%. Under hypotonic conditions there was little change in either parameter.

This effect of hypertonicity might be explained by a decrease in the cell water-rich compartments as the cells become dehydrated. Under such conditions the protein concentration in these compartments could be expected to increase. The fact that the viscosity calculated from D increased much more than that calculated from τ_c, however, implies that longer-range translational motion was affected much more than shorter-range motion, and that barriers rather than an increase in fluid viscosity was most likely responsible.

DISCUSSION

These studies were carried out to determine the relatively short and long-range movements of a small spin-label molecule, comparable in size

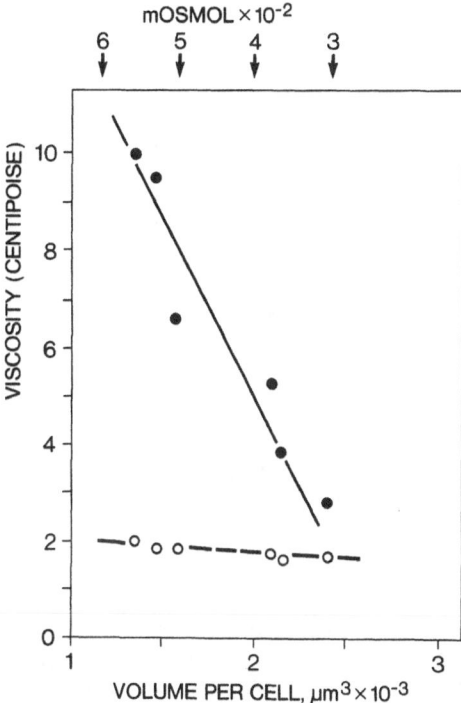

Fig. 10. Apparent viscosities of BHK cells under hypertonic conditions. BHK cells were labeled with deuterated tempone. $NiCl_2$ (75 mM) and increasing concentrations of KCl were added to increase the hypertonicity of the medium. The osmolality ranged from 290 to 550 milliosmols (mosmol). Spin label was used at 10 mM to calculate D and at 1 mM to measure τ_c. Viscosities were calculated from the Stokes-Einstein equation. ($\bullet$) η calculated from D; ($\bigcirc$) η calculated from τ_c. At 290 mosmol, $D = 2.3 \times 10^{-6}$ cm^2/sec; and τ_c was 0.57×10^{-10} sec. At 550 mosmol, $D = 6.7 \times 10^{-6}$ cm^2/sec; and τ_c was 0.68×10^{-10} sec. (Adapted from Mastro et al. 1984; Mastro and Keith, 1984).

to glucose, in the aqueous compartment of a mammalian cell. Moreover, a comparison of diffusion constants obtained from the rotational motion, τ_c with that obtained from translational motion, D, enabled us to differentiate the effects of viscosity from that of microbarriers on movement within the cell cytoplasm. The differential effect of barriers vs viscosity was clearly seen in the case of spin label trapped in polyacrylamide beads of various pore sizes. The internal bead network slowed down translational motion significantly under conditions in which rotational motion was unaffected. It was not until the pore size was made very small did it slow down the rotational motion, and only then to a much lesser degree than the translational motion.

This differential effect between translational and rotational motion was not seen in spin-label measurements made in mammalian cells. Movement calculated from both translational and rotational parameters was similar. That is, the apparent viscosity of the aqueous compartment was calculated to be two to three times that of water (2-3 cp), using either τ_c or D. The value of D for PCAOL averaged 3.3×10^{-6} cm^2/sec for several cell lines at room temperature. This D value corresponds to that found for an aqueous solution of about 15% sucrose. Likewise, calculations of viscosity from τ_c for the same sucrose solutions showed the same distinction.

The cell is clearly not a homogeneous solution of sucrose, so one must ask why the spin-label parameters are thus affected. For one, the cell contains many molecules, ions, organelles, proteins, etc. There is no evidence that the spin label slows down because it binds to these elements. If we assume that the protein content of cytoplasm is about 7% (similar to that of serum), then we can ask if this protein concentration can affect the apparent viscosity. When spin-label diffusion was measured in aqueous solutions of 10% protein, of various sizes, there was little effect on motion (Table 3). At most, D was changed by 20% compared with diffusion in water. In other studies it had also been determined that PCAOL does not bind to DNA nor to synthetic polymers (Keith et al., 1979).

There is also the possibility that spin label is sequestered in bound water. There is considerable debate, as to what extent water is present in a free or bound state in the cytoplasm. In a model system using phospholipid multilayers and a water-soluble spin label, it was seen that the label partitions non-preferentially between the free and bound water (Keith et al., 1977). Although the exchange between the two kinds of water was rapid, the exchange rate was longer than the spin-label relaxation time. Thus, two distinct signals were visible, one from the bound and one from the free water. In a system containing both, the spectrum is an arithmetic mean of the individual spectra of the two signals, but is dominated by that of the free water because of the more narrow line widths. In the mammalian cell system the shape and the intensity of the signal indicated that the spin label is predominantly in free water. Furthermore, based on the intensity of the signal, for a given concentration of spin label, both under isotonic and hypertonic conditions, a calculation of the free water domain suggests that it is about 90% of the water of the cell.

The original hypothesis was that translational motion, D, would be affected by a barrier of cytoskeletal components, while rotational motion would not. However, under normal physiological conditions such a differential effect was not apparent. Both D and τ_c changed to about the same degree. Apparent cytoplasmic viscosity calculated form either measure was 2-3 cp. Thus, over the dimensions detectable with the spin-label concentrations used (about 50-100 Å), there was no evidence that barriers inhibited the motion of the spin label.

When cytochalasin B (a reagent which modifies the microfilaments) was added to cells, there was an affect on D, suggesting that barriers were involved. However, under these conditions, τ_c also changed, which implied that the effect of cytochalasin B may have been indirect.

In contrast, the strongest evidence that cytoplasmic structures can play a role in diffusion comes from studies with cells in hypertonic solutions. Under these conditions the cell volume decreased. The translational diffusion constant also decreased, thereby causing an apparent increase in cytoplasmic viscosity. In contrast, the rotational motion changed very little. This differential effect on rotational _vs_ translational motion was reminiscent of that seen when spin label partitioned into porous beads, and could be explained by the presence of physical barriers to diffusion. What is the source of these barriers under hypertonic conditions? For one, the spacing of the cytomatrix, visualized by Porter and others (Tucker and Porter, 1981) with high-voltage EM, also decreases under hypertonic conditions when the cell volume shrinks. However, a simple decrease in the aqueous compartment is probably not sufficient to explain the barriers with a lattice spacing of between 50 and 100 Å, i.e., those detectable at the spin-label concentrations used. When the cell volume decreases by 50%, it is estimated that the aqueous compartment decreases by about 75% (Mansell and Clegg, 1983). A uniform decrease in the molecular lattice spacing would lead to distances between barriers averaging about 500 Å. However, all of the spaces are not uniform in size; some may shrink to 100 Å, while others remain more than 500 Å. Furthermore, tortuosity increases as the network becomes more compact. Moreover, as the cell water decreases, the concentration of protein molecules and structural elements increases, possibly leading to aggregations (Fulton, 1982). The data here suggest that the barriers formed are, on the average, spaced about 100 Å apart. Schobert and Marsh (1982) and Mansell and Clegg (1983) have reported increases in the density of cytoplasmic ground substance under hypertonic conditions in algae or in L cell fibroblasts, respectively.

Over the years, many workers have measured the translational motion of various solutions in different kinds of cells. However, the movement of large molecules such as BSA and IgG has been determined over relatively long distances (micrometers and millimeters) and for long times (seconds to hours), compared with the measurement using spin labels and ESR. However, to get a better idea of how barriers can affect diffusion, the values of D for PCAOL and tempone were compared with those of other molecules measured by other techniques (Table 3).

Diffusion theory predicts that D for molecules in aqueous solutions varies approximately inversely with the size of the molecule. In spite of the variety of molecules, techniques, and cells, this relationship was seen to hold for the D of small molecules, dextran spheres, and at least one protein in the cell cytoplasm (Table 3, Figure 11). The D in the cytoplasm was, in general, two- to five-fold less than in water.

However, this relationship did not hold for such larger macromolecules as insulin, actin, ovalbumin, BSA, IgG, vinculin, actinin, and apoferritin (Table 3, Figure 11). The movement of these was measured by recovery after fluorescence photo-bleaching (FPR), with fluorescent derivatives injected into cells. On the whole, these

Table 3. Diffusion Constants of Various Molecules in Cytoplasm

Compounds	M_r	Radius (Å)	Diffusion Constant (10^{-7} cm^2/sec)	Viscosity[c] (η)	Cells[d]
PCAOL	170	3.2	33	2.1	Mouse fibroblasts (ref. 1)
Sorbitol	182	2.5[a]	50	1.7	Barnacle muscle fibers (ref. 2)
Methylene blue	320	3.7[a]	15	4.0	Squid axons (ref. 3)
Sucrose	324	4.4[a]	20	2.5	Frog oocytes (ref. 4)
Eosin	648	6.0[a]	8.0	4.5	Squid axons (ref. 3)
Dextran	3,600	12.0	3.5	5.2	Frog oocytes (ref. 4)
Inulin	5,500	13.0[b]	3.0	5.6	Frog oocytes (ref. 5)
Dextran	10,000	23.3	2.5	3.7	Frog oocytes (ref. 4)
Insulin	12,000	16.0	0.09	151.5	Human fibroblasts (ref. 6)
Myoglobin	16,900	17.0[b]	5.1	2.5	Bovine heart muscle (ref. 7)
Dextran	24,000	35.5	1.5	4.1	Frog oocytes (ref. 4)
Actin	43,000	23.2[a]	0.03	179.0	Chicken gizzard fibroblasts (ref. 8)
Ovalbumin	45,000	23.8[b]	0.34	26.7	Mouse "macrophage-like" cells (ref. 9)
Bovine Serum Albumin	68,000	36.0	0.10	61.0	Human fibroblasts (refs. 8, 10)
			0.06	101.0	Chicken gizzard fibroblasts (refs. 8, 10)
			4.0	1.5	Amebae (refs. 8, 10)
IgG[F(ab')2]	100,000	40.6	0.16	33.6	Human fibroblasts (ref. 6)
IgG	153,000	35.0[b]	0.09	50.0	Human fibroblasts (refs. 8, 10)
			0.06	75.0	Chicken gizzard fibroblasts (refs. 8, 10)
Vinculin	130,000	33.3[b]	0.03	218.4	Human fibroblasts (ref. 6)
α-Actinin	200,000	78.2	0.03	93.0	Chicken gizzard fibroblasts (ref. 11)
Apoferritin	440,000	61.0	0.16	22.4	Human fibroblasts (ref. 6)

Table 3 (cont.)

[a]Calculated from structure

[b]Calculated for a sphere of equivalent volume.

[c]Viscosity (η) was calculated from the Stokes-Einstein equation, $\eta = kT/(6 \cdot \pi r Df/f_o)$ using the literature values for D, except in the case of myoglobin where the reported value was η, and D was calculated. For comparison for all the compounds, except vinculin where data were not available, the same calculation of η based on diffusion in aqueous solution was calculated and found to be 1.047 ± 0.14. This table was in part adapted from Mastro and Keith, 1984, and Mastro et al. 1984.

[d]References: (1) Mastro et al., 1984, Mastro and Keith, 1984; (2) Caille and Hinke, 1974; (3) Hodgkin and Keynes, 1956; (4) Horowitz, 1972; (5) Horowitz and Moore, 1974; (6) Jacobson and Wojcieszyn, 1984; (7) Livingston et al., 1983; (8) Kreis et al., 1982; (9) Wang et al., 1982; (10) Wojcieszyn et al., 1981; (11) Suzuki et al., 1976, Geiger et al., 1984.

---

molecules moved about 10- to 100-fold more slowly in cells than the other molecules, including dextran spheres of about the same size and shape as BSA. The apparent deviation is not surprising for actin, because measurements were made under conditions in which it is largely immobilized in the cytoplasm. Actin binds to microfilaments. However, molecules like BSA and IgG were determined to be greater than 90% mobile over the time course of the measurements.

Using the Stokes-Einstein relation, $D = kT/(6 \cdot \pi r \eta \cdot f/f_o)$, the apparent viscosity of each molecule in water and in cytoplasm can be calculated (Figure 11). A plot of this information indicates that the molecules in water experience an apparent viscosity of approximately 1.0 cp. In the cytoplasm, the apparent viscosity is about 2-5 cp for the small molecules, dextran spheres, and myoglobin. However, for all of the larger proteins, it is in the realm of 20-200 (Table 3). Thus, these molecules move much more slowly than predicted on the basis of cytoplasmic viscosity.

One explanation for the slow movement is the encounter of barriers. However, the fact that there is no proportional relationship between size and diffusion suggests that barriers alone are not the explanation. For example, insulin moves more slowly than the larger molecules apoferritin and IgG, as noted by Wojcieszyn and Jacobson (1984). In addition, a dextran sphere of the same approximate dimension as BSA moved 20-50 times faster in the cytoplasm. Furthermore, when the rotational motion of myoglobin was obtained **in situ** by NMR, it behaved as if it were in a solution with a viscosity of about 2.5 cp, in line with the values determined by ESR and other techniques (Table 3).

Thus, it seems likely that the explanation for the apparent slow diffusion of BSA and some of the other proteins is that they bind with low affinity to cytoplasmic structures. If binding were reversible with rate constants in the same order as the rate of diffusion, then the overall effect of continual binding and release would be seen as a decrease in the diffusion constant. Horowitz et al. (1970), in fact,

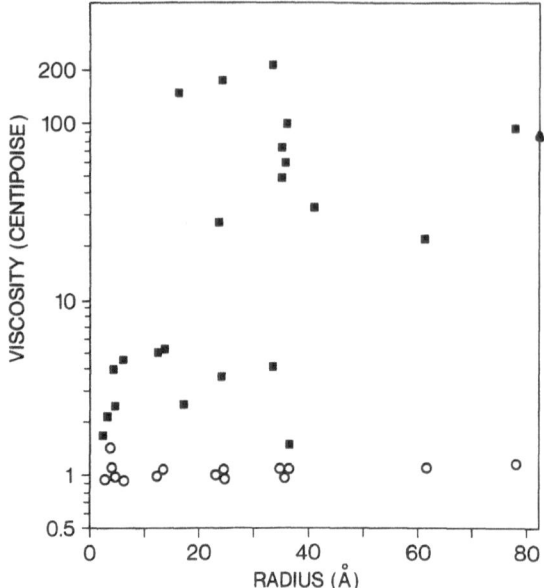

Fig. 11. Comparison of the apparent viscosities of water and cytoplasm
calculated for molecules of different sizes (see data in Table
3.). (■) water; (○) cytoplasm.

described such a chromatographic effect for the movement of cysteamine
phosphate in the oocyte cytoplasm. On the basis of theoretical
considerations, Gershon et al. (1983) also concluded that BSA must bind
to cytoplasmic proteins. Wojcieszyn and Jacobson (1984) came to the same
conclusion, after comparing the diffusion constants of proteins of various
sizes. Either in the case of binding or barriers (or both), one could
reach the mistaken conclusion that a large protein such as BSA were in a
more viscous solution than is really the case.

It is also interesting to note that when the movements of BSA or
ovalbumin were measured by FPR in amebae, they appeared to diffuse much
faster than in mammalian cells (Wang et al., 1982), suggesting the
cytoplasmic structure of amebae may be very different from that of
mammalian cells.

In regard to the possible binding of injected proteins to cytoplasmic
structures, it is important to know if it is of physiological importance
or a possible artifact of the experimental system. The rotational value
for myoglobin in cells implies that not all proteins experience this
binding. However, this measurement is the only one thus far available
for the rotational motion of a protein in situ. The ability to examine
the rotational motion of larger proteins in situ, as well as the
development of more sensitive spin-label techniques which allow the use
of spin-label concentrations in the range 1×10^{-8} to 1×10^{-7} M would
permit the detection of barriers spaced about 100 to 1000 Å. This is the
proposed size of the spacing of the MTL, a possible network of
cytoplasmic proteins and a potential barrier to free diffusion in the
cell cytoplasm.

ACKNOWLEDGEMENTS

We thank Don Ohi and Andrew Shyjan, undergraduate research students, for their contributions to this study. This work was supported in part by a grant from the Office of Naval Research (N00014-79-C-0253) and by a grant from the National Science Foundation (PCM-8309109). A.M.M. is the recipient of a Research Career Development Award CA00705 from the National Cancer Institute, DHHS.

REFERENCES

Caille, J.P., and Hinke, J.A.M., 1974, The volume available to diffusion in the muscle fiber, Can. J. Physiol. Pharmacol., 52:814.

Clegg, J.S., 1984, Properties and metabolism of the aqueous cytoplasm and its boundaries, Am. J. Physiol., 246:R133.

Conklin, E.G., 1940, Cell and protoplasm concepts: historical account, in "The Cell and Protoplasm." F.R. Moulton, ed., The Science Press, Washington, D.C.

Crick, F.H.C., and Hughes, A.F.W., 1950, The physical properties of cytoplasm: a study by means of the magnetic particle method, Exp. Cell Res., 1:37.

Fulton, A.B., 1982, How crowded is the cytoplasm? Cell, 30:345.

Geiger, B., Kreis, T.E., Avnur, Z. and Schlessinger, J., Cell and Muscle Motility, 1985, in press.

Gershon, N., Porter, K. and Trus, B., 1983, The microtrabecular lattice and the cytoskeleton, their volume, surface area and the diffusion of molecules through it, in "Biological Structures and Coupled Flows," A. Oplatka and M. Balaban, eds., Academic Press, New York.

Hammerstedt, R.H., Keith, A.D., Boltz, R.C., and Todd, P.W., 1979, Use of amphiphilic spin labels and whole cell isoelectric focusing to assay charge characteristics of sperm surfaces, Arch. Biochem. Biophys., 194:565.

Hodgkin, A.L., and Keynes, R.D., 1956, Experiments on the injection of substances into squid giant axons by means of a microsyringe, J. Physiol., 131:592.

Horowitz, S.B., 1972, The permeability of the amphibian oocyte nucleus, in situ., J. Cell Biol., 54:609.

Horowitz, S.B., and Moore, L.E., 1974, The nuclear permeability, intracellular distribution, and diffusion of insulin in the amphibian oocyte, J. Cell Biol., 60:405.

Horowitz, S.B., Fenichel, I.R., Hoffman, B., Kollmann, G., and Shapiro, B. 1970, The intracellular transport and distribution of cysteamine phosphate derivatives, Biophys. J., 10:944.

Jacobson, K. and Wojcieszyn, J., 1984, The translational mobility of substances within the cytoplasmic matrix, Proc. Natl. Acad. Sci. USA, 81:6747.

Keith, A.D. and Mastro, A.M., Cell and water viscosity, in "Biomembranes", Methods in Enzymology, L. Packer, ed., in press.

Keith, A.D., and Snipes, W., 1974, Viscosity of cellular protoplasm, Science (Wash. DC), 183:666.

Keith, A.D., Mastro, A.M., and Snipes, W., 1979, Diffusion relationships between cellular plasma membrane and cytoplasm, in "Low Temperature Stress in Crop Plants," J. Lyons and J. Raison, eds., Academic Press, New York.

Keith, A.D., Snipes, W., and Chapman, D., 1977, Spin label studies on the aqueous regions of phospholipid multilayers, Biochemistry, 16:634.

Keith, A.D., Snipes, W.C., Melhorn, R.J., and Gunter, T., 1977, Factors restricting diffusion of water-soluble spin labels, Biophys. J., 19:205.

Kreis, T.E., Geiger, B., and Schlessinger, J., 1982, Mobility of microinjected rhodamine actin within living chicken gizzard cells determined by fluorescence photobleaching recovery, Cell, 29:835.

Mansell, J., and Clegg, J.S., 1983, Cellular and molecular consequences of reduced cell water content, Cryobiology, 20:591.

Mastro, A.M. and Pepin, K.G., 1980, Supressiono of lectin-stimulated DNA synthesis in bovine lymphocytes by the tumor promoter 12-0-tetradecanoyl phorbol-13-acetate, Cancer Res., 40:3307.

Mastro, A.M. and Keith, A.D., 1984, Diffusion in the aqueous compartment, J. Cell Biol. 99:180s.

Mastro, A., and Keith, A., 1981, Spin label viscosity studies of mammalian cell cytoplasm, in "The Transformed Cell," L. Cameron and T. Pool, eds., Academic Press, New York.

Mastro, A.M., Babich, M., Taylor, W., and Keith, A., 1984, Diffusion of a small molecule in the cytoplasm of mammalian cells, Proc. Natl. Acad. Sci. USA, 81:3414.

Paine, P.L. and Horowitz, S.B., 1980, The movement of material between nucleus and cytoplasm, in "Cell Biology: A Comprehensive Treatise," D.M. Prescott and L. Goldstein, eds., Academic Press, New York.

Porter, K.R. and Tucker, J.B., 1981, The ground substance of living cells, Sci. Am., 244(3):57.

Porter, K.R., Boggs, D.P., and Anderson, K.L., 1982, The distribution of water in the cytoplasm, 40th Annual Proceedings of the Electron Microscope Society of America, 4-7.

Schobert, B., and Marsh, D., 1982, Spin label studies on osmotically induced changes in the aqueous cytoplasm of Phaeodacytlom. tricornutum, Biochim. Biophys. Acta., 720:87.

Suzuki, A., Goll, D.E., Singh, I., Allen, R.E., Robson, R.M., and Stromer, M.H., 1976, Some properties of purified skeletal muscle α-actinin, J. Biol. Chem., 251:6860.

Wang, Y., Lanni, F., McNeil, P.L., Ware, B.R., and Taylor, D.L., 1982, Mobility of cytoplasmic and membrane-associated actin in living cells, Proc. Natl. Acad. Sci. USA, 79:4660.

Wolosewick, J.J., and Porter, K.R., 1979, Microtrabecular lattice of the cytoplasmic ground substance, J. Cell Biol., 82:114.

Wojcieszyn, J.W., Schlegel, R.A., Wu, E.S., and Jacobson, K.A., 1981, Diffusion of injected macromolecules within the cytoplasm of living cells, Proc. Natl. Acad. Sci. USA, 78:4407.

RESTRICTED MOTION OF CELLULAR K$^+$ IN

PERMEABILIZED CELLS

Miklos Kellermayer[1] and C. F. Hazlewood[2]

[1]Department of Clinical Chemistry
University Medical School
Pecs, Hungary, and [2]Department of Physiology
Baylor College of Medicine
Houston, Texas 77030, U.S.A.

INTRODUCTION

One of the fundamental characteristics of the living state is the ability of a cell to maintain an asymmetric distribution of solutes relative to the medium in which it is bathed. This characteristic is well recognized in the case of monovalent electrolytes. The failure of cells to maintain an asymmetric distribution of solutes, such as potassium (K$^+$) and sodium (Na$^+$), is associated with the deterioration of cellular and organ function. More fundamental knowledge about the cellular mechanisms that give rise to and substain the asymmetric accumulation of K$^+$ and Na$^+$ is needed.

The various roles of cell membranes in the establishment and maintenance of the asymmetric distribution of solutes in general, and ions in particular, have been well studied (Dean, 1941; Krogh, 1946; Hodgkin, 1951; Bonting, 1970; Knight and Welt, 1974; Glynn and Karlish, 1975). On the other hand, the role that protein structures within the cytoplasm and the nucleus may play in the establishment and maintenance of the asymmetric distribution of solutes is less well studied (Damadian, 1969, 1971; Ling and Cope, 1969; Edzes et al., 1977; Hazlewood, 1979; Ling, 1981a, b; Ling and Negendank, 1980; Ling et al., 1980; Negendank, 1982). The recent discovery of the cytoplasmic and nuclear structures that remain following non-ionic detergent treatment of cells (Schliwa et al., 1981; Penman et al., 1983; Capco et al., 1982), the time dependent release of cytoplasmic and nuclear proteins following detergent treatment, and the time-dependent loss of the asymmetric distribution of potassium and sodium implicate proteins in the maintenance of this asymmetry (Kellermayer et al., 1984).

In view of the findings listed above, and since diffusional equilibrium should be reached within a few seconds after the disruption of membrane integrity, and since the asymmetry of K$^+$/Na$^+$ is maintained for minutes after the disruption of cellular and nuclear membranes, a detailed study of asymmetric distribution of K$^+$ and Na$^+$ in membraneless cellular and subcellular preparations is warranted. In addition, the identification of specific proteins associated with the time-dependent loss of K$^+$/Na$^+$ asymmetry should provide further insights into the selective retention of ions by the cells.

METHODS AND PROCEDURES

The methods and procedures may be found in the literature (Kellermeyer et al., 1984). In general, calf thymus lymphocytes were isolated and suspended in various buffered solutions with and without non-ionic detergents. The lymphocytes were exposed to Triton X-100 or Brij 58 for 5, 10, or 20 minutes and then centrifuged at 40,000xg for 10 minutes. Sodium and potassium concentrations of the digested pellets were determined by flame photometry.

SUMMARY OF RESULTS AND CONCLUSIONS

Non-ionic detergents, Triton X-100 and Brij 58, removed lipoid membranes of suspended thymus lymphocytes within 5 minutes. The mobilization and solubilization of cytoplasmic and nuclear proteins occurred much faster (< 5 minutes) with Triton X-100 treatment than with Brij 58 treatment (<10 minutes). In Triton X-100 treated cells the loss of K^+ was complete within 5 minutes, whereas with Brij 58 treatment the K^+ loss was not complete after 10 minutes. Thus, the high concentration of K^+ and the low concentration of Na^+ in the nuclei can remain near normal for minutes in the absence of membrane structures. If the ions were in free solution within the cells, disruption of membrane integrity should lead to equilibration of the ions with external media within seconds. The time-dependent decrease of K^+ in the Brij 58 treated cells was correlated with the solubilization of the proteins. These results support the view that K^+ and Na^+ are not freely dissolved in the cellular water, but are co-compartmentalized with proteins inside the living cell (Kellermeyer and Hazlewood, 1979; Kellermeyer et al., 1980, 1984).

ACKNOWLEDGMENTS

We are grateful to the United States Office of Naval Research (contract number N00014-76-C-0100), the Robert A. Welch Foundation (Q-390), the National Health Ministry of Hungary (05/3-16/255), and a contribution given in the memory of Stanley T. Weiner for the financial support of this work.

REFERENCES

Bonting, S. L., 1970, in: "Membrane and Ion Transport", E. E. Bittar, ed., Vol. 1, 257.

Capco, D. G., Wan, K. M., and Penman, S., 1982, Cell, 29:847.

Damadian, R., 1969, Science, 165:79.

Damadian, R., 1971, Biophys. J., 11:739.

Dean, R. B., 1941, Biol. Symp., 3:331.

Edzes, H. T., Ginzburg, M., Ginzburg, B. Z., and Berendsen, H. J. C., 1977, Experientia, 33:732.

Glynn, I. M., and Karlish, S. J. D., 1975, Ann. Rev. Physiol., 37:13.

Hazlewood, C. F., 1979, in: "Cell Associated Water," W. Drost-Hansen and J. S. Clegg, eds., Academic Press, New York.

Hodgkin, A. L., 1951, Biol. Rev. (Cambridge), 26:339.

Kellermayer, M. and Hazlewood, C. F., 1979, Cancer Biochem. Biophys., 3:181.

Kellermayer, M., Rouse, D., Gyorkey, F. and Hazlewood, C. F., 1984, Physiol. Chem. and Physics and Med. NMR, 16:503.

Kellermayer, M., Timar, T., Jobst, K., Komaromy, L. and Trombitas, K., 1980, *Eur. J. Cell Biol.*, 23:204.

Knight, A. B., and Welt, L. G., 1974, *J. Gen. Physiol.*, 63:351.

Krogh, A., 1946, *Proc. Roy. Soc. Ser. B.*, 133:140.

Ling, G. N., 1981a, *in*: "International Cell Biology 1980-1981", H. G. Schweiger, ed., Springer-Verlag, New York.

Ling, G. N., 1981b, *Physiol. Chem. Phys.*, 13:356.

Ling, G. N., and Cope, F. W., 1969, *Science*, 163:1335.

Ling, G. N., and Negendank, W., 1980, *Pers. Biol. and Med.*, 23:215.

Ling, G. N., Ochsenfeld, M. M., Walton, C., and Berginser, T. J., 1980, *Physiol. Chem. Phys.*, 12:3.

Negendank, W., 1982, *Biochim. Biophys. Acta*, 694:123.

Penman, S., Capco, D. G., Fey, E. G., Chatterjee, P., Reiter, T., Ermiosh, S. and Wan, K., 1983, *in*: "Modern Cell Biology," 2:385, Alan Liss Inc., New York.

Schliwa, M., van Blerkom, J., and Porter, K. R., 1981, *Proc. Nat. Acad. Sci., U.S.A.*, 78:4329.

THE INTRACELLULAR DISTRIBUTION OF

ADENOSINE TRIPHOSPHATE

Samuel B. Horowitz and David S. Miller

Department of Physiology and Biophysics
Michigan Cancer Foundation
Detroit, Michigan 48201

INTRODUCTION

Eukaryotic cells are not internally homogeneous, but are compartmentalized into cytoplasmic and nuclear ground substances and the numerous organelles embedded therein. Because of spatial heterogeneity, a comprehensive understanding of the cell's metabolism requires not only data on the rates of synthesis and utilization of adenosine triphosphate, but also a description of how ATP is distributed. Unfortunately, it is difficult to measure ATP concentrations regionally within cells, so that very little is known about its distribution. When considering its role in specific metabolic pathways, it is usual to accept ATP concentrations averaged over the entire cell (or tissue) volume as an adequate measure of local concentrations. This is equivalent to assuming that intracellular ATP is uniformly distributed. (Mitochondria, thought to have much lower ATP levels than the surrounding cytoplasm, are usually exempted from this assumption.) With the development of cryomicrodissection (Frank and Horowitz, 1978; Tluczek et al., 1984), it is now possible to measure ATP in regions of a single cell, the amphibian oocyte. We are beginning to develop some notion as to the assumption's validity and, when necessary, the reasons for its failure.

In cryomicrodissection, a cell is rapidly frozen with liquid nitrogen to prevent solute redistributions; and, while the cell remains frozen, it is dissected on the stage of a low-temperature ($45°C$) microdissection apparatus (Fig. 1C). Three specimens were taken from each oocyte: the nucleus and two samples of cytoplasm, one each from the animal and vegetal hemispheres. While still frozen, each cryomicrodissected sample was placed in a tared foil packet, weighed, and then placed in a Potter-type microhomogenization tube containing 80-150 µl of frozen 0.05N NaOH. At this point, each sample was removed from the ultra-low-temperature environment to an ice bath where it was thawed and homogenized. Samples were neutralized with 0.5 N HCl, and ATP determined using a luciferin-luciferase assay (Lust et al., 1981). Concentrations were calculated from pmoles ATP/µg sample and sample water content. The latter was determined from wet and dry weights of parallel cryomicrodissected samples, dried overnight at $60°C$ over P_2O_5. Water contents (% wet weight) were: nucleus 87.8 ± 0.2 (SE); animal cytoplasm 44.1 ± 0.3; vegetal cytoplasm 40.4 ± 0.5. For the purposes of this report, data for

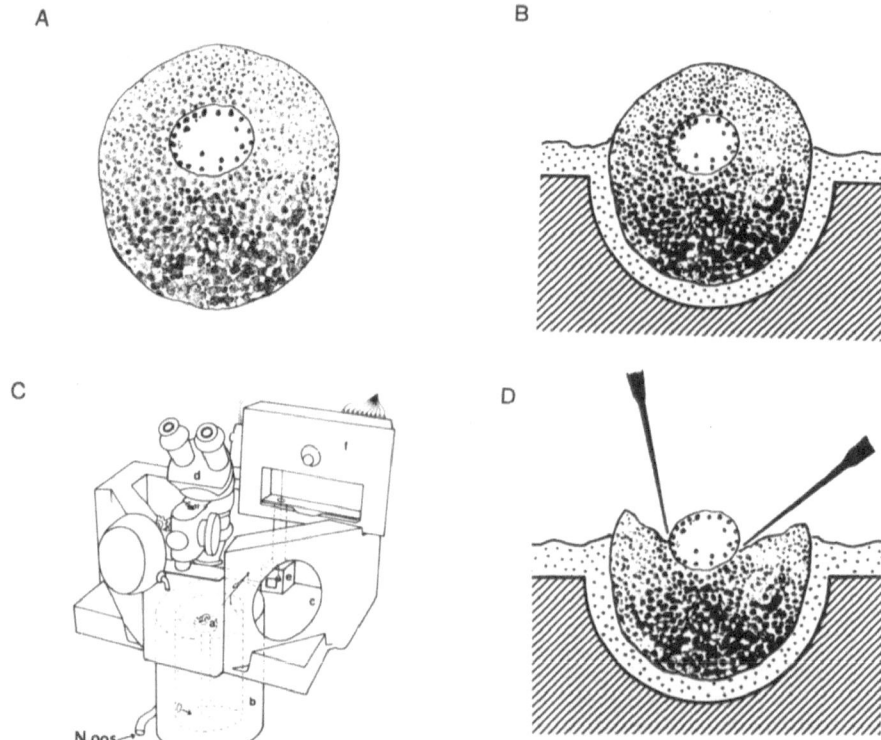

Fig. 1.   Cryomicrodissection of an amphibian oocyte. (A) Schematic of
          the full-grown ovarian oocyte. In Rana pipiens, the cell is
          ~1500 µm in diameter and contains an eccentrically located ~400
          µm nucleus or germinal vesicle. The nucleus consists largely of
          nuclear ground substance, with less than 1% of its volume
          occupied by nucleoli and a contracted chromosome frame.
          Cytoplasm contains a ground substance which is densely packed
          with organelles, including yolk platelets, mitochondria,
          endoplasmic reticulum, ribosomes, etc. Every cytoplasmic
          component exhibits some form of animal-vegetal gradient in
          distribution. (B and D) For cryomicrodissection, the oocyte is
          placed on a metal mounting and frozen rapidly in liquid
          nitrogen. Cell and mounting are placed on the low-temperature
          (-45°C) stage of a microdissection apparatus (C), following which
          samples of cytoplasm and the nucleus are isolated by free-hand
          dissection. The apparatus is an insulated box with glove ports
          (c) to allow access to the low-temperature stage (a), which is
          mounted in a Dewar flask (b). Temperature is maintained by the
          thermostatically regulated flow of nitrogen gas. The dissection
          is observed through a stereoscopic microscope (d), whose focus is
          remotely controlled by a foot-switch. Isolated nuclei and
          cytoplasm are weighed at -20°C to -10°C in a compartment (e)
          mounted below an electrobalance weighing chamber (f).   The
          procedure used for reference-phase microdissection closely
          resembles that used for the nucleus.

animal and vegetal cytoplasm were combined to give an average cytoplasmic value for each cell.

ATP is not uniformly distributed intracellularly but is more concentrated in the nucleus than in cytoplasm (Table 1). The unequal distribution is due partially to differences in nuclear and cytoplasmic water content. However, even after normalization for regional water content, the nuclear ATP concentration is still 2.6 times the cytoplasmic concentration (Table 1).

Additional information is disclosed when ATP concentration data are displayed as an isotherm (Fig. 2). Cytoplasm/nucleus partition ratios are not constant, but vary with concentration. In control cells (i.e., not cyanide-poisoned), the isotherm is adequately represented by a straight line with a slope of 0.23. When extrapolated to the ordinate, this line intercepts at an ATP concentration of ~0.9 mM. We interpret these data to mean that the isotherm is biphasic, exhibiting both a linear (albeit, not equimolar) component and a non-zero intercept; the latter is suggestive of a "bound" ATP fraction. This interpretation is supported by data for cyanide-poisoned cells (Fig. 2, squares), which show that when ATP concentrations are low, the isotherm curves toward the origin in a manner typical of bound fractions.

Many equilibrium and kinetic mechanisms, alone or in combination, might account for the complex nuclear/cytoplasmic isotherm. Equilibrium mechanisms might include, for example, (1) ATP adsorption by nondiffusive nuclear and/or cytoplasmic macromolecules and (2) solvent differences between nuclear and cytoplasmic water. Kinetic mechanisms include (1) impermeability of membrane-enclosed cytoplasmic organelles, (2) active transport through the nuclear envelope or in cytoplasmic vesicles, and (3) diffusion gradients arising because of spatially separated ATP sources and sinks.

Faced with a choice of several mechanisms as explanations, the problem is to determine which are in fact operative. Indeed, it is not even apparent whether mechanisms localized in the cytoplasm, nuclear envelope or nucleus should be the focus of our attention. Perhaps something could be learned by observing changes in the cytoplasm/nucleus

Table 1. ATP in Rana pipiens oocytes[*]

Sample	Content mmoles/kg. wet wt.	Concentration mmoles/l H_2O
Cytoplasm	1.0 ± 0.1	2.4 ± 0.2
Nucleus	5.5 ± 0.6	6.2 ± 0.7
Whole Cell	1.2 ± 0.1	2.7 ± 0.2

[*]Oocytes are about 96% cytoplasm by volume, which accounts for the similar cytoplasmic and whole cell ATP values. Data given as mean $\pm$ SE.

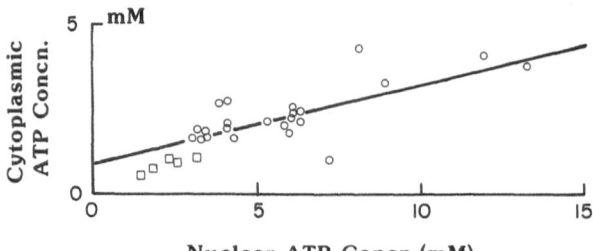

Nuclear ATP Concn.(mM)

Fig. 2. ATP isotherm showing cytoplasmic concentration as a function of
nuclear concentration in control (circles) and cyanide-treated
(squares) Rana pipiens oocytes. [We show elsewhere that the
presence of an iRP does not affect how ATP partitions between
cytoplasm and nucleus (Miller, D. and Horowitz, S.B., in
preparation).] The line was calculated for control cells by the
method of least squares (r = 0.75, P < 0.001). The line
intercepts the ordinate at 0.90 ± 0.07 mM and has a slope of
0.23 ± 0.02.

ATP isotherm as a function of various cell treatments (e.g., introduction
of exogenous ATP and the use of metabolic poisons, precursors or
transport inhibitors). However, the interpretation of such data is often
ambiguous. This is because each compartment in the tripartite cytoplasm-
nuclear envelope-nucleoplasm system is complex, and none can be assumed
to remain invariant during treatment. Thus, none can serve as a control
which would provide the normalizing function necessary to assess changes
that take place in the other compartments. We address this problem by
introducing into the cell a simpler artificial organelle or internal
reference phase, that can serve this normalizing function (Horowitz and
Miller, 1984; Horowitz and Paine, 1974; Horowitz et al., 1979).

An internal reference phase (iRP) is a bolus of gelatin located in
the oocyte's vegetal hemisphere. Dilute gelatin gels are physiologically
inert and do not bind ATP, nor do they interact with water in a manner
that alters its solvent properties (Miller, D. and Horowitz., S.B., in
preparation).

To create a reference phase, 11% gelatin, a gel at ambient temperatures (20°C), is liquified by warming to 30°C and a 30nl droplet is
injected into an oocyte. Following this, the cell is briefly cooled to
hasten gelation and then returned to the ambient incubation temperature.
The resultant intracytoplasmic gelatin gel droplet, or iRP, is an
ordinary aqueous solution which possesses a fibrous protein mesh fine-
structure. The iRP gelatin mesh excludes large formed-structures, such
as cellular organelles and the cytoskeleton, but permits the free entry
of diffusive metabolites and proteins. The iRP, like other (naturally-
occurring) cytoplasmic organelles, is embedded in ground substance.
Because it lacks a membrane it comes to diffusional equilibrium quickly.
Because the iRP is water-like in solvent properties, at equilibrium it
accurately reflects the solute concentrations of the cell's "soluble"
space (Siebert and Humphrey, 1965).

When the iRP reaches diffusional equilibrium with the cell interior
(about 2 hrs in the present experiments), the oocyte is frozen with
liquid nitrogen; and the cytoplasmic samples, nucleus and iRP are

isolated by cryomicrodissection. ATP and water contents are determined and the resultant concentration data plotted as nucleus/iRP and cytoplasm/iRP isotherms (Figs. 3 and 4). The two isotherms are dissimilar, the nuclear isotherm being much simpler.

The nucleus/iRP isotherm for ATP is an equimolar line (Fig. 3); at every concentration, this solute partitions uniformly between the water of the two compartments. Equimolarity demonstrates that the nuclear envelope is not actively transporting ATP and that nucleoplasm neither binds nor excludes the solute. Like the iRP, the nucleus must be diffusionally contiguous with cytoplasm's soluble compartment (at least for smaller solutes) and water-like in its solvent properties. The unequal subcellular distribution of ATP (Table 1) and the complexity of the cytoplasm/nucleus isotherm (Fig. 2) are apparently not due to nucleoplasm or to the nuclear envelope, but rather to cytoplasm's properties.

Nucleus/iRP isomolarity implies that the cytoplasm/iRP isotherm will closely resemble the cytoplasm/nucleus isotherm. This can be confirmed by comparing Figs. 2 and 4. Both isotherms are biphasic, consisting of a linear component with a slope of about 0.2 which dominates at high ATP concentrations and a second, saturable component which accounts for the ordinate intercept of the extrapolated linear component at a concentration significantly above zero.

A striking point is made by the linear component. Despite the fact that diffusional equilibrium exists among the three regions, ATP concentrations in cytoplasm are, on average, only a third as great as

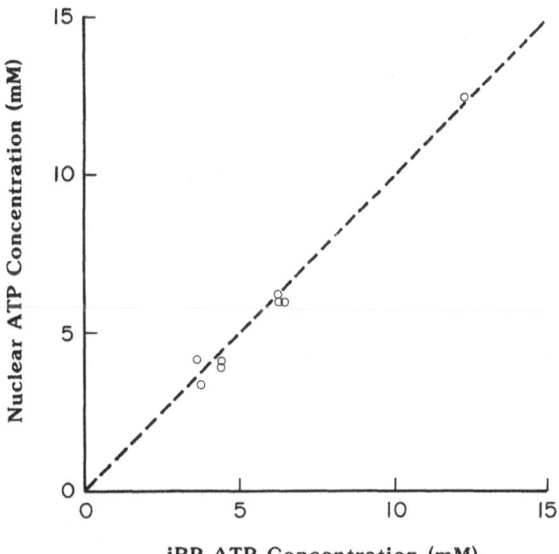

Fig. 3. ATP isotherm showing nuclear concentration as a function of internal reference phase (iRP) concentration in control oocytes. Isomolarity is indicated by the dashed line. Slope and intercept, determined by the least squares method, were 1.02 ± 0.03 and 0.13 ± 0.21, respectively, and do not differ significantly from the isomolar line.

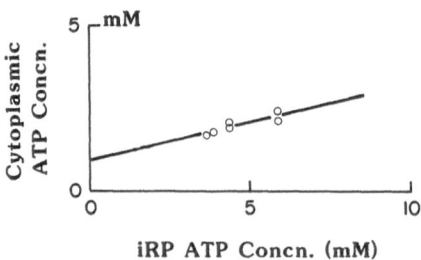

Fig. 4. ATP isotherm showing cytoplasmic concentration as a function of
internal reference phase (iRP) concentration in control oocytes.
The line intercept (0.89 ± 0.09 mM) and slope (0.24 ± 0.02) were
determined by the least squares method (r = 0.95; P < 0.01).
These variables do not differ significantly from those which
describe the cytoplasm/nucleus isotherm (Fig. 2).

those in nucleus and iRP. The slope of the linear component indicates
that approximately 80% of cytoplasmic water is unavailable to ATP.
Apparently there are cytoplasmic regions which effectively exclude ATP
from entry. Several organelles can be suggested. First are the
mitochondria which selectively extrude ATP (Erecinska and Wilson, 1982)
and are abundant in oocyte cytoplasm. But mitochondrial compartmentalization cannot account for the bulk of water unavailable to ATP. A
more abundant candidate organelle is the yolk platelet. Stereographic
measurements show that platelets occupy about 70% of the cytoplasmic
volume, while other data show that platelets contain about 60% of
cytoplasm's water (Tluczek et al., 1984). Yolk platelets are membrane-
enclosed and contain a protein crystal whose water of hydration is known
to have a reduced capacity to dissolve sucrose (Tluczek et al., 1984;
Wallace, 1963). Together, yolk platelets and mitochondria can account
for most of cytoplasm's ATP-excluding water. Other organelles may also
play this role. In fact, because of ATP's size and polar nature, every
membrane-bounded organelle and every organelle that possesses a
crystalline or semicrystalline hydrate region is suspect.

The second significant feature of cytoplasmic ATP isotherms is the
nonzero intercepts of the extrapolated linear component. These imply
that about 0.9 mM of cytoplasmic ATP is "bound" in a manner that
restricts the translational diffusion required to enter the nucleus or
iRP. Furthermore, the data available for cyanide-poisoned cells (Fig.
2) suggest that this fraction is in a mass action-like equilibrium with
the cytoplasm's diffusive or "free" ATP. We do not know where in
cytoplasm the bound ATP fraction is located. NMR experiments, currently
underway, may help determine whether this ATP is in vesicle-like
structures (and therefore free to undergo rotational motion) or is
adsorbed to macromolecules.

Whether sequestered in vesicles or adsorbed, the presence in cytoplasm of a dissociable-bound fraction raises the interesting possibility
that ATP "activity" (i.e. availability to enter into metabolic reactions)
may be buffered. In other words, in metabolically deficient cells free
ATP may not be determined solely by rates of synthesis and utilization,
but may be kept from decreasing to unacceptable levels by the buffering
capacity of bound reserves. Considering ATP's importance to the cell,
some buffering capacity would seem to have survival value, not only in
the oocyte but in other cells as well.

ACKNOWLEDGEMENTS

This work was supported by NIH grants GM 19548 and HD 12512 and by an institutional grant from the United Foundation of Greater Detroit. We thank P. Paine and Y.T. Lau for useful discussions and S. Harmon, L.J.M. Tluczek, and J. White for technical assistance.

REFERENCES

Erecinska, M. and Wilson, D.F., 1982, J. Membr. Biol., 70:1.
Frank, M. and Horowitz, S.B., 1978, J. Cell. Sci., 19:127.
Horowitz, S.B. and Miller, D.S., 1984, J. Cell. Biol. 99:172s.
Horowitz, S.B. and Paine, P.L., 1974, Biophys. J., 25:45.
Horowitz, S.B., Paine, P.L., Tluczek, L. and Reynhout, J.K., 1979,
    Biophys. J., 25:33.
Lust, W.D., Feussner, G.K., Barbehehn, E.K. and Passoneau, J.V., 1981,
    Anal. Biochem., 110:258.
Siebert, G. and Humphrey, G.B., 1965, Adv. Enzymol., 239.
Tluczek, L., Lau, Y.T. and Horowitz, S.B., 1984, Dev. Biol., 104:97.
Wallace, R., 1963, Biochem. Biophys. Acta, 74:495.

ORGANIZATION OF MACROMOLECULAR SYNTHESIS

NEW VIEWS OF CELL AND TISSUE CYTOARCHITECTURE:

EMBEDMENT-FREE ELECTRON MICROSCOPY AND BIOCHEMICAL ANALYSIS

Edward G. Fey and Sheldon Penman

Department of Biology
Massachusetts Institute of Technology
Cambridge, Massachusetts 02139

Our perception of the cell interior is undergoing a fundamental
change. What once appeared to be formless "plasms" are now seen to
contain highly organized structural scaffolds. Even the "soluble"
proteins, those not a part of skeletal structures, are no longer
considered free, but associated with the skeletal framework. The
implications of these studies of cell structure are beginning to emerge.
The end result will likely be a change in many of the paradigms of our
science. As with most changes in paradigm, progress is halting and not
without conflict.

The images observed in micrographs of cytoskeletal networks are often
reminiscent of the architecture of Buckminster Fuller. The resemblance
is not fortuitous since, like spider webs, cytoskeletal fibers are
organized into "tensegrity" structures in which there are no bending
moments. These complex space-filling structures are continuous from the
nuclear interior to the cell surface. The cytoskeletal networks are
extensive in tissues, especially epithelial sheets, where the filament
networks interconnect at the cell borders forming a tissue-wide skeletal
framework. This macrostructure may serve as a scaffold for the
establishment of differentiated tissue morphology.

Using new techniques for electron microscopy and cell fractionation,
two different views of the cell interior have emerged. The dense "microtrabeculae" described by Porter are observed when intact cells are
prepared as unembedded sections. These images of microtrabeculae suggest
that all cellular components are organized into structural domains. The
new imagine technique (more correctly an old technique resurrected
[Anderson, 1951]) consists of viewing samples in the electron microscope
with no embedding material present. This technique is possible because
the cell and its structural networks are self-supporting. Proteinaceous
structures, previously masked by the embedding substance, are viewed with
great clarity in these unembedded specimens. Applying the technique to
intact cells reveals a microtrabecular lattice not observed in embedded
specimens (Figure 1a) (Guatelli et al., 1982; Wolosewick and Porter,
1976, 1979).

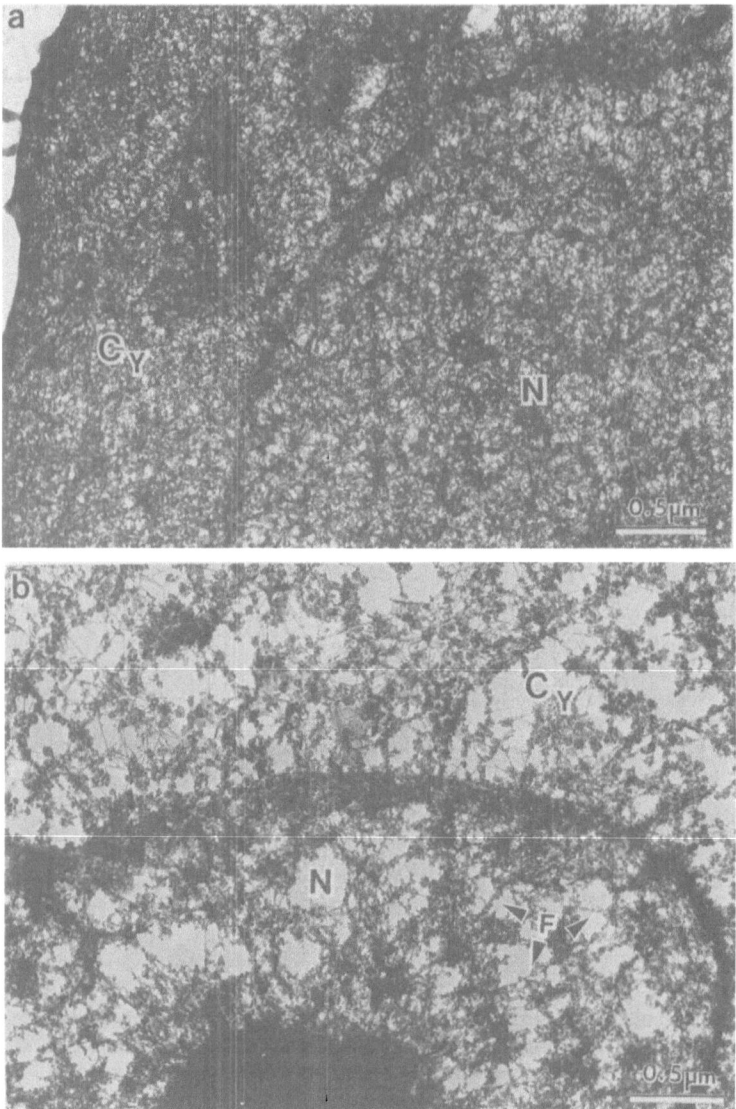

Fig. 1. Transmission electron micrographs comparing an intact HeLa nucleus
and a HeLa nucleus after extraction with Triton X-100. Intact
cells (a) and cells extracted with Triton X-100 (b) were prepared
as unembedded sections (Capco et al., 1984). The intact cell (a),
seen in a 0.2 μm section, appears as a "microtrabecular lattice"
as described by Porter (Guatelli et al., 1982; Wolosewick and
Porter, 1976, 1979). The extracted cell (b) is a skeletal framework, the structure that remains after phospholipid and >70% of
the protein is removed (Ben-Ze'ev et al., 1979; Lenk et al., 1977;
Herman et al., 1978; Fulton et al., 1980; Fey et al., 1984). The
extracted cell is less dense than the intact cell, yet many aspects
of cellular morphology are retained in the skeletal framework.

Underlying the microtrabecular lattice are the structural elements of the cell, the skeletal framework and the nuclear matrix. These elements are obscured by the microtrabecular lattice, but are seen clearly when the phospholipids and soluble proteins (greater than 70% of the cellular proteins) are removed by extraction with non-ionic detergents. The skeletal framework, consisting of 20-30% of the cellular protein, is seen to be qualitatively different from the trabecular lattice (Figure 1). These images of cytoskeletal elements require the concerted use of embedment-free microscopy and a reliable means of removing phospholipids and soluble proteins. Fortunately the fractionation and embedment-free techniques are simple and require no special equipment or skills not already present in most laboratories.

The clarity of images of protein networks obtained from embedment-free sections is seen in Figure 2. A 0.2% collagen gel was prepared for electron microscopy in both epon-embedded (Figure 2a) and unembedded sections (Figure 2b). The image obtained from the embedded section shows the collagen as a series of unconnected fibers and black dots, presumably due to the filaments at grazing incidence to the section surface and filaments intersecting the section obliquely. In contrast, the image from the unembedded section (Figure 2b) shows the collagen gel to be an interconnected filament network. Figure 2c and 2d compare images of a 2%

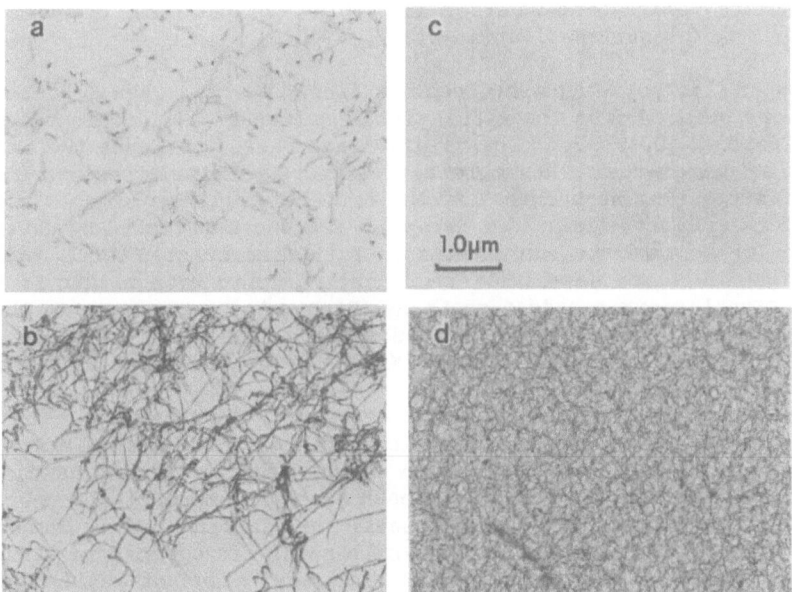

Fig. 2. Comparisons between TEM images of epoxy sections and embedment-free sections of collagen and agar gels. All sections are of equivalent thickness (gold interference). The Epon-Araldite sections (a&c) were post-stained with uranyl acetate and lead citrate. The embedding medium was removed from the DGD sections (b&d) with n-butyl alcohol as described by Capco et al. (1984). The epon-embedded 0.2% collagen gel (a) contains many short discontinuous filaments. The embedment-free collagen section (b) shows a network of collagen fibers. A 2% agar gel embedded in epon (c) did not produce an identifiable image, while the same gel viewed in an unembedded section (d) shows a complex agarose network in high contrast.

agarose gel in embedded and unembedded sections of equal thickness. Essentially no structure is visible in the embedded section (Figure 2c), although care was taken to insure that a sample was present. Presumably, the heavy metal stains used in this preparation have little affinity for the polysaccharides of the agar gel and thus do not enhance the image. In contrast, the unembedded section (Figure 2d) shows the polymeric agar network with great clarity.

Why does the view of the protein fiber networks afforded by transmission electron micrographs of unembedded samples (Figure 1) appear so different from that obtained using embedded samples? Conventional electron microscopy, employing embedded sections, is affected by the presence of the embedding material to a degree not often appreciated. The embedding resins, especially those based on epoxy, scatter electrons to essentially the same degree as the embedded specimen. Thus, the specimen is rendered invisible except where heavy metal stains are bound (Guatelli et al., 1982). Since the heavy metals do not penetrate the plastic interior of the section, the stain is concentrated at the surface of the section (Meek, 1976; Shalla et al., 1964). The familiar electron micrograph of an intact cell is essentially a two-dimensional slice a few angstroms thick. The three-dimensional organization of cellular morphology is poorly represented in embedded sections. Such images, formed of exceedingly thin planes, are excellent for visualizing many biological structures, such as membrane cross-sections. However, they are unsuitable for three-dimensional networks of protein filaments (Capco et al., 1982) such as those constituting the cytoskeletal framework (i.e. the surface lamina [Ben-Ze'ev et al., 1979] and protein filament networks surrounding the nucleus [Penman et al., 1983]).

The utility of embedment-free microscopy emerges when soluble proteins are removed from the cytoskeletal networks. Non-ionic detergents like Triton X-100 remove phospholipids and thereby release the soluble cellular components. The skeletal framework remains after extraction as a self-supporting structure. If the extraction buffer closely mimics the interior cellular milieu, the skeletal framework revealed by detergent extraction retains the configuration of the unextracted cell. Such preparations appear devoid of structural elements when viewed in conventional epoxy-embedded sections (Figure 3a). However, in the unembedded whole mount or resin-free section, the dense three-dimensional protein networks are made visible (Figure 3b-d) (Capco et al., 1982).

The whole-mount preparation has important technical limitations. Cells must be relatively thin, and the filament networks relatively sparse for the images to be interpretable. Also, structures such as desmosomes, centrioles and nuclear pores have been seen predominantly as cross-sections in conventional embedded sections. The three-dimensional appearance of these structures in whole-mount micrographs is almost unrecognizable. Perhaps the most serious limitation of whole-mount microscopy is the inability to image the interior of the nucleus because of the surrounding nuclear lamina.

Unembedded sections for electron microscopy can be prepared quite easily. The method was pioneered by John Wolosewick (1980) who used polyethylene glycol as a removable embedding compound. Our own contribution has been to employ diethylene glycol distearate as the removable embedding medium (Capco et al., 1984) and to apply the technique to fractionated cytoskeletal structures.

We have been analyzing the organization and composition of

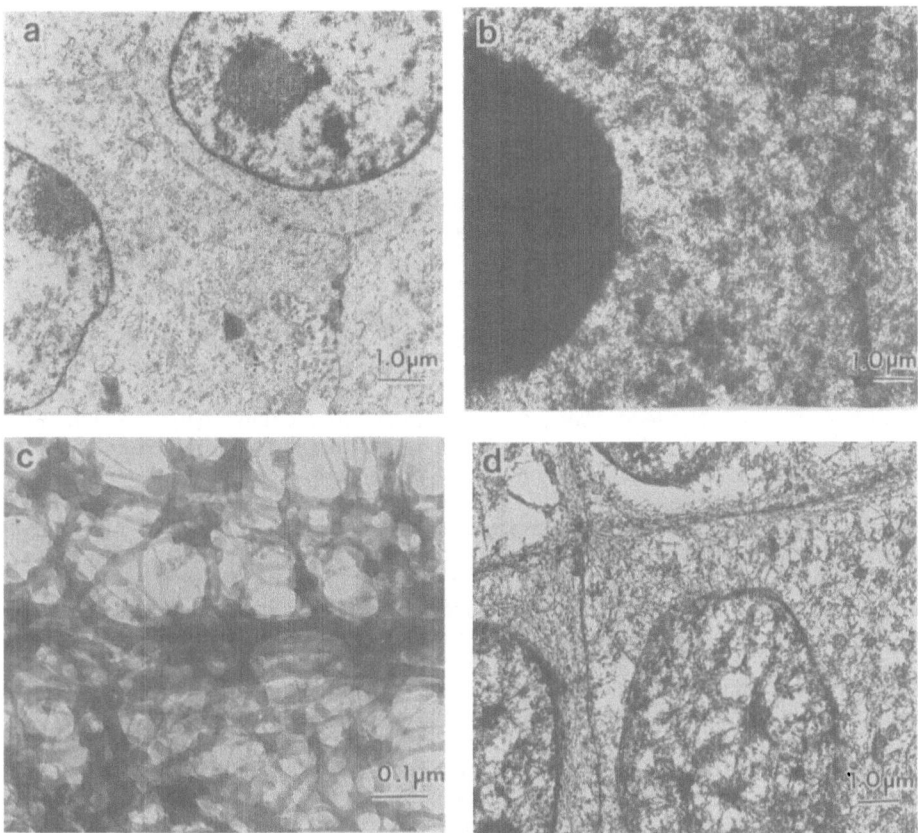

Fig. 3.   TEM images of an epoxy-embedded thin section (a), unembedded
          whole mounts (b&c), and an embedment-free section of MDCK cells
          (d) all after extraction with Triton X-100.   The sections are
          transverse and correspond geometrically to the apical view of
          the whole-mount preparation (b&c).   The cytoplasm of the epoxy-
          embedded section (a) appears as a sparse region with little
          indication of cytoskeletal structure.   The whole-mount images
          of identical cells (b) show the cytoplasm to be densely
          organized with a complex network of overlapping filaments.   At
          higher magnification (c) these filaments are seen to display
          heterogeneous diameters.   The embedment-free section (d) allows
          observation of the filament organization in more detail.
          Filaments are observed as a continuous network throughout the
          cytoplasm.   There are also filaments that are in direct contact
          with the nuclear lamina.

cytoskeletal structures for a number of years and have developed
procedures that optimize both morphological preservation and biochemical
fractionation of elements involved in cytoarchitecture.   These studies
initially employed single cells grown in culture (Capco et al., 1982;
Ben-Ze'ev et al., 1979; Lenk et al., 1977; Herman et al., 1978; Fulton et
al., 1980).   More recently, the analysis involved model cell systems
having tissue-like organization (Fey and Penman, 1984; Fey et al., 1984).
The preparation of cytoskeletal structures with Triton X-100 in
physiological buffer has been very carefully characterized, and the

separation of soluble and structural fractions has been shown to have a firm biochemical basis. Further dissection of cells with nuclease digestion followed by elution with ammonium sulfate reveals a more fundamental structure, the nuclear matrix-intermediate filament (NM-IF) scaffold. This network of nuclear and cytoplasmic filaments is composed of a small fraction (usually 5% less) of total cell proteins, yet retains many aspects of differentiated morphology. When epithelia are fractionated in this manner the differentiated morphology of the tissue type is retained (Figure 4) (Fey et al., 1984). The preparative procedures also yield fractions that are distinct in composition; each fraction retains a unique subset of proteins.

Much of our recent work has employed a line of epithelial Madin Darby canine kidney (MDCK) cells. These cells were derived from distal kidney tubules and retain, to a remarkable degree, the simple cuboidal phenotype characteristic of these cells in vivo (Madin and Darby, 1975; McRoberts et al., 1981). Figure 3 shows electron micrographs of the skeletal framework from an MDCK colony using an epoxy-embedded section (Figure 3a), an unembedded whole mount (Figure 3b and 3c), and an embedment-free section (Figure 3d). The cells were extracted with 0.5% Triton X-100 in a physiological buffer and prepared for TEM (Capco et al., 1984). The sections (Figure 3a and 3d) were cut parallel to the surface of the culture dish and all preparations observed at the same magnification. The epon thin section (Figure 3a) shows well preserved junctional complexes, but cell interiors show a random array of dots and some filament segments surrounding a nucleus. There is little to suggest the organization of the detergent-resistant cytoskeleton. The whole mount in Figure 3b shows the existence and organization of the filament network. At this magnification, details of filament organization are obscured by the density of superimposed filaments. When the cytoskeletal network of this whole mount is viewed at a higher magnification (Figure 3c) considerable detail of filament-filament interaction becomes apparent. The embedment-free section, Figure 3d, shows the nucleus with nucleoli and cytoskeletal filaments as an anastomosing network throughout the cell interior. This image provides a clear indication of the filamentous perichromatin networks in the nuclear interior. The nuclear lamina is seen to be a structural interface between chromatin inside the nucleus and cytoplasmic filaments that appear to associate directly with the nuclear surface.

Methods of further cell dissection have been developed which remove chromatin from the nucleus and the majority of all structural elements from the cytoplasm. The resulting structure, consisting of the nuclear matrix, intermediate filaments and junctional complexes including desmosomal proteins appears to comprise a fundamental structural entity found in all cells. We have suggested the term nuclear matrix-intermediate filament (NM-IF) scaffold, a designation whose rationale is most apparent in studies of tissue (Fey et al., 1984).

The NM-IF scaffold preparation, consisting of only 5% of the total cellular protein, affords striking images when examined as a whole mount by transmission electron microscopy. Apical views of the NM-IF scaffold derived from an MDCK epithelial sheet are shown in Figure 4. The filament network is much simpler in composition and overall organization than the skeletal framework (Figure 3). However, the structure remaining after removal of phospholipid, nucleic acids, and 95% of the total cellular protein still clearly retains the organization characteristic of an epithelium. The polygonal outline of the epithelial cells is still identifiable from the remnant cell-boundary structures. Included in

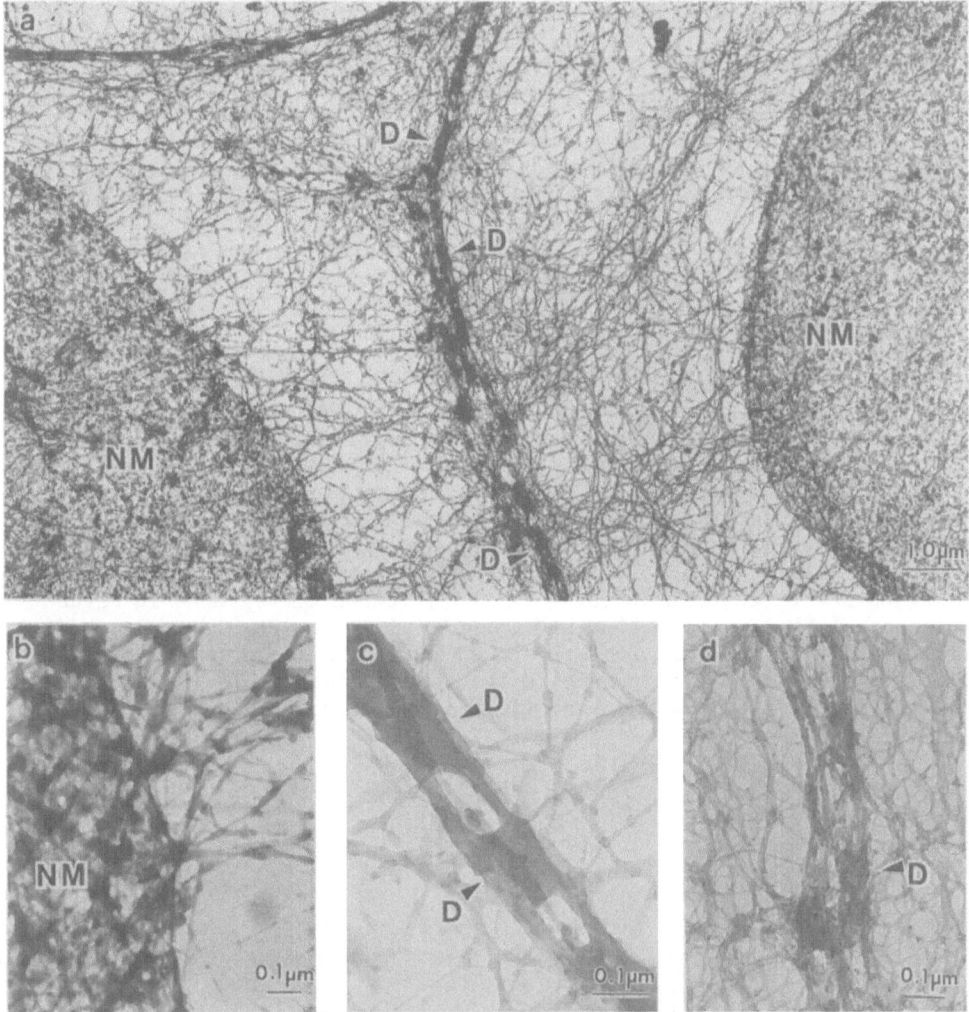

Fig. 4.   Nuclear matrix-intermediate filament (NM-IF) scaffold
structures of MDCK epithelial colonies. MDCK epithelial
colonies were fractionated as described by Fey et al. (1984).
The NM-IF scaffolds viewed in whole mount transmission electron
micrographs (a-d) retain only 5% of the total cell protein and
virtually no phospholipid or nucleic acid. The chromatin-
depleted nuclear matrices (NM) (a) are associated with
cytoplasmic filaments largely consisting of cytokeratins.
These cytokeratin networks often terminate in residual
desmosome structures (D). Higher magnifications of this
structure provide further details of the nuclear matrix (b) and
desmosomal complex (c&d) structures. NM-IF core structures
were fixed in formaldehyde, and antibody reactions were carried
out with desmosomal antiserum followed by a second antibody
conjugated to gold beads as described. Greater than 75% of the
gold bead staining from this antibody is observed in the dense
structures tentatively identified as residual desmosomal cores
(d).

these boundaries are dense plaques that serve as termini for bundles of filaments. These dense plaques have been identified as remnant desmosomal structures by immuno-electron microscopy (Figure 4d).

A web of filaments, most of 10 nm diameter, with many originating at the nuclear surface, extend throughout most of the cytoplasmic space and form a continuous network throughout the epithelial colony. These are the cytokeratins whose organization is revealed here with striking clarity. In the NM-IF network, numerous filaments appear to originate on the surface of the nuclear matrix (Figure 4a and 4b), and many terminate at the dense plaques or desmosomes of the remnant junctional complexes (Figure 4c). Some of these fibers form direct connections between the desmosomes and the nuclear matrix.

Aside from the obvious significance for fundamental cell biology, these procedures may have profound implications for pathological studies that make use of cell structure and its anomalies as a diagnostic tool. Our experience in this regard is quite striking. Whenever a change in cell morphology is detectable, either with the light microscope or by conventional electron microscopy, an enormous alteration is observed in the cytostructure seen by the extraction-based, embedment-free microscopy (Fey and Penman, 1984). The retention of phenotypic alterations in the purified NM-IF scaffold is shown in the electron micrographs in Figure 5. The control NM-IF preparation in Figure 5a shows the regular polygonal organization of the cell borders. When the cells are given a temporary malignant phenotype by exposure to the tumor promoter TPA (Figure 5b), the organization of the NM-IF scaffold is altered and shows the formation

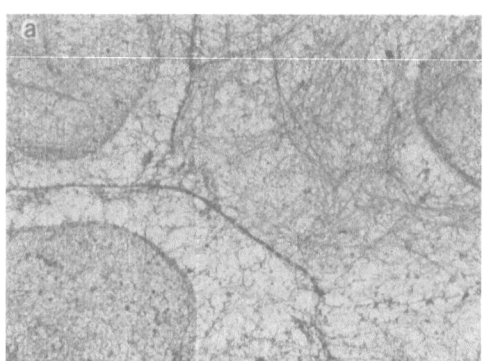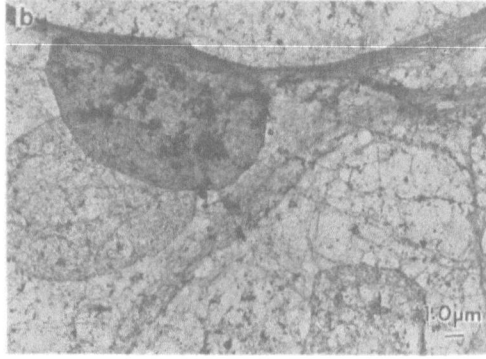

Fig. 5.   Nuclear matrix-intermediate filament (NM-IF) scaffold structures of MDCK epithelial colonies: efffects of TPA on epithelial organization. MDCK colonies were treated with 0.02% DMSO (a) or the potent tumor promoter 12-0-tetradecanoyl phorbol-13-acetate (TPA) in DMSO for 4 hours at 37°C. The cells were fractionated and viewed as whole mounts in transmission electron micrographs as described (Fey et al., 1984). Control MDCK colonies displayed characteristic epithelial morphology (a) while the TPA treated colonies were observed to flatten, become motile and extend numerous processes. These aspects of morphological alteration were retained in the NM-IF scaffold preparations of TPA-treated cells (b) (Fey and Penman, 1984). The nuclear matrices were observed to elongate and dense bundles of cytoplasmic filaments are often in close tangential association with the nuclear matrix.

of filament bundles and a breakdown of junctional complexes.

The power of the concerted approach is apparent in the biochemical analysis of structures seen in the electron micrographs above. The proteins obtained in the cytoskeleton, chromatin, and NM-IF fractions, together with the soluble protein from the initial Triton X-100 extract, represent four distinct protein subsets of the cell, as is shown in the two-dimensional gel electropherograms in Figure 6. The soluble fraction contains the majority of cellular proteins, and its electropherogram shows a complex, dense pattern of major proteins. Each of the structural fractions (the cytoskeleton, chromatin and NM-IF fractions) has a characteristic protein pattern whose major proteins are found predominately in only one of the three fractions. Some of the proteins characteristic of each fraction are identified by arrows (Figure 6). Careful examination of these electropherograms shows few proteins in more than one fraction. Thus, the sequential extraction technique produces subfractions that represent unique populations of cellular proteins, as well as morphologically distinct structures.

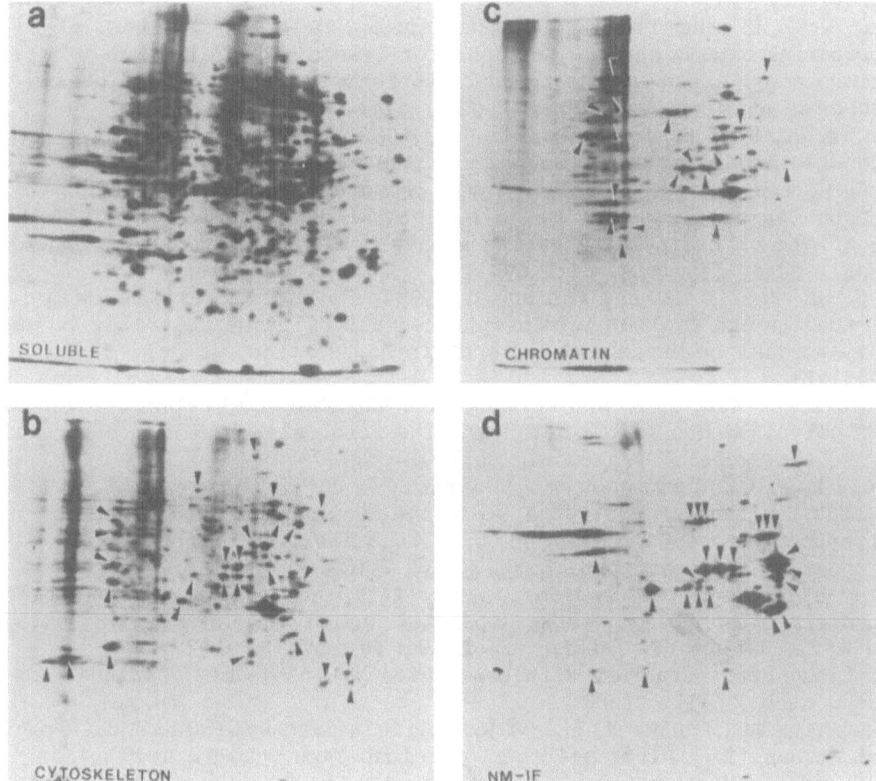

Fig. 6.   Two-dimensional gel profiles of proteins obtained after fractionation of MDCK colonies. Fractionation and two-dimensional gel electrophoresis were carried out as described by Fey et al. (1984). The first dimension ranges from pH 10 to pH 3 (left to right). The patterns of the cytoskeleton, chromatin and NM-IF fractions are individually characteristic with little overlap of protein from one fraction to another. Major proteins that are found predominately in only one of these fractions are indicated by arrowheads.

The combined morphological and biochemical approach to cell and tissue structure has so far been applied to individual cultured cells and cultured cell models of tissue. The next obvious technical development will apply these methods to tissues in general. The extension will not be trivial, but the potential benefits to pathology are enormous. A replacement for the frozen tissue section with far higher resolution is possible using these techniques. The use of diagnostic antibodies is greatly enhanced by the extraction techniques. Since the methodologies are simple and well within the capability of most laboratories, a powerful method for the analysis of tissue and organ structure is now available.

REFERENCES

Anderson, T. F., 1951, Techniques for the preservation of three-dimensional structure in preparing specimens for the electron microscope, Trans. N.Y. Acad. Sci., Ser. II, 13:130.

Ben-Ze'ev, A., Duerr, A., Solomon, F. and Penman, S., 1979, The outer boundary of the cytoskeleton: A lamina derived from plasma membrane proteins, Cell 17: 859.

Capco, D. G., Krochmalnic, G. and Penman, S., 1984, A new method of preparing embedment-free sections for transmission electron microscopy: Applications to the cytoskeletal framework and other three-dimensional networks, J. Cell Biol., 98:1878.

Capco, D. G., Wan, K. M. and Penman, S., 1982, The nuclear matrix: Three-dimensional architecture and protein composition, Cell, 29:847.

Fey, E. G. and Penman, S., 1984, Tumor promoters induce a specific morphological signature in the nuclear matrix-intermediate filament scaffold of Madin-Darby canine kidney (MDCK) cell colonies, Proc. Natl. Acad. Sci. USA, 81:4409.

Fey, E. G., Wan, K. M. and Penman, S., 1984, Epithelial cytoskeletal framework and nuclear matrix-intermediate filament scaffold: Three-dimensional organization and protein composition, J. Cell Biol., 98:1973.

Fulton, A. B., Wan, K. M. and Penman, S., 1980, The spatial distribution of polyribosomes in 3T3 cells and the associated assembly of proteins into the skeletal framework, Cell, 20:849.

Guatelli, J. C., Porter, K. R., Anderson, K. L. and Boggs, D. P., 1982, Ultrastructure of the cytoplasmic and nuclear matrices of human lymphocytes observed using high voltage electron microscopy of embedment-free sections, Biol. Cell, 43:69.

Herman, R., Weymouth, L. and Penman, S., 1978, Heterogeneous nuclear RNA-protein fibers in chromatin depleted nuclei, J. Cell Biol., 78:663.

Lenk, R. L., Ransom, Y., Kaufmann, Y. and Penman, S., 1977, A cytoskeletal structure with associated polyribosomes obtained from HeLa cells, Cell, 10:67.

Madin, S. A. and Darby, N. B., 1975, American Type Collection Catalogue of Strains II. First edition. American Type Culture Collection, Rockville, MD.

McRoberts, J. A., Taub, M. and Saier, M. H., 1981, The Madin Darby canine kidney (MDCK) cell line, in: "Functionally Differentiated Cell Lines," G. Sato, ed., Alan R. Liss, New York.

Meek, G. A., 1976, "Practical Electron Microscopy for Biologists," Wiley, New York.

Penman, S., Capco, D. G., Fey, E. G., Chatterjee, P., Reiter, T., Ermisch, S. and Wan, K. M., 1983, The three-dimensional structural networks of the cytoplasm and nucleus, in: "Spatial Organization of Eukaryotic Cells," J. R. McIntosh, ed., Alan R. Liss, Inc., New York.

Shalla, T. A., Carroll, T. W., DeZoeten, G. A., 1964, Penetration of stain into ultrathin sections of tobacco mosaic virus, Stain Technol., 39:257.

Wolosewick, J., 1980, The application of polyethylene glycol to electron microscopy, J. Cell Biol., 86:675.

Wolosewick, J. J. and Porter, K. R, 1976, Stereo high-voltage electron microscopy of whole cells of the human diploid line WI-38, Am. J. Anat., 147:303.

Wolosewick, J. J. and Porter, K. R, 1979, Microtrabecular lattice of the ground cytoplasmic substance: artifact or reality, J. Cell Biol., 82:114.

UNDERSTANDING THE ORGANIZATION OF CELL METABOLISM IN

EARLY EMBRYONIC SYSTEMS: DEVELOPMENTAL IMPLICATIONS

Roberto Marco, Beatriz Batuecas, Manuel Calleja,
Mario Carratalá, Margarita Cervera,
Alberto Domingo, Carmen Ferreiro,
Rafael Garesse, Carlos Urquía
and Isabel Vernós

Instituto de Investigaciones Biomédicas del CSIC

and

Departamento de Bioquímica de la Facultad de Medicina
Universidad Autónoma de Madrid
Madrid   28029   SPAIN

INTRODUCTION

Although early in this century the study of cell metabolism in developmental systems was a major concern (e.g., see Needham, 1931, 1942; Willier et al., 1955), the current surge of molecular biology has shifted the interest of the majority of developmental biochemists to the study of the regulation of gene expression during development. Without discussing the obvious conceptual and historical reasons behind such change, this virtual elimination of a formerly fashionable subject has left many unanswered questions, if not for other reasons, as a consequence of the limited technical and conceptual potentiality of biochemistry half a century ago.

Two major points seem particularly relevant in the light of the topics of this Workshop: 1. The peculiar metabolic properties and organization of early embryonic systems and 2. The potential implications that these peculiar properties may have in our understanding of one of the major unsolved problems of current biology, the emergence of diversity during embryonic development.

An important principle to remember in this regard is, that biological systems make multiple uses of any element or process which is a part thereof. A corrolary of this principle is that, recognition of the function or role of a particular component or process does not necessarily preclude additional functions. In fact, when a system is amenable to genetic studies, this principle is often recognized in the pleiotropy shown by the elimination through mutation of defined components of the system.

THE PECULIAR PROPERTIES OF THE CELLULAR AND METABOLIC

ORGANIZATION DURING EARLY EMBRYONIC DEVELOPMENT

In Table 1, a series of properties, peculiar in one way or another to early developmental systems, is summarized. They can be simply

Table 1. Peculiar Properties of the Cellular and Metabolic Organization of Early Eucaryotic Developmental Systems*

A. Huge cells : complexity unknown?

B. Cytoskeletal organization : practically unknown

C. Different overall cytoplasmic organization : bound water, diffusion constants, cytoplasmic pH, others?

D. Presence of specific, relatively poorly characterized subcellular organelles: protein and lipid yolk spheres, cortical granules, polar granules, etc.

E. Self-contained, practically isolated from the environment

F. Many components (maternal) stored in the cytoplasm in an ill-defined way: ions (Ca^{++}....), metabolites, enzymes, mRNA, tRNAs, ribosomes, nuclear and cell membrane components, subcellular organelles, etc.

G. Relative inhibition and/or repression of many (if not all) cellular and metabolic activities:

Early released after activation (fertilization):

    1. Membrane potential (very fast)
    2. Intracellular pH (very fast)
    3. Protein translation (5 min)
    4. Transport of external components (different in different organisms: see E)
    5. DNA replication and nuclear cell division (slower)

Later released after activation:

    1. RNA transcription (mid-blastula transition)
    2. Cellular respiration (progressive during early development)

H. Coupling of the different levels of regulation : Positional information and plasticity of the whole system (gap junctions, extracellular matrix, cell-cell interactions, etc.)

---

*Compiled from Balinsky and Fabian (1981), Browder (1984), and Davidson (1976).

considered the consequence of the need of dissociating two basic biological processes, normally tightly linked: a) the increase in cellular components needed for cell division and b) the actual event of cell division, preceding cell differentiation. This dissociation, which shifts most of process a) from early development to oogenesis, is required if early developmental systems, which have to pass through the single haploid genome stage to undergo sexual mating, are to proceed fast enough to become self-sufficient, independent, multicellular, differentiated organisms with as few requirements as possible from the external environment.

This seemingly contradictory situation, of being unicellular from the point of view of the genomic constitution of the egg, but harboring enough cellular components to become pluricellular (sometimes very extensively pluricellular), in a record period of time is solved by the mature oocyte by developing special mechanisms of inhibition and repression of the function of its huge cytoplasmic machinery, which remain mostly poorly characterized. For example, it has been known since the studies of Warburg (1908) at the beginning of the century, that respiration of the mature oocytes and early embryos is much lower than what could be expected from its huge cytoplasm, and that it increases markedly during early development to achieve normal levels at some point after cleavage has been completed and differentiation has started. On the other hand, molecular biological studies (Dawid, 1970) have shown that the mature oocyte contains a mitochondrial DNA supply sufficient for the developed animal, and that very little (or none) of this DNA synthesis occurs during early development.

Recently, we have reexamined this problem in two biological systems, the early development of Drosophila melanogaster and Artemia, a crustacean, and found that, at least in part, it can be explained by the storage in a masked form of a significant population of mitochondria inside the yolk granules--from which they are released in an active form at specific times during development (Marco et al., 1981, 1983). This finding suggests that the yolk granule, besides its well known role of providing the early embryo with amino acids (the principal precursors of cell metabolism in these systems), can have subsidiary roles of storage of other preformed components, e.g., enzymes (like proteases [Ezquieta and Vallejo, 1986] and DNases [Domingo, A., Cervera, M., Martin, E., and Marco, R., submitted]) or organelles like mitochondria, which can be released in an active form at the appropriate time of development and, perhaps, in the required region of the embryo.

POTENTIAL IMPLICATIONS FOR EARLY DEVELOPMENT

One way of considering the above described properties of early developmental systems is that they constitute adaptations to the peculiar situation encountered by these systems, to lack of efficient external feeding sources, to uncertainties when the actual switch-on of development will occur (normally controlled by insemination), etc. If this is the case, many of the events occurring during early embryonic development can be simply interpreted as the substitution of this peculiar "metabolism" by the normal cellular metabolism proper of the developed organism. On the other hand, one must not forget that this substitution occurs while the process of development itself is actually taking place. The precise origin of diversity in the developing embryo is still one of the major unsolved questions in biology. Although not yet proven, it is widely agreed that the multiplication and distribution of equivalent genomes across the huge cytoplasmic domain of the fertilized oocyte leads to their differential activation by inhomogeneities produced in the cytoplasm once it has become compartmentalized after cleavage. These inhomogeneities may have arisen a) from the preformed inhomogeneous cytoplasmic organization of the oocyte, or b) de novo from an initially homogeneous cytoplasm through self-organizing processes involving both cytoplasmic diffusion-driven dissipative structures and/or cell-cell interactions.

As suggested more than half a century ago (Child, 1941), it is possible that metabolic processes may be involved in the production of these inhomogeneities. In fact, the above discussed desinhibition and/or derepression of different metabolic processes (Table I) may be one of the specific ways in which these inhomogeneities may be produced, particularly if the desinhibition and/or derepression does not occur uniformly across the embryo.

In principle, the minimal kinetic and diffusion properties, of a self-organizing system driven from a homogeneous initial system by random fluctuations, have been defined (Gierer, 1981; Lacalli and Harrison, 1979). Some of the metabolic processes specified above may fullfil many of them.

Several arguments suggest, although do not prove, that, indeed, metabolic and/or ionic gradients may be involved in the production of morphogenetic fields and in the generation of positional information in higher organisms. The genetic evidence obtained using homeotic transformations in Drosophila (Postlethwait and Sheneiderman, 1974; Morata, 1975) and the experimental evidence obtained in grafting homotopic and heterotopic, homologous and heterologous structures in different organisms (Fallon and Crosby, 1977) suggest that the elements involved in the generation of these morphogenetic gradients, if they exist, have to be of a very fundamental, general and widespread nature. Furthermore, the difficulty of finding evidence of the chemical nature of morphogenetic gradients by direct or indirect (genetic) means also suggests that they must be of a very simple and basic nature. In systematic genetic analyses (Nusslein-Volhard et al., 1984; Jurgens, 1984) of mutations affecting Drosophila development, mutations with properties suggestive of affecting directly the production of these hypothetical morphogenetic gradients failed to appear. This finding suggests either that they are not the actual mechanism of generation of diversity in the embryo (that they do not have the postulated properties of current models), or that they correspond to very basic cell functions. In the latter case, mutations would be lethal, making it very difficult to study them either in the whole organism or even in isolated cells (for instance, by clonal analysis).

If the state of water in the cytoplasm of cells (Clegg, 1982) is different from that traditionally accepted, particularly in the case of these early developmental cells like oocytes and early embryos, it may have implications through its direct or indirect effects on the diffusion constants of the metabolites involved in the formation of gradients (Kauffman et al., 1978).

Finally, the cytoskeletal organization of the oocyte and early embryo may also be of crucial importance, a) by stabilizing any preformed or self-generated inhomogeneities during oogenesis or early embryogenesis or b) by providing alternatives to simple diffusion models for directional transport of the components involved in the establishment of the inhomogeneities in the embryo, particularly in the case of Drosophila and other insects where the initial development occurs in a syncytial embryo.

ACKNOWLEDGEMENTS

The expert typing of the manuscript by Pilar Ortega and the financial support of the CAICYT, FIS, and CONIE are gratefully acknowledged.

REFERENCES

Balinsky, B.I. and Fabian, B.C., 1981, "An Introduction to Embryology" (5th ed.), Saunders, Philadelphia.

Browder, L.W., 1984, "Developmental Biology" (2nd ed.), Saunders, Philadelphia.

Child, C.M., 1941, "Patterns and Problems of Development", University of Chicago Press, Chicago.

Clegg, J.S., 1982, Interrelationships between water and cell metabolism in Artemia cyts. IX. Evidence for organization of soluble cytoplasmic enzymes, Cold Spring Harbor Symposium on Quantitative Biology, 46:23.

Davidson, E.H., 1976, "Gene Activity in Early Development", Academic, New York.

Dawid, I.B., 1970, Cytoplasmic DNA, in: "Oogenesis", J.D. Biggers and A.W. Schuetz, eds., University Park Press, Baltimore.

Ezquieta, B. and Vallejo, C.G., 1986, The trypsin-like proteinase of Artemia. Yolk localization and developmental activation, Comp. Biochem. Physiol. B, in press.

Fallon, J.F. and Crosby, G.M., 1977, Polarizing zone activity in limb buds of amniotes, in: "Vertebrate Limb and Somite Morphogenesis", D.A. Ede, J.R. Hinchcliff, and M. Balls, eds., Cambridge University Press, Cambridge.

Gierer, A., 1981, Generation of biological patterns and form: Some physical, mathematical and logical aspects, Prog. Biophys. Molec. Biol., 37:1.

Jurgens, G., Wieschaus, E., Nusslein-Volhard, C. and Kluding, H., 1984, Mutations affecting the pattern of the larval cuticle in Drosophila melanogaster. II Zygotic loci on the third chromosome, Wilhelm Roux Archives, 193:283.

Kauffman, S.A., Shymko, R.M. and Trabert, K., 1978, Control of sequential compartment formation in Drosophila, Science, 199:259.

Lacalli, T.C. and Harrison, L.G., 1979, Turing's conditions and the analysis of morphogenetic models, J. Theoret. Biol., 76:419.

Marco, R., Batuecas, B. and Garesse, R., 1983, Mitochondrial storage in Drosophila melanogaster yolk granules, Eighth European Drosophila Research Conference, Cambridge.

Marco, R., Garesse, R. and Vallejo, C.G., 1981, Storage of mitochondria in the yolk platelets of Artemia dormant gastrulae, Cell. Mol. Biol., 27:515.

Morata, G., 1975, Analysis of gene expression during development in the homeotic mutant contrabithorax of Drosophila melanogaster, J. Embryol. Exp. Morph., 34:19.

Needham, J., 1931, "Chemical Embryology", Cambridge University Press, Cambridge.

Needham, J., 1942, "Biochemistry and Morphogenesis", Cambridge University Press, Cambridge.

Nusslein-Volhard, C., Wieschaus, E. and Kluding, H., 1984, Mutations affecting the pattern of the larval cuticle in Drosophila melanogaster. I. Zygotic loci in the second chromosome, Wilhelm

  Roux's Archives, 193:267.

Postlethwait, J.H. and Sheneiderman, H.A., 1974, Developmental genetics
  of Drosophila imaginal discs, Ann. Rev. Genet., 7:381.

Warburg, O., 1908, Beobechtungen uber die Oxidationsprozesse in
  Seeigelei, Hoppe-Seyler's Z. Physiol. Chem., 57:1.

Willier, B.H., Weiss, P.A. and Hamburger, V., eds., 1955, "Analysis of
  Development", Saunders, Philadelphia.

ORGANIZATION OF BIOSYNTHETIC AND
BIODEGRADATIVE PROCESSES

ORGANIZATION OF ENZYMES IN THE

TRYPTOPHAN PATHWAY

John A. DeMoss

Department of Biochemistry and Molecular Biology
University of Texas Health Science Center at Houston
Houston, Texas   77005, U.S.A.

INTRODUCTION

The reactions which constitute the pathway for tryptophan biosynthesis exhibit an unusual degree of organizational diversity
throughout the biological world.   The component reactions of this
pathway, including those involved in general aromatic biosynthesis, are
organized in different patterns of multifunctional complexes; and this
diversity is reflected in specific patterns of gene-enzyme relationships.
Some of the first reported examples of multifunctional proteins and
channelling of intermediates between active sites in multifunctional
proteins were uncovered by the extensive biochemical and genetic studies
of this pathway.   To illustrate the organization of active sites in
multifunctional protein complexes, I will review several well-defined
examples from this pathway.   In this review my emphasis will be on the
current status of our understanding of the structure of multifunctional
enzymes and the possible functional role of enzyme organization within
this pathway.

The biosynthesis of the tryptophan occurs by the reactions diagrammed
in Fig. 1.   Beginning with erythrose-4-P and PEP, the first seven
reactions leading to the syntheseis of chorismate constitute a common
pathway for synthesis of all aromatic amino acids; while the last five
reactions are specifically involved in the synthesis of tryptophan.   All
organisms which synthesize their own tryptophan, including diverse
species among the procaryotes and lower eucaryotes, utilize this same
sequence of reactions.   However, unique patterns of gene-enzyme
relationships and of organization of the catalytic components exist in
various organisms.

TRYPTOPHAN SYNTHASE: INTERACTION OF ACTIVE SITE DOMAINS
IN A MULTIFUNCTIONAL PROTEIN

In one of the first cases of multifunctional proteins discovered,
David Bonner and his colleagues (Bonner et al., 1960) demonstrated that
the terminal enzyme in tryptophan biosynthesis in Neurospora crassa,

A. COMMON AROMATIC AMINO ACID PATHWAY

Erythrose-y-P

1. $\downarrow$ + PEP

3-Deoxy-D-arabino-heptulosonate 7-phosphate

2. $\downarrow$ + NAD$^+$

5-Dehydroquinate

3. $\downarrow$

5-Dehydroshikimate

4. $\downarrow$ + NADPH

Shikimate

5. $\downarrow$ + ATP

Shikimate 5-phosphate

6. $\downarrow$ + PEP

3-Enoylpyruvylshikimate 5-phosphate

7. $\downarrow$

Chorismate

B. TRYPOTPHAN-SPECIFIC PATHWAY

Chorismate

1. $\downarrow$ + glutamine   Anthranilate   synthase

Anthranilate

2. $\downarrow$ + PRPP   PR-transferase

N-5'-Phosphoribosylanthranilate (PRA)

3. $\downarrow$ PRA isomerase

1(0-Carboxyphenylamino)-

1-deoxyribulose 5-phosphate (CdRP)

4. $\downarrow$ InGP synthase

Indole-3-glycerol phosphate (InGP)

5. $\downarrow$ + Serine   Tryptophan synthase

Tryptophan

Fig. 1. The pathway for tryptophan biosynthesis. Abbreviations used
for intermediates are shown in parentheses; PRP is 5-
Phosphoribosyl-1-pyrophosphate; PEP is Phosphoenolpyruvate. **A.**
Reactions 1-7 are involved in the synthesis of chorismate,
which is a precursor of phenylalanine, tyrosine, and
tryptophan. **B.** Reactions 1-5 are specifically involved in
tryptophan biosynthesis from chorismate. The names of the
enzymes catalyzing the tryptophan-specific reactions are shown
on the right.

apparently encoded by a single gene, catalyzes the following three
reactions.

reaction 1: InGP + serine $\xrightarrow{B_6P}$ tryptophan + glyceraldehyde-3-P

reaction 2: indole + serine $\xrightarrow{B_6P}$ tryptophan

reaction 3: InGP $\rightleftharpoons$ indole + glyceraldehyde-3-P

Reaction 1 is the overall reaction catalyzed by tryptophan synthase,
while reactions 2 and 3 are reactions which together account for this
overall reaction. Analysis of a large number of mutants deficient in the
overall reaction suggested that reactions 2 and 3 are catalyzed at
distinct sites on the polypeptide chain. One class of mutations led to
the formation of altered proteins which were deficient in the overall
reaction but retained the capacity to catalyze either reaction 2 or 3.
These mutations mapped in segregated clusters in different regions of the
linear map of the gene (Bonner et al., 1960).

On the basis of studies with the partially purified enzyme, I
proposed a model for the organization of active sites for reactions 2 and
3 on the enzyme (Fig. 2) which permits channelling of the indole formed
by reaction 3f to the active site of reaction 2 without mixing with the
surrounding medium (DeMoss, 1962). In these studies, I showed that the
indole binding sites for reaction 3r (the back reaction of reaction 3)
and reaction 2 had quite distinct kinetic properties (Table 1). Along
with the observation that either reaction 2 or 3 was selectively altered
by mutations which mapped in different regions of the gene, these data
provided strong evidence that reactions 2 and 3 are catalyzed by distinct
regions of the polypeptide chain.

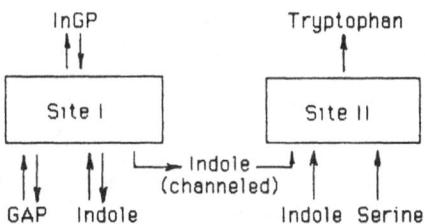

Fig. 2. Two-site model for the tryptophan synthase reaction (DeMoss,
1962). Site 1 catalyzes reactions 3f and 3r. Site 2 catalyzes
reaction 2. Reaction 1 is catalyzed when the two sites
interact.

Table 1. Kinetic Properties of Tryptophan Synthase from N. crassa[a]

Reaction	Conditions[b]	Product	V_{max}	K_m(Indole)
1	InGP + Serine + B_6P	Tryptophan	6.4	-
2	Indole + Serine + B_6P	Tryptophan	13.8	7×10^{-5}
3f	InGP	Indole	0.12	-
3f	InGP + Serine + B_6P	Indole	0.00	-
3r	Indole + GAP	InGP	4.9	2×10^{-3}
3r	Indole + GAP + Serine + B_6P	InGP	27.6	3×10^{-3}

[a]From DeMoss, 1962.
[b]Concentrations of substrate were saturating for V_{max} measurements.
Abbreviation: B_6P, pyridoxal phosphate; GAP, glyceraldehyde 3-phosphate; InGP, indole-3-glycerol phosphate.

The studies with the partially purified enzyme also showed that the two active sites interacted in some direct way. When radioactive InGP was converted to tryptophan in the presence of serine and pyridoxal phosphate, intermediate indole formed by reaction 3f was converted to tryptophan without mixing with a pool of unlabeled indole in the surrounding medium (DeMoss, 1962). This provided one of the first direct demonstrations of channelling of an intermediate between active sites on a single protein.

The kinetic studies also showed that binding of substrates to the active site for reaction 2 had a direct stimulatory effect on the V_{max} for reaction 3. When the partially purified enzyme was saturated with serine and pyridoxal phosphate, reaction 3r was stimulated 4-fold as shown in Table 1 (DeMoss, 1962). In addition it had also been shown that in many indole-accumulating mutants, which are deficient in reaction 2 but retain reaction 3, indole accumulation is stimulated by pyridoxal phosphate plus serine (DeMoss and Bonner, 1959). Thus the two active sites appear to be organized on the enzyme surface in a way that permits a direct interaction and results in efficient catalysis of the overall reaction by channelling of the indole between the two sites.

The tryptophan synthase of Neurospora has subsequently been purified to homogeneity and shown to be a homodimer of a 75 kDa subunit (Matchett and DeMoss, 1975), thereby demonstrating that, as inferred from the earlier genetic studies (Bonner et al., 1960), the two active sites reside on the same polypeptide chain. Using the purified enzyme, Matchett (1974) showed that indole formed from InGP was physically trapped on the enzyme. As previously shown, during the conversion of the InGP to tryptophan this sequestered indole did not equilibrate with exogenous indole when added indole was present at a concentration (0.5 mM) which just saturated reaction 2. However, when the exogenous indole concentration was increased to 2.5 mM, the sequestered indole completely equilibrated with the exogenous indole before it was converted to tryptophan. These experiments demonstrated that the intermediate indole exists free on the enzyme surface in a "compartment" which does not normally equilibrate with the surrounding medium; however, high concentrations of exogenous indole penetrate this compartment and equilibrate with the indole formed by the enzyme. On the basis of these observations, Matchett (1974) proposed that the two active sites are organized on the enzyme surface so that indole formed by reaction 3f tends to remain in an unmixed solvent layer which provides direct access to the active site for reaction 2.

The tryptophan synthase from Escherichia coli provides an interesting contrast to the Neurospora enzyme. The enzyme from E. coli, which has been studied in great detail both genetically and biochemically (Crawford, 1974; Yanofsky and Crawford, 1972), is composed of two distinct subunits, A and B, in an A_2B_2 structure. Isolated A subunit catalyzes only reaction 3 and isolated B subunit (as B_2) catalyzes only reaction 2. The reaction catalyzed by each subunit is significantly stimulated when the two subunits are mixed and the heterotetramer reconstituted. As shown for the Neurospora enzyme, indole does not appear to be a freely diffusible intermediate in the conversion of InGP to tryptophan (Yanofsky and Rachmeler, 1958), and loading the active site with serine and pyridoxal phosphate appears to stimulate reaction 3 in the intact heterotetramer (Crawford, 1975). These observations suggest that the active sites for reactions 2 and 3 are organized in the tetrameric E. coli enzyme in a manner similar to that in the dimeric Neurospora enzyme.

Studies on a broad range of organisms (Crawford, 1975) suggest that all procaryotic organisms produce tryptophan synthase composed of two distinct subunits as in E. coli, while studies on a more limited range of fungi suggest that the lower eucaryotes generally produce the enzyme as a homodimer as in the case of Neurospora (Hutter et al., 1986). In Saccharomyces cerevisiae (Zalkin and Yanofsky, 1981), as well as in Neurospora, the monomeric subunit is approximately 75 kDa, which is approximately equal to the sum of the A (28 kDa) and B (45 kDa) subunits from E. coli. The amino acid sequence of the S. cerevisiae monomer, deduced from the nucleotide sequence of the cloned gene, exhibits a significant degree of homology with the sequences of both the A and B subunits from E. coli, with the regions of homology arranged in the order, amino-terminal:A(reaction 3):B(reaction 2):carboxy-terminal (Zalkin and Yanofsky, 1981). Recent studies with chain-terminating tryptophan synthase mutants of Neurospora suggest the same arrangement of activity domains in the Neurospora tryptophan synthase (Matchett, Lacy and DeMoss, unpublished data). These observations provide strong support for the early postulation by Bonner et al. (1965) that the gene which encodes tryptophan synthase in lower eucaryotes was evolved by fusion of the two genes which encode the A and B subunits in bacteria. Significantly, the organization and interaction between the active sites for reaction 2 and 3 have been preserved in the evolution of a single chain carrying both active sites.

In considering the possible organization of the two active sites on the enzyme, it is of some significance that in both Neurospora and S. cerevisiae tryptophan synthase is a homodimer. It is theoretically possible that the channelling of indole between the active sites for reactions 2 and 3 involves an interaction between sites on the same polypeptide chain or sites on different subunits. I shall return to this question for another case of enzyme channelling later.

ANTHRANILATE SYNTHASE: THE CASE OF THE PROMISCUOUS DOMAIN

Anthranilate synthase, which catalyzes the first committed step in tryptophan biosynthesis, provides an example of a complex reaction site that involves the interaction of active site domains from two different and distinct polypeptide chains. In all organisms studied to date, this enzyme is composed of two distinct subunits and catalyzes the following two reactions:

reaction a: chorismate + NH_3 $\longrightarrow$ anthranilate

reaction b: chorismate + glutamine $\longrightarrow$ anthranilate + glutamate

One of the subunits, α (Hulett and DeMoss, 1975) or component I (Ito and Yanofsky, 1969), by itself catalyzes reaction a, in which ammonia serves as the amino-group donor. When α is combined with the second subunit, β (Hulett and DeMoss, 1975) or component II (Ito and Yanofsky, 1969), the complex can utilize glutamine as the amino group donor (reaction b). Very high, nonphysiological concentrations of ammonium ions are required for reaction a (Arroyo-Begovich and DeMoss, 1973); furthermore, mutants which lack only the β component cannot synthesize their own tryptophan, indicating that reaction b is the physiologically important reaction catalyzed by anthranilate synthase (Chalmer and DeMoss, 1970).

In the $\alpha_2 \beta_2$ complex of N. crassa, the β subunit apparently provides a glutamine binding site and glutamine amido-transferase activity to the active site for the anthranilate synthase reaction (Keesey et al., 1981). In the complex, glutamine degradation or the covalent binding of glutamine analogues to the β chain requires the presence of chorismate; neither glutamine degradation nor the specific binding of glutamine analogues occurs with isolated β subunit alone. These observations suggest that glutamine binding and the amido-transferase activity are expressed only when the specific β glutamine-domain reacts directly with the active site domain for anthranilate synthase.

During the course of evolution, the glutamine amido-transferase domain (G-domain) has fused with different enzymatic components of the tryptophan pathway. In these cases the other enzymes become part of a multifunctional complex when active anthranilate synthase is formed by interaction of the two specific domains, thus creating the wide variation of enzyme organization observed for the tryptophan pathway. In a number of bacteria the G-domain is provided by a subunit which expresses no other known activity (Crawford, 1975). In most of the enterobacteria, the G-domain is the amino-terminal domain of a polypeptide chain that also catalyzes the next step in tryptophan biosynthesis, the anthranilate-PRPP phosphoribosyl transferase activity. The assembled complex, therefore, catalyzes the first two reactions of the tryptophan pathway (Ito and Yanofsky, 1969). In a broad range of fungi, or lower eucaryotes, the G-domain is the amino-terminal domain of a polypeptide chain which catalyzes the third and fourth reactions of tryptophan biosynthesis, the PRA isomerase and InGP synthase reactions; in this case the assembled complex catalyzes the first, third and fourth reactions of the tryptophan pathway (Hutter et al., 1986).

The selective advantage of these associations is not clear. Do these complexes represent stages of evolution towards an organized complex of the tryptophan enzymes? Or do they reflect more stable components of an existing, higher level of organization which has gone undetected in the biochemical and genetic studies on the tryptophan pathway of these organisms? While it is difficult to devise experiments which answer these questions, these complexes offer excellent experimental systems for the study of domain interactions in multifunctional proteins and multisubunit complexes.

PRA ISOMERASE AND InGP SYNTHASE: A CASE OF EVOLUTION OF ENZYME ORGANIZATION?

The proteins which catalyze the two sequential reactions leading to the formation of InGP (Fig. 1B, reactions 3 and 4), exhibit a still greater diversity of association and functional organization. These two reactions involve an Amidori rearrangement of the ribose side-chain of PRA to produce the intermediate CdRP (reaction 3), followed by the closure of the indole ring and elimination of the carboxyl group to produce InGP (reaction 4).

In many bacteria these two reactions are catalyzed by separate proteins encoded by separate genes (Crawford, 1975). In the enterobacteria, including E. coli, these two reactions are catalyzed by a monomeric 45 kDa polypeptide chain which is encoded by one gene (Creighton, 1970). While many mutations within this gene lead to the loss of both activities, some mutations affect specifically one or the other reaction; and these mutations map in clusters toward opposite ends of the gene (Smith, 1967). Based on kinetic studies (Creighton, 1970), limited proteolysis studies (Kirschner et al., 1980), and substrate-analogue binding studies (Bisswanger et al., 1979), the InGP synthase and PRA isomerase reactions appear to be catalyzed by separate domains located, respectively, at the amino and carboxyl regions of this bifunctional protein. In a kinetic study of the two reactions in E. coli, Creighton (1970) found no evidence for interaction between the two active sites; CdRP accumulation and utilization, as well as the measured transition time for the overall reaction, corresponded to a sequence which involved independent sites and complete equilibration of CdRP with the surrounding medium.

In Neurospora the same two activities appear to be structurally and functionally more highly organized. As indicated above, the PRA isomerase and InGP synthase reactions are catalyzed by a multifunctional enzyme complex, which also catalyzes the anthranilate synthase reaction (Kirschner et al., 1980). This complex is composed of distinct α (75 kDa) and β (85 kDa) subunits in an $\alpha_2\beta_2$ structure (Keesey et al., 1981). The β subunit is a trifunctional protein which includes the active sites for the PRA isomerase and InGP synthase reactions, as well as the G-domain discussed above. Based on limited proteolysis studies, which permitted separation of active domains (Walker and DeMoss, 1983), and on amino acid sequence homologies with proteins from bacteria and yeast which catalyze the same reactions (Shechtman and Yanofsky, 1983), the three domains appear to be arranged within the polypeptide chain as shown in Fig. 3A. When the G-domain was separated from InGP synthase and PRA isomerase by limited proteolysis with elastase (Walker and DeMoss, 1983; M. Walker and J.A. DeMoss, unpublished data), active glutamine-dependent anthranilate synthase was separated as an α:G-domain structure, while InGP synthase and PRA isomerase were recovered as a fully active, intact dimer as shown diagrammatically in Fig. 3B. Therefore, the G-domain is functionally independent of the activity domains for InGP synthase and PRA isomerase; and dimer formation is a property of the polypeptide region which defines the latter two domains, while recognition and interaction with α are properties of the G-domain.

Aspergillus (Hutter and DeMoss, 1967a) and a number of other fungi (Hutter and DeMoss, 1967b) produce anthranilate synthase complexes with

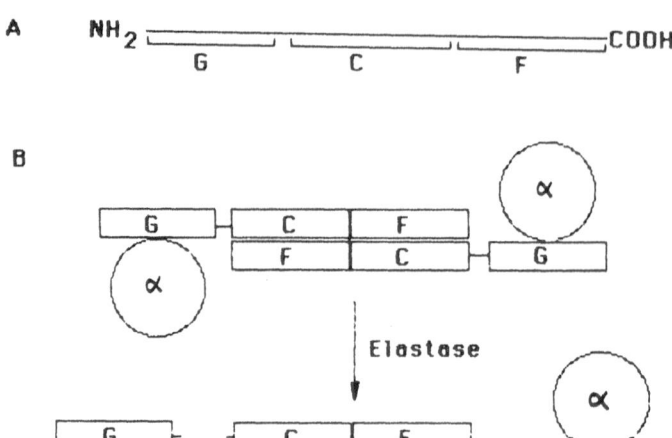

Fig. 3.   Proposed structure for the anthranilate synthase complex from
          N. crassa. A. Domain arrangement in the β subunit as deduced
          from amino-acid sequence homologies. B. Separation of
          activity domains by limited proteolysis with elastase.

structures similar to that of N. crassa which catalyze the same three
reactions. Saccharomyces cerevisiae, which like Neurospora is an
ascomycete, has the three activities organized in a way which is unique
among the fungi. In this organism PRA isomerase is a separate monomeric
protein which is encoded by a separate gene (DeMoss, 1962; Tschumper and
Carbon, 1980), and the anthranilate synthase complex is a heterodimer
containing one α subunit and one bifunctional β subunit composed of a G-
domain and an InGP synthase domain (Prantl et al., 1985). Although the
direction of evolution of these complexes cannot be established with
certainty, it seems likely that genes encoding the separate PRA isomerase
and InGP synthase in S. cerevisiae evolved from the trifunctional β gene
found in other fungi by a translocation of the region encoding the
carboxyl terminal PRA isomerase domain to another location in the genome.

     In the anthranilate synthase complex from N. crassa, the active sites
for the PRA isomerase and InGP synthase reactions appear to be organized
so that they interact more directly than was found for the bifunctional
protein from E. coli. In the N. crassa case, when the two reactions were
measured independently with added substrates the ratio of the V_{max}'s for
the two reactions, PRA isomerase and InGP synthase, was approximately 8 to
1 at all stages of purification (Gaertner and DeMoss, 1969), seemingly
eliminating any need for a channelling mechanism, since the predominant
PRA isomerase activity should produce more than enough CdRP to accumulate
and saturate the InGP synthase reaction. Furthermore, several years
earlier we had found with relatively crude extracts that when radioactive
PRA was converted to InGP, the radioactive CdRP formed equilibrated
completely with an exogenous pool of unlabeled, exogenous CdRP before it
was converted to InGP (Wegman and DeMoss, 1965). However, kinetic
studies with a relatively purified preparation of the complex suggested
that some kind of channelling of CdRP does occur between the two active
sites (Gaertner et al., 1970). With relatively purified preparations of

the complex, the V_{max} for the conversion of PRA to InGP (the overall reaction) was approximately two-fold higher than the V_{max} for the conversion of exogenous CdRP to InGP. These observations with the more purified complex indicate that CdRP formed by the PRA isomerase reaction is channelled to InGP synthase and appear to contradict our results of the radioactive channelling experiment. This apparent paradox might be resolved by a detailed examination of isotope channelling at different substrate concentrations and during the initial as well as steady-state stages of the reaction. It is possible, for example, that a physiologically significant channelling effect may occur only at steady-state subsaturating substrate concentrations which occur within a cell.

In any case, from the kinetic experiments we conclude that the active sites for PRA isomerase and InGP synthase are organized in the dimeric β subunit of the Neurospora complex, so that enzyme-generated CdRP is used more effectively than exogenously-added CdRP for InGP synthesis. In principle, interaction could occur between sites on the same polypeptide chain or between sites on different polypeptide chains as diagrammed in Fig. 4. The fact that the bifunctional enzyme from E. coli is a monomer in which there is no apparent channelling of CdRP between the sites could be used to argue that channelling must result from an interaction between sites on different chains. However, it is also possible to argue that the bifunctional, monomeric enzyme represents an intermediate stage in the evolution of the organized complex in which specific domain interaction within a single polypeptide chain has not yet been evolved. In either case, to understand the structural basis for CdRP channelling in the Neurospora complex, it will be important to establish whether the interaction occurs between sites on the same or different chains.

The picture, which emerges from the variations observed in structure and functional association of PRA isomerase and InGP synthase, is an evolution towards fusion of separate genes to produce a bifunctional protein, followed by changes which promote interaction and substrate channelling between active sites. The existence of the two activities as two enzymes encoded by separate genes in S. cerevisiae and other closely related yeasts (Braus et al., 1985) is apparently unique among the lower eucaryotes studied to date, and it probably resulted from a specific translocation event which became stabilized during the evolution of the yeast.

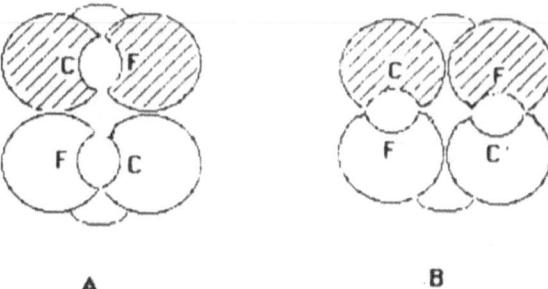

A   B

Fig. 4. Alternate modes of interaction between the active sites for PRA isomerase and InGP synthase. A. Interaction of active sites within the same subunit of the β subunit dimer. B. Interaction of active sites of different subunits within the β subunit dimer.

THE AROM COMPLEX: A CASE OF SUPER-ORGANIZATION

The enzymes which comprise the common aromatic pathway (Fig. 1A) in E. coli are separate enzymes (Berlyn and Giles, 1969). In contrast, in N. crassa and S. cerevisiae, five sequential enzymes of this pathway (reactions 2-6, Fig. 1A) are catalyzed by a single polypeptide chain (Gaertner and Cole, 1977) encoded by a single gene (Giles et al., 1967). Gaertner et al. (1970) found that this multifunctional protein catalyzes the overall reaction with a V_{max} which is greater than some of the individual reactions measured with exogenous substrates, and Welch and Gaertner (1975) demonstrated that the initial kinetics for the overall reaction exhibited a transition time shorter than that expected for a simple sum of the individual reactions. Although the order of the activities within the polypeptide chain can be deduced from the linear genetic map of the gene (Giles et al., 1967), little is known about the organization or interactions of the five active sites within the active dimer (Gaertner and Cole, 1977). It is not clear whether this catalytic facilitation represents an activation by the binding of the initial substrate or the channelling of intermediates to one or more of the limiting reaction sites.

To date, these activities have been observed to be either catalyzed by separate enzymes or by the pentafunctional protein. It is, therefore, difficult to speculate how this multifunctional protein might have evolved, or what selective value intermediates in the evolutionary process might have conferred. What is clear is that the lower eucaryotes produce a multifunctional protein which catalyzes a sequence of five reactions by a mechanism which is more efficient than a simple sum of the reactions. This complex system provides another excellent experimental system for probing the role of structure and interaction of domains in the organization of enzyme activities.

SUMMARY AND CONCLUSION

The pathway involved in the biosynthesis of tryptophan is an unusually rich source of examples of multifunctional proteins and variations in enzyme organization. When this pathway was studied in diverse organisms throughout the procaryotic and eucaryotic world, several different gene fusions appear to have led to multifunctional proteins and multisubunit complexes which catalyze multiple reactions in the tryptophan pathway. In several of these complexes, the domains appear to be organized so that intermediates are channelled between sequential active sites. In each case of intermediate channelling which has been documented, the enzyme complex involves multiple copies of the same subunit, suggesting that the interaction of active sites may involve interchain rather than intrachain domain interactions. In the best studied case, tryptophan synthase, channelling appears to involve a noncovalently-bound intermediate which is sequestered in a normally unmixed layer on the enzyme surface. It remains to be determined, what the structural basis is for this phenomenon and whether the same general mechanism applies to other cases of intermediate channelling in multifunctional proteins.

REFERENCES

Arroyo-Begovitch A. and DeMoss, J.A., 1973, J. Biol. Chem. 248:1262.

Berlyn, M.B. and Giles, N.H., 1969, J. Bacteriol., 99:222.

Bisswanger, H., Kirschner, K., Cohn, W., Hager, V., and Hansen, E., 1979, Biochemistry, 18:5946.

Bonner, D.M., DeMoss, J.A. and Mills, S.E., 1965, in: "Evolving Genes and Proteins", V. Bryson and H.J. Vogel, eds., Academic Press, New York.

Bonner, D.M., Suyama, Y. and DeMoss, J.A., 1960, Fed. Proc., 19:926.

Braus, G., Furter, R., Prantl, F., Niederberger, P., and Hutter, R., 1985, Arch. Microbiol., 142:383.

Chalmer, J.H. and DeMoss, J.A., 1970, Genetics, 65:213.

Crawford, I.P., 1974, in: "Subunit Enzymes", K. Ebner, ed., Dekker, New York.

Crawford, I.P., 1975, Bacteriol. Rev., 39:87.

Creighton, T., 1970, Biochem. J., 120:699.

DeMoss, J.A., 1962, Biochim. Biophys. Acta, 62:279.

DeMoss, J.A. and Bonner, D.M., 1959, Proc. Nat. Acad. Sci. U.S.A., 45:1405.

Gaertner, F.H. and Cole, K.W., 1977, Biochem. Biophys. Res. Commun., 75:259.

Gaertner, F.H. and DeMoss, J.A., 1969, J. Biol. Chem., 244:2716.

Gaertner, F.H., Ericson, M.C. and DeMoss, J.A., 1970, J. Biol. Chem., 245:595.

Giles, N.H., Case, M.E., Partridge, C., and Ahmed, I., 1967, Proc. Natl. Acad. Sci. U.S.A., 58:1453.

Hulett, F.M. and DeMoss, J.A., 1975, J. Biol. Chem., 250:6648.

Hütter, R. and DeMoss, J.A., 1967a, Genetics, 55:241.

Hütter, R. and DeMoss, J.A., 1967b, J. Bacteriol., 94:1896.

Hütter, R., Niederberger, P. and DeMoss, J.A., 1986, Ann. Rev. Microbiol., in press.

Ito, J. and Yanofsky, C., 1969, J. Bacteriol., 97:734.

Keesey, J., Paukert, J., and DeMoss, J.A., 1981, Arch. Biochem. Biophys., 207:103.

Kirschner, K., Szadkowski, H., Henschen, A., and Lottspeich, F., 1980, J. Mol. Biol., 143:395.

Matchett, W.H., 1974, J. Biol. Chem., 249:4041.

Matchett, W.H. and DeMoss, J.A., 1975, J. Biol. Chem., 250:2941.

Prantl, F., Strasser, A., Aebi, M., Furter, R., Niederberger, P., Kirschner, K., and Hütter, R., 1985, Eur. J. Biochem., 146:95.

Shechtman, M. and Yanofsky, C., 1983, J. Mol. Appl. Gent., 2:83.

Smith, O., 1967, Genetics, 57:95.

Tschumper, G. and Carbon, J., 1980, Gene, 10:157.

Walker, M. and DeMoss, J.A., 1983, J. Biol. Chem., 258:3571.

Wegman, J. and DeMoss, J.A., 1965, J. Biol. Chem., 240:3781.

Welch, G.R. and Gaertner, F., 1975, Proc. Natl. Acad. Sci. U.S.A., 72:4218.

Yanofsky, C. and Crawford, I.P., 1972, in: "The Enzymes", Vol. 7, P.D. Boyer, ed., Academic Press, New York.

Yanofsky, C. and Rachmeler, M., 1958, Biochim. Biophys. Acta, 28:640.

Zalkin, H. and Yanofsky, C., 1981, J. Biol. Chem., 257:1491.

IMPLICATIONS OF METABOLIC COMPARTMENTATION

IN PROKARYOTIC CELLS

V. Moses

School of Biological Sciences
Queen Mary College (University of London)
London E1 4NS, UK

INTRODUCTION

Contemplation of the sheer complexity of cellular metabolism, coupled with the realization that in bacterial cells it all takes place in a minute volume exhibiting a minimum of cellular architectural detail, soon begins to prompt questions about the nature of intracellular organization in prokaryotic cells. The very absence of membrane-bounded organelles suggests that, aside from reactions localized in or about the cell membrane, organizational structure is likely to be based on direct macromolecular interactions. These would yield macrostructures which in some cases might also act in a more-or-less compartmentalized fashion, leading to the separate channelling of whole series of sequential reactions. There is some variation in the way in which the terms "compartment" and "channel" are used. In the present contribution a "channel" is defined as a structure or a facet of organization which acts as a constraint to free chemical diffusion. A "compartment" is a sequestered volume; it may be similar to a channel. Any defined aspect of biochemical organization may represent a compartment so long as it remains chemically inactive. For example, a ligand bound to an enzyme is in a compartment until it undergoes a reaction.

The number of studies to explore compartmentation and channelling of metabolites and metabolism in bacteria is limited; a common view has been that in organisms so small there is ample time and opportunity for reacting molecules to encounter one another with a frequency quite sufficient to account for the rates of reaction observed in practice. That may in a theoretical sense be true but, by itself, does not eliminate the possibility of a more elaborate and subtle intracellular organization developed for purposes other than securing the most rapid reaction rates. Furthermore, diffusion rates have usually been measured in dilute aqueous solution, clearly different from cytoplasm. Bacteria by their very simplicity and versatility offer ideal systems for studying multienzyme complexes and structures at their most fundamental level.

For the past two decades a group of us has been exploring channelling phenomena as evinced by glycolysis and two amino acid biosynthetic pathways in _Escherichia coli_ and a few closely related species. Apart

from the present author, five members of the group were postdoctoral
fellows, five were graduate students and one was an undergraduate project
student at the time of his involvement. At no time were more than three
people working together. Their names are recorded at the end of this
paper.

E. coli was chosen as the object of study because it is
undifferentiated and simple in structure, because it is robust and easy
to handle, and because its very well developed genetics offered
(especially at the beginning of this work) a great deal of flexibility
and opportunity in the use of mutants to explore particular points of
interest. Glycolysis has played the most prominent role in this extended
study. Exploration of the glycolytic pathway in E. coli followed an
early attempt to examine glycolytic compartmentation in Ehrlich ascites
tumor cells of mice (Moses and Lonberg-Holm, 1965), a study which, while
suggesting certain conclusions, demonstrated clearly the desirability of
working with simpler cells. The experimental technique which was first
adopted, that of following the fates of labelled carbon atoms from
several mono- and di-saccharides and related compounds fed simultaneously
to the bacteria, was well suited to E. coli, an organism possessing a
variety of inducible sugar uptake mechanisms many of which can be
manipulated at will. The two amino acid biosynthetic pathways, those for
histidine and proline, were selected deliberately as representing, on the
one hand a long pathway with many enzymic steps encoded in bacteria by
contiguous genes transcribed into a polycistronic messenger RNA, and on
the other a short metabolic sequence specified by three non-contiguous
genes. Much of the work has been published and will be reviewed in
general outline, but some is new and will be presented in greater detail.

GLYCOLYSIS

The first indication that glycolysis is channelled, at least to some
degree, in E. coli came from a study (McBrien and Moses, 1968) in which
resting cultures, treated with chloramphenicol to prevent additional
enzyme synthesis, were divided into several identical subcultures each of
which was supplied with the same mixture of glucose, galactose and
lactose. Each culture also received a small quantity of a sugar
generally labelled with ^{14}C. These were, for the four subcultures,
glucose, galactose, lactose labelled in the glucose moiety and lactose
labelled in the galactose moiety. Sampling of the cultures during the
ensuing 45 minutes was followed by an analysis of the flow of ^{14}C from
the external substrate into intracellular glycolytic metabolites, as well
as into those participating in the tricarboxylic acid cycle. The results
suggested that carbon atoms from the differently labelled substrates
contributed unequally to the accumulation of ^{14}C in metabolites. These
data signified the differential metabolism of sugar carbon atoms supplied
to the cell; all the cultures were, of course, identical as far as both
their history and the range of available substrates were concerned; they
differed only in respect of the ^{14}C labelling patterns.

These studies were technically exacting. Not only was it difficult
to sample simultaneously from several reaction vessels, but the
glycolytic and tricarboxylic acid cycle intermediates are present at
rather low intracellular concentrations, detracting from the accuracy of
^{14}C incorporation determinations. A second study (Macnab et al., 1973)
sought to overcome these problems by employing cultures of growing cells
and measuring the incorporation of ^{14}C from the substrates into protein
amino acids. The product yields were consequently much higher, the

sampling protocol more extended, and the analyses simpler. Two competition experiments were undertaken. In the first, replicate cultures were supplied with a mixture of galactose and lactose, with the label either in the free galactose or in the galactose moiety of lactose. This experiment served partly to check the galactose/lactose data of the earlier study. The second experiment used maltose and glycerol with either one or the other generally labelled. The purpose here was to investigate the possibility of preferential metabolism of intermediates derived from maltose carbon (entering glycolysis at the hexose level) compared with glycerol carbon (joining glycolysis at the point of triosephosphate isomerase). Maltose rather than glucose was used as the hexose substrate in order to minimize catabolite repression effects on glycerol metabolism. In both cases the data confirmed the conclusions derived from the first study and indicated a substantial measure of channelling. Figure 1 summarizes these conclusions.

Interest then turned to a physical explanation for these phenomena. At a time when a number of workers were reporting the existence of presumptive multienzyme complexes as evidenced by the co-migration of catalytic activities in a variety of separatory procedures, as well as the dispersion and spontaneous reaggregation of the pyruvate dehydrogenase and other complexes, the concept of a specific aggregate of glycolytic enzymes offered the possibility of insight into the mechanism of channelling. By disrupting E. coli cells, using osmotic lysis with a minimum dilution of the released intracellular contents and separating the soluble proteins by gel-exclusion chromatography, it was possible to show the presence of a number of glycolytic enzymes in two distinct molecular weight ranges: one corresponded to the mono-dispersed enzymes (in each case consistent with the known molecular weight for the enzyme in this organism), the other apparently to a comigrating complex of all

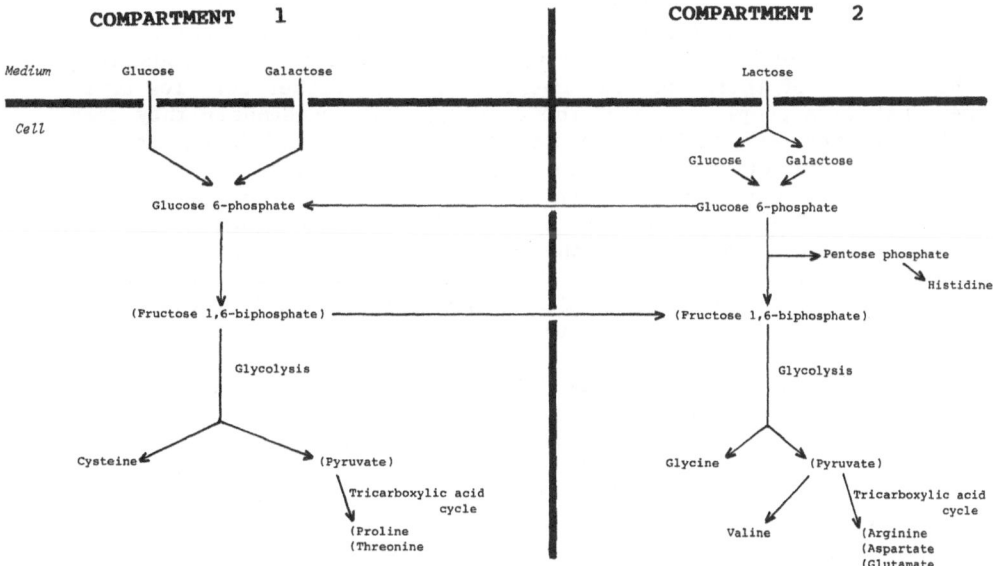

Fig. 1. Summary of metabolic compartmentation model for E. coli based on the labelled products derived from substrate galactose, glucose and lactose.

of them together, perhaps with further unidentified components (Mowbray and Moses, 1976; Gorringe and Moses, 1980).

The pooled enzymes from the low molecular-weight range, although individually present at concentrations several fold greater than those same enzymes from the high molecular-weight eluate range, failed almost entirely to glycolyze; i.e. no pyruvate (measured after conversion, with added glutamate and aminotransferase, to alanine) was detected. By contrast, the putative high molecular-weight complex (approximately 7.5 million daltons) did convert glucose to pyruvate. The integrity of the complex was sensitive to dilution, as witnessed by a fall in the specific activity of pyruvate formation from glucose. After dilution of the complex, further chromatography demonstrated glycolytic enzymes only in the low-molecular weight fractions; the complex itself had vanished. Pooling the enzymes from the low molecular-weight fractions, concentrating the mixture by ultrafiltration, and rechromatographing the concentrate showed that while the greater part of each enzyme continued to migrate as a low molecular-weight component, a minor proportion co-chromatographed with minor proportions of the other glycolytic enzymes in eluate fractions characteristic of the complex. Electron microscopic examination of solutions containing the complex revealed a collection of fairly uniform diameter particles, of a size commensurate with a molecular weight of 7.5 million, and with indistinct traces of fine structure.

Competition studies have provided evidence for channelling within the complex. Using labelled glucose or glucose 6-phosphate as a primary substrate for the complex, added equimolar concentrations of unlabelled fructose 1,6-bisphosphate, 3-phosphoglyceric acid or phosphoenolpyruvic acid produced no more than a limited reduction of ^{14}C incorporation into pyruvate. This finding suggested a failure of the unlabelled intermediates to equilibrate with and consequently dilute the labelled intermediates lying between glucose and pyruvate, each of which must certainly have been present in the whole reaction mixture but at a much lower concentration than the same intermediate added unlabelled to the medium. It was also found that the activities of phosphoglucose isomerase and of triosephosphate isomerase, determined individually by the addition of their specific substrates, were either very low or not measurable in preparations of the complex. Yet, presumably they were active in the complex both because pyruvate was formed from glucose and because of the fact that when the very same preparation of complex was diluted and rechromatographed those two enzymes could indeed be detected in the low molecular-weight fractions. The conclusion was drawn from these data that glycolysis by the complex is at least partly channelled, and that while they remain part of the complex the active sites of some of the enzymes are not accessible to substrates present in the surrounding medium but only to those generated by the complex itself.

Some of these findings have been confirmed by S. Barnes and P.D.J. Weitzman (personal communication) who, in addition, have recently reported evidence for an apparent complex involving most of the tricarboxylic acid cycle enzymes of E. coli (Barnes and Weitzman, 1985). S. Mowbray and G.A. Petsko (personal communication) also found the glycolytic complex in yeast, which they were able to stabilize in vivo with dimethylsuberimidate. Recent work at Queen Mary College (Kingdon, 1983) has supported the notion of stabilization of binding by dimethylsuberimidate, especially between triosephosphate isomerase and 3-phosphoglyceraldehyde dehydrogenase, with some evidence for the presence of an additional component in the complex. Measurement of the diffusion coefficients of glycolytic enzymes with respect to gels showing with a range of exclusion properties has indicated binding between 3-phosphoglyceraldehyde dehydrogenase and triosephosphate isomerase, but not as yet between other glycolytic enzymes.

AMINO ACID BIOSYNTHETIC PATHWAYS

Two sequences were chosen for study, those leading to histidine and proline.

Histidine Biosynthesis

Histidine biosynthesis in Salmonella typhimurium requires the sequential action of ten catalytic steps, two of which may be catalyzed by the same enzyme. The genes coding for the nine proteins are contiguous, though not in the same order as the enzymes in the sequence. There are no branches on the pathway, but one intermediate (5'-phosphoribosyl-4-carboxamide-5-aminoimidazole) is also an intermediate in purine nucleotide synthesis. Such a complex system as the histidine pathway, with nine contiguous genes arranged as an operon and transcribed into a polycistronic messenger RNA, and leading as it does to a single product, might well be expected to function as a multienzyme system, perhaps with channelled intermediates.

The proposition was partially investigated in S. typhimurium by Bearden and Moses (1972). They used whole cells to investigate the incorporation of ^{14}C from glucose into histidine in the presence and absence of an unlabelled pathway intermediate added to the reaction medium. All of the metabolic intermediates of histidine biosynthesis with the exception of the last two, histidinol and histidinal, are phosphorylated. None except these two would be likely to be taken up by intact bacteria. Histidinol met the requirement for histidine by auxotrophs blocked earlier in the pathway, though a concentration higher than that for histidine itself was required to achieve the maximal growth rate. Unlabelled histidinol present in the medium totally abolished the incorporation of ^{14}C from glucose into histidine, but, as anticipated, had no effect on incorporation into alanine. Clearly no evidence for channelling of histidine biosynthesis was obtained from this study. Conceivably, the use of other intermediates in a cell-free synthetic system might yield contrary data, but none have been reported. We conclude that a gene cluster coding for a multi-step single-product pathway does not necessarily imply either a multienzyme complex or metabolic channelling.

Although not strictly relevant to a review of channelling in prokaryotes, it is worth noting the differences in histidine biosynthesis between Salmonella and yeast. In the latter the genes are mostly unlinked, although one gene cluster, HIS-4, appears to code for a single polypeptide chain with a molecular weight of 95,000 which carries three enzymic activities. Significantly, perhaps, one of these is histidinol dehydrogenase. In yeast histidinol is able to meet the histidine requirement shown by an auxotroph, and measurements both of the cell volume available for histidinol penetration and directly of the uptake of [^{14}C]histidinol show that it is taken up into the cell. Yet unlabelled histidinol in the medium failed to influence the incorporation in growing cells of ^{14}C from glucose into histidine. The biosynthesis of histidine in yeast may therefore be channelled, but nothing is known of any physical structure to account for it.

Proline Biosynthesis

The proline pathway starting from glutamate is believed to proceed through four reactions, only three of which are enzyme-catalyzed. The cyclization of glutamic-γ-semialdehyde to Δ^1-pyrroline 5-carboxylate (P5C) takes place spontaneously. The three enzymes (A, B and C) are specified by three genes which are not clustered: pro$_A$ and pro$_B$ appear to be contiguous, but pro$_C$ is separated from the first two genes by the lac operon and other markers. E. coli pro$_A$ or pro$_B$ auxotrophs are able to grow in the presence of P5C; pro$_C$ auxotrophs require proline itself for growth. Competition experiments with the proline pathway were carried out with E. coli using protocols analogous to those described above for exploring channelling in the histidine pathway of S. typhimurium: unlabelled P5C present in the medium totally abolished the incorporation of ^{14}C from glucose into proline but had no effect on incorporation into valine (Gamper and Moses, 1974).

However, an interesting aspect of proline biosynthesis is the lability in an aqueous medium of the first pathway intermediate, γ-glutamyl phosphate (Katchalsky and Paecht, 1954), with the formation of 2-pyrrolidone 5-carboxylate (Strecker, 1957). γ-Glutamyl phosphate is formed by activation of the glutamate γ-carboxyl with ATP. In a number of similar carboxyl-activation reactions there is evidence for the involvement of an intermediate acyl phosphate, and γ-glutamyl phosphate has been proposed by analogy as the intermediate in proline biosynthesis; see Gamper and Moses (1974) for references supporting these arguments.

In order to study the involvement of γ-glutamyl phosphate, Gamper and Moses (1974) developed a cell-free proline biosynthesis system. Because of the sensitivity of glutamyl kinase (the first enzyme in the sequence) to retroinhibition by proline, it was important to prevent an accumulation of proline in the reaction mixture. This was achieved by additionally incorporating into the mixture living proline-starved cells of a proline auxotroph, together with [^{14}C]phenylalanine as an indicator of protein synthesis and hence of proline availability. Using pro$_B$ or pro$_C$ mutants it was possible to determine separately the quantities of either proline alone, or of proline + P5C, synthesized by the cell-free system. Although specific rates of synthesis were low, they were directly proportional to the concentration of cell extract in the reaction mixture, suggesting the presence of a multifunctional enzyme complex. N-Acetylglutamate was not a substrate for proline synthesis. Ammonia and imidazole had little or no effect on cell-free proline synthesis, although, as nucleophiles, both would be expected to react with free γ-glutamyl phosphate. The fact that they did not do so suggests that the intermediate is protected in the pathway, perhaps by complex formation between the first two enzymes.

No proline synthesis from glutamate was observed in cell-free preparations from proline auxotrophs. Gel filtration analyses of the enzymic activities of these extracts with respect to proline synthesis were carried out by Hayzer and Moses (1978) and indirect evidence was obtained for the first enzyme (glutamyl kinase) forming part of an enzyme complex.

In vitro complementation between enzymes in extracts from different proline auxotrophs, attempted without success by Gamper and Moses (1974), was later reported by Garrib (1983). Let $\underline{A}$ and $\underline{B}$ represent the active forms of the complexing enzymes, $\underline{a}$ and $\underline{b}$ the inactive forms, $\underline{AB}$ the wild-type complex and $\underline{Ab}$ or $\underline{aB}$ the complexes from the auxotrophs. Garrib proposed that mixing $\underline{Ab}$ and $\underline{aB}$ extracts, diluting them to give the separate enzymes $\underline{A}$, $\underline{a}$, $\underline{B}$ and $\underline{b}$, and reconcentrating them to promote complex formation should yield a mixture of $\underline{AB}$, $\underline{ab}$, $\underline{Ab}$ and $\underline{aB}$. Thus 25% of wild-type specific activity would be expected; a value of 18% was actually obtained experimentally.

Garrib (1983) went on to explore _in vivo_ complementation between genotypes. Transfer of an _E. coli_ episome F' $\underline{lac^+}$ $\underline{pro_A^-}$ $\underline{pro_B^+}$ into a female recipient $\underline{F^- lac^-}$ $\underline{pro_A^+}$ $\underline{pro_B^-}$ $\underline{pro_C^+}$ allowed growth without proline, but transfer into $\underline{F^-}$ $\underline{lac^-}$ $\underline{pro_A^-}$ $\underline{pro_B^-}$ $\underline{pro_C^+}$ did not. Conversely, $\underline{F^-}$ $\underline{lac^-}$ $\underline{pro_A^-}$ $\underline{pro_B^-}$ $\underline{pro_C^-}$ grew prototrophically with respect to proline after transfer of F' $\underline{lac^+}$ $\underline{pro_A^-}$ $\underline{pro_B^-}$, but $\underline{F^-}$ $\underline{lac^-}$ $\underline{pro_A^+}$ $\underline{pro_B^-}$ $\underline{pro_C^+}$ did not.

Very similar results were obtained with interspecific crosses using wild type and proline auxotrophs of _Proteus mirabilis_. Combined extracts of _E. coli_ $\underline{Ab}$ and _P. mirabilis_ $\underline{aB}$ gave specific rates of proline synthesis equivalent to 12% and 17%, respectively, of the wild-type rates for _E. coli_ and _P. mirabilis_. Combining $\underline{Ab/aB}$ extracts from _P. mirabilis_ mutants showed a surprisingly high specific activity of nearly 79%. Heteromerodiploids between _P. mirabilis_ and _E. coli_ are easy to construct by F-duction using $\underline{lac^+}$ as a selective marker, since _P. mirabilis_ is $\underline{lac^-}$. Such a heteromerodiploid with the structure _P. mirabilis_ F^- $\underline{lac^-}$ $\underline{pro_A^-}$ $\underline{pro_B^+}$/_E. coli_ F' $\underline{lac^+}$ $\underline{pro_A^-}$ $\underline{pro_B^-}$ was prototrophic for proline, and cell-free extracts prepared from it demonstrated a specific activity for proline formation 30% and 44%, respectively, of the _E. coli_ and _P. mirabilis_ wild-type activities (Garrib, 1983). There seems little doubt that interspecific recognition can be fairly effective in promoting the formation of the glutamyl kinase/γ-glutamyl phosphate reductase complex.

CONCLUSIONS

The argument is sometimes levelled against proposals for compartmentalized metabolism that such mechanisms are not necessary because, particularly in small bacterial cells, the rates of free diffusion are sufficiently high to account for observed rates of reactions, although Ling and Cope (1969) concluded otherwise. It should nevertheless be noted that cytoplasm may not act as a dilute solution (Atkinson, 1969), and that under cytoplasmic conditions many metabolites may be protein-bound (Sols and Marco, 1970). One role of a compartmentalized or channelled system may be to raise the metabolite concentrations locally in the pathway with which it is associated to sufficiently high levels to allow saturation of enzyme binding sites and rapid reaction rates. The global concentration taken in the cell as a whole may not permit that.

A second function for channelling might be directly related to regulation and metabolic economy, ensuring for major pathways that all (or the bulk) of material entering the sequence is converted to the primary product without being diverted elsewhere. Intermediates confined to such a channel would presumably be more susceptible to metabolic control exerted at one or two critical points, than would be those same metabolites able to participate freely and competitively in a wide range of cellular biochemistry. From such a standpoint, channelling might be more valuable for a major sequence than for a minor one. Thus, for an organism such as E. coli, in which hexoses are major primary substrates, glycolysis is one of the main pathways for conversion to a large number of products via the tricarboxylic acid cycle and other routes. Channelling metabolic intermediates of glycolysis may be more important for control than channelling those of histidine biosynthesis, which have nowhere else to go. Our own results, summarized in this paper, offer support to such a conclusion.

Finally, we may recognize the role of enzyme complexes for the stabilization of valuable intermediates; γ-glutamyl phosphate is a case in point. Reduction of the γ-carboxyl of glutamic acid to an aldehyde requires prior activation, and there may be few alternative ways to do this other than by phosphorylation with ATP as the phosphate donor. The resulting unstable product is protected from rapid degradation in the aqueous environment by remaining attached to the enzyme complex. This, however, is not the only way to protect γ-glutamyl phosphate. The arginine pathway also starts from glutamic acid and proceeds via phosphorylation of the γ-carboxyl to the γ-semialdehyde. But in that sequence glutamic acid is N-acetylated before phosphorylation, and N-acetyl-γ-glutamyl phosphate is not rapidly degraded in water. Furthermore, retention of the N-acetyl group through the next steps of the arginine pathway ensures that the γ-semialdehyde does not cyclize to P5C and effectively prevents any diversion of intermediates towards proline. It would be interesting to know whether any enzyme complexes participate in the arginine pathway, but no studies appear to have been done.

The very concept of specific enzyme-enzyme interactions to yield complexes is presumably dependent on enzyme structure as determined genetically. In some cases, at least, the recognition site for the protein-protein interaction will be sufficiently distinct from the catalytic site for the former to be inactivated without influencing the latter. It should therefore be possible to identify a class of metabolically defective mutants in which the lesion results from a failure of organization rather than a failure of any particular catalytic step. Manney (1970) reported a case in yeast which might be of this type. In a study of tryptophan synthetase, he found that indole-accumulating and indole-requiring lesions could complement each other in vivo in diploid recombinants, but such strains showed long growth lags related to the initial cell density in the culture. He suggested that the lesion in the mutants interrupted the association of the two enzymic reactions firstly converting indole glycerophosphate (IGP) to indole plus 3-phosphoglyceraldehyde, the second step then catalyzing the formation of tryptophan from the indole so produced. In the wild type, indole is believed to be channelled from the first enzyme to the second. In the mutants, however, channelling does not take place; indole is released into the cytoplasm and some leaks out of the cells into the medium. At high cell densities enough indole is leaked to the medium to generate an intracellular level permitting rapid growth by cross-feeding between cells. These strains have not been studied in greater detail, and we know of no work to explore the biochemistry and physiology of protein recognition in relation to multienzyme channelled systems. They would be interesting problems to explore.

At a recent Harden Conference, Pflugfelder, Adams and Kirschner (1985) reported that the catalytic domains of phosphoribosyl anthranilate (PRA) isomerase:IGP synthase support each other. The bifunctional monomeric enzyme unfolds reversibly. In the partially folded intermediate, the IGP synthase domain is still folded but both catalytic activities are strongly reduced. Unfolding the PRA isomerase domain may destabilize the IGP synthase domain. The appropriate DNA coding sequences, when cloned separately in expression plasmids, lead to the production of separate, stable, functional polypeptide fragments but with specific activities only 10% those of the wild type enzyme. The authors suggest that each domain needs the presence of the other for maximal activity.

ACKNOWLEDGEMENTS

It is with very great pleasure that I record the names of colleagues who have worked with me during the past two decades on the various problems discussed in this paper: Drs. Lydia Bearden, Deborah Galbraith, Howard Gamper, Alam Garrib, Diana Gorringe, David Hayzer, Charles Kingdon, Karl Lonberg-Holm, Robert Macnab, David McBrien, and John Mowbray. Some of the concepts which were developed during these studies originated even earlier in the course of a collaboration with Professor Melvin Calvin and Dr. Osmund Holm-Hansen.

REFERENCES

Atkinson, D.E., 1969, Curr. Top. Cell. Regul., 1:29.
Barnes, S. and Weitzman, P.D.J., 1985, This volume.
Bearden, L. and Moses, V., 1972, Biochim. Biophys. Acta, 279:513.
Gamper, H. and Moses, V., 1974, Biochim. Biophys. Acta, 354:75.
Garrib, A., 1983, Ph.D. Thesis, University of London: "The enzymes of proline biosynthesis in Escherichia coli and Proteus mirabilis".
Gorringe, D.M. and Moses, V., 1980, Int. J. Biol. Macromol., 2:161.
Hayzer, D.J. and Moses, V., 1978, Biochem. J., 173:219.
Katchalsky, A. and Paecht, M., 1954, J. Amer. Chem. Soc., 76:6042.
Kingdon, C.F.M., 1983, Ph.D. Thesis, University of London: "The status of glycolytic organization in Escherichia coli".
Ling, G.N. and Cope, F.W., 1969, Science, 163:1335.
Macnab, R., Moses, V. and Mowbray, J., 1973, Eur. J. Biochem., 34:15.
Manney, T.R., 1970, J. Bacteriol., 102:483.
McBrien, D.C.H. and Moses, V., 1968, J. Gen. Microbiol., 51:159.
Moses, V. and Lonberg-Holm, K.K., 1968, J. Theor. Biol., 10:336.
Mowbray, J. and Moses, V., 1976, Eur. J. Biochem., 66:25.
Pflugfelder, M., Adams, B. and Kirschner, K., 1985, Abstracts of the 24th Harden Conference, 19.
Sols, A. and Marco, R., 1970, Curr. Top. Cell. Regul., 2:227.
Strecker, H.J., 1960, J. Biol. Chem., 235:2045.

ORGANIZATION OF GLUCOSE METABOLISM:

A MODEL OF COMPARTMENTS BY POLY-ISOZYMIC COMPLEXES

Tito Ureta and Jasna Radojković

Departamento de Biologia
Facultad de Ciencias
Universidad de Chile
Casilla 653
Santiago, Chile

"Adso," William said, "solving a mystery is not the same as deducing from first principles. Nor does it amount simply to collecting a number of particular data from which to infer a general law. It means, rather, facing one or two or three particular data apparently with nothing in common, and trying to imagine whether they could represent so many instances of a general law you don't yet know, and which perhaps has never been pronounced." ...

... *"And this is what I am doing now. I line up so many disjointed elements and I venture some hypotheses. I have to venture many, and many of them are so absurd that I would be ashamed to tell them to you."*

(Umberto Eco, The Name of the Rose, Fourth Day, Vespers)

BRANCH-POINTS IN METABOLISM

A metabolite being released from the active site of an enzyme may find itself in the predicament of deciding which pathway to choose among several possibilities. For instance, once glucose-6-P has been synthesized several alternative paths are possible, e.g., glycogen synthesis via phosphoglucomutase, glycolysis via phosphoglucose-isomerase, pentose-P pathway via glucose-6-P dehydrogenase, or back again to glucose through glucose-6-phosphatase (Fig. 1). Furthermore, the same situation will happen again at every branch-point along the metabolic maze until a committed step is reached.

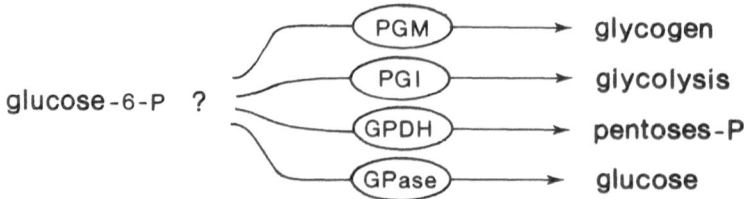

Fig. 1. Alternative pathways for glucose-6-P metabolism. PGM,
         phosphoglucomutase; PGI, phosphoglucose isomerase; GPDH,
         glucose-6-P dehydrogenase; GPase, glucose-6-phosphatase.

The decision to turn **somewhere**, of course, does not belong to the
puzzled intermediate. According to current thinking, once the needs of
the cell are perceived, instructions to modify enzyme rates in order to
control the traffic of intermediates at the branching points are issued
by a hitherto undefined molecular controller. Concentrations of key
metabolites and/or cofactors, the activity levels of the enzymes
involved, and their substrate affinities, are thought to be of importance
for the regulation of flux through competing branch-points (Walsh and
Koshland, 1984; Finkelstein and Martin, 1984).

ISOZYMES AND BRANCH-POINTS

The dilemma is further compounded because most enzymes are in fact
components of isozymic systems, i.e., multiple forms catalyzing the same
reaction in the same cell or organism (Markert, 1975; Markert and Moller,
1959). The presence of isozymes constitutes again a branching dilemma
(Fig. 2).

It is known that members of isozymic systems may or may not be
simultaneously present in the same cell. If they do coexist, they may be
located in different organelles and the dilemma does not apply. However,
a number of isozymes **do** coexist in the same compartment, the so-called
cytosol (Ureta, 1978).

Let us consider the situation of glucose-6-P produced by hexokinase
isozymes (Fig. 3).

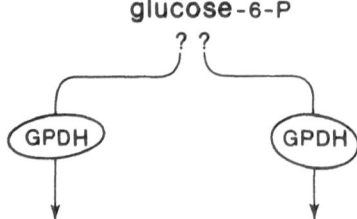

Fig. 2. Glucose-6-P may be converted to 6-phosphogluconate by more than
         one isozyme.

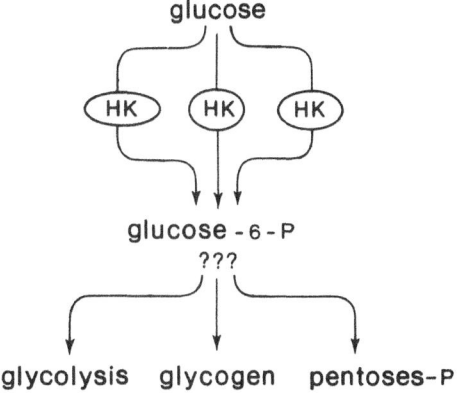

Fig. 3. If one pool only of glucose-6-P exists, then the history of
ester production is irrelevant to its fate.

If one pool only of glucose-6-P exists, then the hexokinase isozyme
catalyzing its production will not specify which pathway to follow, since
either phosphoglucomutase, phosphoglucose isomerase or glucose-6-P
dehydrogenase will be able to draw upon that pool.

Consider now a situation in which the history of glucose-6-P is
important because it determines its fate (Fig. 4).

In this case the hexokinase isozyme used for the synthesis of
glucose-6-P is indeed relevant for the dilemma of metabolic branching
points because, according to this view, branch-points do not exist, i.e.,
no decision has to be made as to where to go since the end-product of the
pathway is specified by the isozymes involved. However, this holds true
only if the product of one isozyme is kept isolated from those formed by
the other isozymes, i.e., if compartmentation of the product exists.

MULTIENZYME COMPLEXES FORMED BY SPECIFIC ISOZYMES: A HYPOTHESIS

It is becoming increasingly clear that compartmentation of metabolic
events is achieved in the cell (apart from the classic organelles)

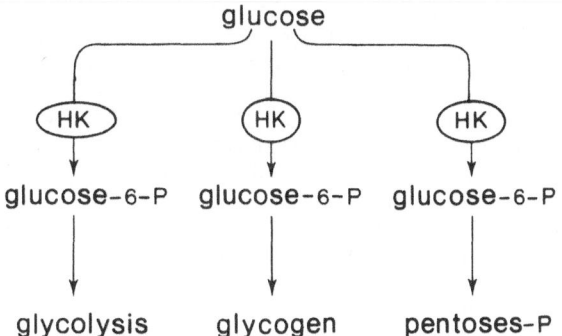

Fig. 4. The origin of glucose-6-P formation is relevant to its fate if
separate pools of the ester are maintained.

through multienzyme complexes (Srere and Mosbach, 1974; Welch, 1977; Stebbing, 1980; Wombacher, 1983), and we therefore propose that isozymes associate in such a way to produce specific channels. This poly-isozymic complex model of metabolism may be formally stated as follows (Ureta, 1978):

> *"Metabolic pathways are unidirectional chain reactions catalyzed by specific isozymes associated as polyisozymic complexes."*

For the specific case of carbohydrate metabolism the model may be represented as shown in Fig. 5.

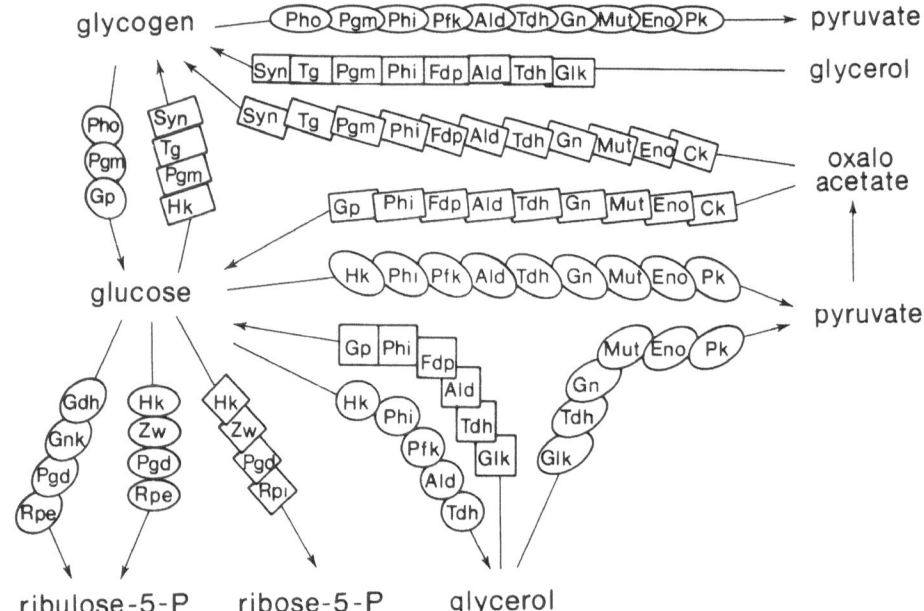

Fig. 5.   Hypothetical poly-isozymic complexes of carbohydrate metabolism. Main pathways are represented as unidirectional reactions. Ovals and rectangles showing abbreviated enzyme names indicate, respectively, catabolic or biosynthetic pathways. Any given enzyme is represented by differently shaped symbols according to the complexes in which they participate in order to indicate different isozymes.   Gp, glucose-6-phosphatase; Pgm, phosphoglucomutase; Tg, UDP-glucose pyrophosphorylase; Syn, glycogen synthase; Pho, glycogen phosphorylase; Zw, glucose-6-P dehydrogenase; Pgd, 6-P-gluconate dehydrogenase; Rpe, ribulose 5-P epimerase; Rpi, ribose 5-P isomerase; Gdh, glucose dehydrogenase; Gnk, gluconate kinase; Phi, glucose-P isomerase; Pfk, phosphofructokinase; Ald, fructose-1,6-bisP aldolase; Tdh, triose-P dehydrogenase; Gn, P-glycerate kinase; Mut, P-glycerate mutase; Eno, enolase; Pk, pyruvate kinase; Fdp, fructose-1,6-bisphosphatase; Glk, glycerol kinase; Ck, phosphoenolpyruvate carboxykinase. A few substrates and products are indicated.

TESTS OF THE HYPOTHESIS

The proposal that metabolic pathways should be considered as unidirectional chain reactions catalyzed by specific isozymes has received attention by several investigators (Masters, 1981; Clegg, 1984 a,b; Wilson, 1985). The hypothesis was developed primarily as an attempt to explain the presence of isozymes in most cells and in practically every metabolic step, but it has obvious pertinence to the quest of the spatio-temporal organization of metabolism (Ureta, 1985; Ureta and Radojkovic, 1985a).

The poly-isozymic complex model has not, as far as we know, been proved wrong. It has not been validated either, but several observations suggest at least its plausibility. A few of them will be briefly discussed.

A multienzyme complex involved in CO_2 fixation was isolated from _Euglena_ by Wolpert and Ernst-Fonberg (1975). The complex contains equimolar amounts of phosphoenolpyruvate carboxylase, malate dehydrogenase, and acetyl-coenzyme A carboxylase. The malate dehydrogenase component was shown to be electrophoretically distinct from other malate dehydrogenases present in the cell. This finding suggests that one, and only one, isozyme (from several alternatives) is used for complex formation.

Several authors have put forward the hypothesis, and in some cases reported data, that _in vivo_ the enzymes of glucose utilization are organized as multienzyme complexes or somehow lumped together in close association with cyto-architectural elements (for reviews, see Ottaway and Mowbray, 1977; Masters, 1981; Ureta, 1985). Nonetheless, the evidence for actual isolation of such complexes has been found wanting. This may be due to lability of the putative enzyme clusters, but, as briefly discussed elsewhere (Ureta, 1985), it may also reflect the transient nature of those associations. Be that as it may, one objection to multienzyme complexes of glucose utilization is the difficulty to reconstitute those complexes _in vitro_ using purified "neighbor" enzymes.

A report by MacGregor et al. (1980) is very illustrative in this respect. It was shown that enzyme-enzyme interaction between aldolase and fructose-1,6-bisphosphatase could be observed only when enzymes purified from the same tissue (liver) were mixed. Association was not observed if one of the enzymes was replaced by an isozyme from another tissue (muscle). These observations suggest that enzyme-enzyme interactions are possible only when the proper "neighbor" _isozymes_ are mixed together and may explain why other investigators failed to observe those associations.

The idea, that very similar pathways may operate in separate functional compartments of the type shown in Fig. 5, has been recently supported by experiments performed in carotid artery rings (Lynch and Paul, 1983). These authors followed specific activities of glucose, lactate, and glycogen under conditions which result in changes of glucose and glycogen utilization. They concluded that glucosyl units from glycogen do not mix with intermediates derived from glucose utilization, and therefore that separate pathways for glycogenolysis and glycolysis are operating under the conditions used (see also Clegg, 1984b).

Studies with bacterial or yeast mutant strains may shed light on this important problem. Breitenbach-Schmitt et al. (1984) have reported data purporting to show the presence of a parallel glycolytic pathway in yeast mutants lacking phosphofructokinase activity. Space limitations do not allow more than a passing reference to that work, but their conclusion that intermediates of the pentose-P pathway may play a role in glucose catabolism implies again that extensive compartmentation of metabolites must exist in yeast.

The immunocytochemical demonstration of differential location of isozymes within the same anatomical compartment should help to validate our hypothesis. Only a few investigators have used specific antisera against members of one isozymic system to study their subcellular distribution. Hatzfeld et al. (1978) observed diffuse reaction in the cytoplasm of hepatoma cells using three specific sera, anti-aldolase A, B, or C. On the other hand, differential distribution of rat liver lactate dehydrogenases M_4 and H_4 was ascertained by Yamashita et al. (1979). Lactate dehydrogenase M_4 was found preponderantly associated to glycogen-rich zones, whereas the H_4 isozyme was recognized in the amorphous cytoplasm near the cell membrane.

In myogenic cells derived from chicken skeletal muscle, which contain three creatine kinase isozymes, only the MM isozyme was found associated to the M-band structure, whereas the BB isozyme was observed at the Z-band (Wallimann et al., 1983).

Rigorous proof of our hypothesis must await, at least, the following: a) the isolation of complexes catalyzing unidirectional overall reactions; and b) evidence that upon dissociation of one such complex one, and only one, isozyme (of a particular reaction) is present, and that in vitro reconstitution of the complex is not possible if one component is replaced by a different isozyme.

The above-mentioned research program, although attractive, seems to be plagued by difficulties. Because the components of the complexes should be very similar, the task of assay and isolation becomes very discouraging. Instead, we have focused on the study of metabolic compartmentation which is, of course, an important corollary of the proposal.

COMPARTMENTS IN GLUCOSE METABOLISM

The study of functional metabolic compartments is fraught with uncertainties. In the case of glucose metabolism, most of the evidence in favor of compartmentation comes from experiments in which cells were incubated with radioactive substrates and specific activities of intermediates and/or end-products measured. In several cases the results suggested the presence of two or more unconnected compartments (for references, see the review by Ottaway and Mowbray, 1977). Besides the various traps involved in the use of radioactive tracers for the study of metabolism and its control, the following should be mentioned: a) the presence of several pathways of glucose utilization and synthesis, which may be acting simultaneously, thus obscuring the results; b) the coexistence of several cell types with different metabolic properties in the tissue or organ under study, e.,g., hepatocytes and sinusoidal cells in liver; c) metabolic zonation, i.e., the coexistence of cell types having different enzyme repertories in a seemingly homogeneous cell preparation (Lowry et al., 1978; Jungermann and Katz, 1982); and d) the destruction of the cells for the determination of specific activities.

THE MICROINJECTED FROG OOCYTE SYSTEM

Some of the limitations listed above can be overcome by the use of microinjected frog oocytes. The advantages of this system for studies on metabolism and its control have been summarized (Ureta and Radojkovic, 1985b). In short, only one cell is used, which is amenable, by way of microinjection, to the modification of intracellular levels of substrates, intermediates, cofactors, and enzymes. Further, from the point of view of glucose utilization, oocytes are simple cells: glucose is primarily directed to glycogen synthesis, and a minor portion is metabolized by the pentose-P pathway. Phosphofructokinase, fructose-1,6-bisphosphatase, and glycogen phosphorylase activities are absent (Radojkovic and Ureta, 1982; Ureta and Radojkovic, 1979). Glycolysis beyond fructose-6-P, gluconeogenesis from fructose bisphosphate, and glycogenolysis are not operative at this stage of oocyte maturation. Thus, the simplicity of oocytes makes them suitable for the search of compartments.

We have previously reported (Ureta and Radojkovic, 1982, 1985a) that in Stage VI oocytes glucose transport is a rate-limiting step for glycogen synthesis, since microinjection of glucose results in a 30-fold increase of glucose carbon incorporation into glycogen when compared to the extent of labeling when glucose is supplied in the external medium. On the other hand, $^{14}CO_2$ production is similar in either condition, suggesting that glucose transport does not limit the operation of the pentose-P pathway, and therefore that the pool of glucose 6-P for glycogen synthesis does not mix with that which will be converted into pentoses-P (Ureta and Radojkovic, 1985a).

Microinjected labeled intermediates are readily incorporated into glycogen and CO_2. Time curves are shown in Fig. 6. Glucose-1-P and fructose-6-P are incorporated significantly faster than glucose or glucose-6-P, the conventional wisdom notwithstanding that fructose-6-P must be first converted into glucose-6-P prior to incorporation into glycogen or to its entry in the pentose-P pathway. Also, glucose-1-P should be first converted into glucose-6-P to undergo utilization in the pentose-P pathway.

Further indications that pools of phosphoryl-hexoses are segregated in oocytes come from experiments in which unlabeled intermediates were coinjected at high concentrations with $[^{14}C]$-glucose at low concentrations (Fig. 7). Under these conditions no effect on $[^{14}C]$ glucose incorporation into glycogen was observed in the presence of intermediates except by glucose-1,6-bisphosphate. On the other hand, carbon incorporation into CO_2 was markedly affected by most intermediates except UDP-glucose.

A different pattern was observed if labeled glucose was supplied to the cells in the incubating medium instead of being microinjected. Under these conditions (Fig. 8), microinjection of unlabeled glucose-6-P, glucose-1-P, UDP-glucose, or 6-phosphogluconate significantly diminished $[^{14}C]$-glucose incorporation into glycogen, although the effect is rather modest. Fructose-6-P or fructose-1,6-bisphosphate did not affect glucose incorporation into the polysaccharide. Some inhibition of $^{14}CO_2$ release was observed, except with UDP-glucose or fructose-1,6-bisphosphate in which cases no effect was seen.

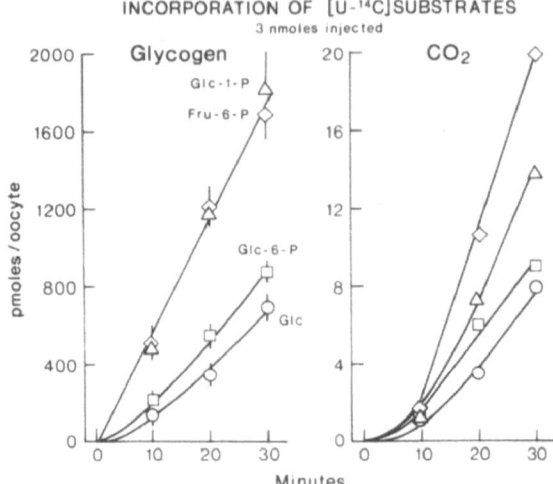

INCORPORATION OF [U-¹⁴C]SUBSTRATES

Fig. 6. Incorporation of labeled glucose, glucose-6-P, glucose-1-P or fructose-6-P into glycogen and CO_2. About 3 nmoles (60,000 to 70,000 cpm per oocyte) of [U-¹⁴C] substrates were microinjected. Calculated intracellular concentrations were about 1 mM. Groups of three oocytes in duplicate tubes were incubated during the indicated times in 50 μL of saline medium at room temperature. ¹⁴CO_2 evolution was monitored every 10 min. After incubation each oocyte was individually digested with 30% KOH at 100° and glycogen isolated by ethanol precipitation and counted. Results are given as means ± S.E. of six individual observations.

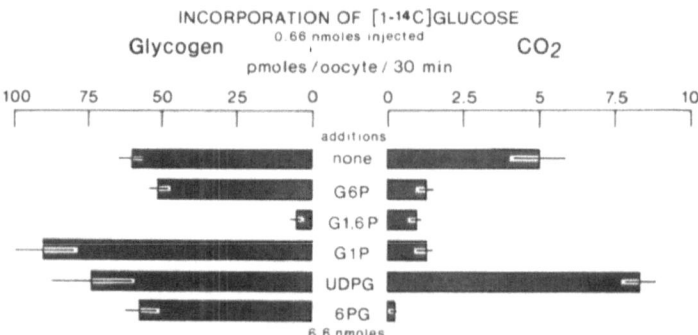

INCORPORATION OF [1-¹⁴C]GLUCOSE

Fig. 7. The influence of co-injected unlabeled intermediates on the incorporation of [1-¹⁴C] glucose into glycogen or CO_2 by frog oocytes. Oocytes were injected with about 0.66 nmoles of [1-¹⁴C]-glucose (injection volume was 50 nL; 70,000 cpm per oocyte). Other oocytes were injected with the same solution plus unlabeled compounds as indicated (about 6.6 nmoles). Oocytes were incubated during 30 min at room temperature. Measurements of radioactivity in glycogen and CO_2 were performed as described in the legend to Fig. 6. The results are given as means ± S.E. of five to six observations.

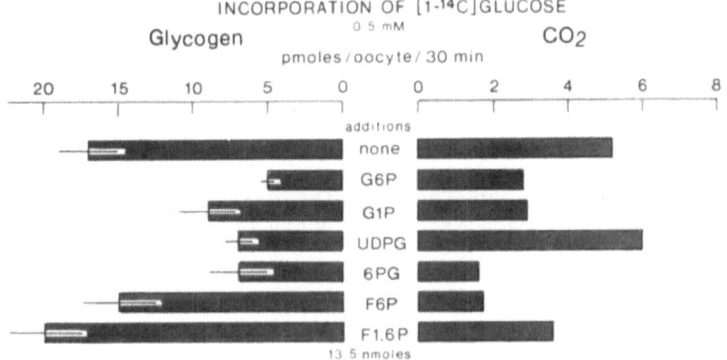

Fig. 8. The effect of injected unlabeled intermediates on the incorporation of $[1-^{14}C]$-glucose present in the incubation medium. Oocytes were microinjected with 45 nL of either saline or the indicated unlabeled intermediates in saline (13.5 nmoles per oocyte). Afterwards, groups of three oocytes in duplicate tubes were incubated in 60 μL 0.5 mM $[1-^{14}C]$-glucose (831,000 cpm) in saline. Incubation time was 30 min at room temperature. Radioactivity measurements in glycogen and CO_2 were performed as described in the legend to Fig. 6.

Taken together, the results of Figs. 7 and 8 suggest that the pathways available to glucose differ according to whether the sugar was supplied to the oocyte by microinjection or through the normal transport system.

CONCLUDING REMARKS

The results presented in this chapter are best understood in terms of extensive compartmentation of phosphoryl-hexoses. Alternative explanations may be suggested. Thus, the lack of isotopic dilution by unlabeled intermediates (Fig. 7) might indicate that the pathway from glucose to glycogen in frog oocytes is not the direct classical one, i.e., glucose-->glucose-6-P-->glucose-1-P-->UDP-glucose-->glycogen. Not a few authors have speculated that indirect pathways for glycogen synthesis from glucose exist (for references, see Hue and Hers [1969], in which a refutation of such proposals was reported). Very recently, it has been claimed that in rat liver glucose must be extensively catabolized prior to incorporation into glycogen (Newgard et al., 1983, 1984). This very interesting "glucose paradox" (Katz and McGarry, 1984) may prove difficult to investigate in a complex system such as the whole liver or even in isolated hepatocytes (see, for instance, the commentary by Pilkis et al. [1985]). Our present data, and unpublished results from our laboratory, may be interpreted as consistent with an indirect pathway which becomes operative when transport is circumvented; but then, again, compartmentation of intermediates must also be invoked.

The underlying hypothesis being tested in our studies of carbohydrate metabolism in frog oocytes is that extensive compartmentation of intermediates of glucose utilization occurs. Our data allow us to suggest that glucose and phosphoryl-hexoses committed to glycogen synthesis are spatially separated from those intermediates committed to the pentose-P pathway. While the results obtained help define the problem of metabolic organization, they do not address the question

whether poly-isozymic complexes do exist. Nonetheless, the fact that
only two routes of glucose metabolism are operative in oocytes correlates
very well with the presence of just two hexokinase isozymes. Thus, it is
tempting to speculate that one of them is uniquely involved in the
production of glucose-6-P available only to phosphoglucomutase, while
the other produces glucose-6-P available only to glucose-6-P
dehydrogenase. Being aware that no degree of correlation can establish a
cause-effect relationship, we feel nonetheless that, in the absence of
direct answers, our approach is a reasonable alternative.

> *"And you," I said with childish impertinence,*
> *"never commit errors?"*
>
> *"Often," he answered. "But instead of conceiving*
> *only one, I imagine many, so I become the slave*
> *of none."*
>
> (Umberto Eco, The Name of the Rose, Fourth Day,
> Vespers)

ACKNOWLEDGEMENTS

We thank members of the Laboratory of Biochemistry and Molecular
Biology, Faculty of Sciences, University of Chile, for their continued
encouragement and advice. Special thanks are due to Dr. Jorge Babul for
his comments and critical review of this manuscript. Experimental work
from our laboratory has been supported by grants from the Departamento de
Investigación y Bibliotecas (B-1989) and from the Fondo Nacional de
Ciencia y Tecnología (1139-84).

REFERENCES

Breitenbach-Schmitt, L, Schmitt, H.D., Heinisch, J. and Zimmermann,
    F.K., 1984, Genetic and physiological evidence for the existence of a
    second glycolytic pathway in yeast parallel to the phospho-
    fructokinase-aldolase sequence, Mol. Gen. Genet., 195:536.
Clegg, J.S., 1984a, Properties and metabolism of the aqueous cytoplasm
    and its boundaries, Am. J. Physiol., 246:R133.
Clegg, J.S., 1984b, Metabolic compartmentation and "soluble" metabolic
    pathways, BioEssays, 1:129.
Finkelstein, J.D. and Martin, J.J., 1984, Methionine metabolism in
    mammals. Distribution of homocysteine between competing pathways, J.
    Biol. Chem., 259:9508.
Hatzfeld, A., Feldmann, G., Guesnon, J., Frayssinet, C. and Schapira,
    F., 1978, Location of adult and fetal aldolases A, B, and C by
    immunoperoxidase technique in LF fast-growing rat hepatomas, Cancer
    Res., 38:16.
Hue, L. and Hers, H.-G., 1969, A reevaluation of the pathway by which
    glucose is converted into glycogen in a liver homogenate, FEBS Lett.,
    3:41.
Jungermann, K. and Katz, N., 1982, Metabolic heterogeneity of liver
    parenchyma, in: "Metabolic Compartmentation", H. Sies, ed., Academic
    Press, New York.
Katz, J. and McGarry, J.D., 1984, The glucose paradox. Is glucose a
    substrate for liver metabolism?, J. Clin. Invest., 74:1901.
Lowry, C.V., Kimmey, J.S., Felder, S., Chi, M.M.-Y., Kaiser, K.K.,

Passonneau, P.N., Kirk, K.A. and Lowry, O.H., 1978, Enzyme patterns in single human muscle fibers, J. Biol. Chem., 253:8269.

Lynch, R.M. and Paul, R.J., 1983, Compartmentation of glycolytic and glycogenolytic metabolism in vascular smooth muscle, Science, 222:1344.

MacGregor, J.S., Singh, V.N., Davoust, S., Melloni, E., Pontremoli, S. and Horecker, B.L., 1980, Evidence for formation of a rabbit liver aldolase-rabbit liver fructose-1,6-bisphosphatase complex, Proc. Natl. Acad. Sci. U.S.A., 77:3889.

Markert, C.L., 1975, Biology of isozymes, in: "Isozymes. I. Molecular Structure," C.L. Markert, ed., Academic Press, New York.

Markert, C.L. and Moller, F., 1959, Multiple forms of enzymes: tissue, ontogenetic, and species specific patterns, Proc. Natl. Acad. Sci. U.S.A., 45:753.

Masters, C.J., 1981, Interactions between soluble enzymes and subcellular structure, CRC Crit. Rev. Biochem., 11:105.

Newgard, C.B., Kirsch, L.J., Foster, D.W. and McGarry, K.D., 1983, Studies on the mechanism by which exogenous glucose is converted into liver glycogen in the rat. A direct or an indirect pathway?, J. Biol. Chem., 258:8046.

Newgard, C.B., Moore, S.V., Foster, D.W. and McGarry, K.D., 1984, Efficient hepatic glycogen synthesis in refeeding rats requires continued carbon flow through the gluconeogenic pathway, J. Biol. Chem., 259:6948.

Ottaway, J.H. and Mowbray, J., 1977, The role of compartmentation in the control of glycolysis, Curr. Top. Cell. Reg., 12:107.

Pilkis, S.J., Regen, D.M., Claus, T.H. and Cherrington, A.D., 1985, Role of hepatic glycolysis and gluconeogenesis in glycogen synthesis, BioEssays, 2:273.

Radojković, J. and Ureta, T., 1982, Regulation of carbohydrate metabolism in microinjected frog oocytes, Arch. Biol. Med. Exp., 15:395.

Srere, P.A. and Mosbach, K., 1974, Metabolic compartmentation: symbiotic, organellar, multienzymic, and microenvironmental, Annu. Rev. Microbiol., 28:61.

Stebbing, N., 1980, Evolution of compartmentation, metabolic channelling and control of biosynthetic pathways, in: "Cell Compartmentation and Metabolic Channelling", L. Nover, F. Lynen, and K. Mothes, eds., Elsevier/North-Holland Biomedical Press, Amsterdam.

Ureta, T., 1978, The role of isozymes in metabolism: a model of metabolic pathways as the basis for the biological role of isozymes, Curr. Top. Cell. Reg., 13:233.

Ureta, T., 1985, Organizacion del metabolismo: localizacion subcelular de enzimas glicoliticas, Arch. Biol. Med. Exp., 18:9.

Ureta, T. and Radojković, J., 1979, Frog oocytes: a model system for in vivo studies on the regulation of glucose metabolism, Acta Cient. Venez., 30:396.

Ureta, T. and Radojković, J., 1985a, Search for compartments of glucose metabolism in the microinjected frog oocyte, Arch. Biol. Med. Exp., 18, in the press.

Ureta, T. and Radojković, J., 1985b, Microinjected frog oocytes: a first-rate test tube for studies on metabolism and its control, BioEssays, 2:221.

Wallimann, T., Moser, H. and Eppenberger, H.M., 1983, Isoenzyme-specific localization of M-line bound creatine kinase in myogenic cells, J. Muscle Res. Cell Mot., 4:429.

Walsh, K. and Koshland, D.E., Jr., 1984, Determination of flux through the branch point of two metabolic cycles. The tricarboxylic acid cycle and the glyoxylate shunt, J. Biol. Chem., 259:9646.

Welch, G.R., 1977, Role of organization of multienzyme systems in cellular metabolism: a general synthesis, Prog. Biophys. Mol. Biol., 32:103.

Wilson, J.E., 1985, Regulation of mammalian hexokinase activity, in: "Regulation of Carbohydrate Metabolism," Vol. 1, R. Beitner, ed., CRC Press, Boca Raton, FL.

Wolpert, J.S. and Ernst-Fonberg, M.L., 1975, Dissociation and characterization of enzymes from a multienzyme complex involved in CO_2 fixation, Biochemistry, 14:1103.

Wombacher, H., 1983, Molecular compartmentation by enzyme cluster formation. A view over current investigations, Mol. Cell. Biochem., 56:155.

Yamashita, S., Yamamoto, N. and Yasuda, K., 1979, Immunohistochemical study of lactate dehydrogenase isozymes in rat liver, Acta Histochem. Cytochem., 12:125.

DIRECT TRANSFER OF METABOLITES VIA ENZYME-ENZYME

COMPLEXES: EVIDENCE AND PHYSIOLOGICAL SIGNIFICANCE

S.A. Bernhard and D.K. Srivastava

Institute of Molecular Biology
University of Oregon
Eugene, Oregon 97403 U.S.A.

INTRODUCTION

This paper deals with the transfer of metabolites (M) from one enzyme site (E_1), the site of synthesis, to another enzyme site (E_2), the site of utilization. Some such metabolite transfers have already been demonstrated to proceed by the direct interaction of a pair of enzymes involved in sequential steps of a metabolic pathway (Weber and Bernhard, 1982; Srivastava and Bernhard, 1984, 1985, 1986a,b). In such transfers the common metabolite is not exposed to an intermediate "solvent" environment. This direct transfer process is in contrast to the usually implicit model of metabolite transfer via dissociation from E_1 and subsequent arrival at E_2 via random diffusion through the aqueous environment (Eq. 1).

$$E_1\text{-}M + E_2 \rightleftharpoons E_1\text{-}M\text{-}E_2 \longrightarrow \text{Products}$$

$$\text{(Direct transfer)} \tag{1}$$

$$E_1\text{-}M \rightleftharpoons E_1 + M + E_2 \rightleftharpoons E_2\text{-}M \longrightarrow \text{Products}$$

$$\text{(Random diffusion transfer)}$$

Prior to our kinetic investigations, the evidence for metabolic channelling has involved either the failure of an isotopically labeled intermediary metabolite to effectively compete with precursor for the formation of final product (Eq. 2), or the demonstration that the kinetics of a multi-step sequence of reactions proceeds more rapidly than would be predicted on the basis of summation of the individual enzyme catalyzed reaction velocities (Welch, 1977; Srivastava and Bernhard, 1986a).

$$\text{Precursor(s)} \rightleftharpoons \boxed{M_1^E \rightleftharpoons M_2^E \rightleftharpoons M_3^E} \rightleftharpoons \text{Product(s)} \tag{2}$$

$$M_2^{aq}$$

$$M^E = \text{Enzyme-bound M}$$
$$M^{aq} = \text{Aqueous M}$$

Such experiments are complicated by the following two possibilities: (1) The inaccessibility of intermediary metabolites to their affine sites in multienzyme structural complexes. (2) The presence of unforeseen enzyme effectors in the multicomponent reaction system.

In contrast to our limited number of studies of the direct transfer of metabolite between enzymes, a large body of information is available regarding specific enzyme-enzyme interactions, particularly among enzymes which are sequential in a metabolic pathway (Ovadi et al., 1978; Halper and Srere, 1977; Fahien and Kmiotek, 1978, 1979, 1983; MacGregor et al., 1980). In the case of glycolytic enzymes, there have been many reports concerning the interaction among glycolytic enzymes in "glycosomes" and particularly on the interaction of specific glycolytic enzymes with membranous and cytoskeletal structural elements (Gorringe and Moses, 1980; Sols and Marco, 1970; Ottaway and Mowbray, 1977; Srivastava and Bernhard, 1986a). Evidence has been presented in addition supporting the view that these particular enzymes interact with other enzymes of the glycolytic pathway. In several instances, direct physical evidence for complex formation has been reported for the enzymes of the glycolytic system (Ovádi et al., 1978; Salerno and Ovádi, 1982; Batke et al., 1980; MacGregor et al., 1980). Still more physical evidence for enzyme-enzyme interactions has been demonstrated among the enzymes of the TCA cycle, and in the enzymic connections between the TCA cycle and enzymes of amino acid and fatty acid biosynthesis and degradation (Srere et al., 1978; Beeckmans and Kanarek, 1981; Sumegi and Alkonyi, 1983; Sumegi et al., 1985; Fahien and Kmiotek, 1978, 1979, 1983). We present herein what we believe to be more convincing experimental demonstrations that the direct transfer mechanism (Eq. 1) is often operative in _in vitro_ systems, and that these demonstrations are physiologically relevant.

EXPERIMENTAL EVIDENCE FOR THE DIRECT TRANSFER OF METABOLITES

It first became apparent to us that something more than the dissociation-random diffusion mechanism was appropriate to understand the phosphoglycerate kinase (PGK) turnover, in considering the rate of dissociation of the metabolite 1,3-diphosphoglycerate (DPG) from the enzyme phosphoglycerate kinase (PGK) (Huskins, 1979; Huskins et al., 1982). ^{32}P-NMR studies of the binding of reaction components to this enzyme demonstrated that one reaction component (DPG) is exceedingly tightly bound to the enzyme (Huskins, 1982; Nageshwara Rao et al., 1978). This tight binding is not substantially affected by the presence or absence of other reaction components (ATP, ADP and 3-phosphoglycerate) (Huskins, 1979). Our NMR experiments with the enzyme from halibut muscle, as well as the experiments of Nageshwara Rao et al. (1978) with yeast PGK, place a maximal value on the magnitude of the dissociation constant, K_d, at $< 10^{-8}$ M. Since the rate constant for diffusion of the metabolite to the enzyme (k_{on}) is maximally of the order of 10^8 M^{-1} s^{-1}, the specific unimolecular dissociation rate constant (k_{off}) cannot exceed 1 s^{-1} (Eq. 3).

$$
\begin{array}{c}
\text{3PGA} \\
+ \ \text{PGK} \rightleftharpoons \text{PGK} \Big\langle {}^{\text{ATP}}_{\text{3PGA}} \rightleftharpoons \text{PGK} \Big\langle {}^{\text{ADP}}_{\text{DPG}} \\
\text{ATP}
\end{array}
$$

$$
\text{PGK} + \text{DPG} \xrightarrow{\ k_{on}\ } \text{PGK-DPG} \qquad (3)
$$

$$k_{off} = K_d \times k_{on} = 10^{-8} \text{ M} \times 1 \times 10^8 \text{ M}^{-1}\text{sec}^{-1}$$

$$= 1 \text{ sec}^{-1}$$

This specific "off" rate constant is probably in reality substantially smaller than that estimated on the basis of equilibrium dissociation constant.

A classic assay procedure either for PGK activity or for ATP concentration involves the coupling of the reaction of ATP with 3-phosphoglycerate catalyzed by PGK to the oxidation of NADH by DPG catalyzed by the enzyme glyceraldehyde-3-PO$_4$ dehydrogenase (GPDH) as is indicated in Eq. 4.

$$
\begin{array}{c}
\text{3PGA} \\
\text{ATP}
\end{array}
\xrightleftharpoons{\qquad} \text{ADP + DPG}
\xrightarrow[\text{NADH}]{\text{GPDH}}
\begin{array}{c}
\text{G-3-P} \\
\text{NAD}^+ \\
\text{P}_i
\end{array}
\qquad (4)
$$

Under conditions of large excess of GPDH over PGK, the coupled reaction assay becomes independent of GPDH concentration (Weber and Bernhard, 1982; Huskins, 1979). Under such conditions and in sufficient excess of both PGK substrates (3-phosphoglycerate and ATP) and GPDH substrate (NADH), a turnover number (k_{cat}) for the coupled reaction limited by PGK concentration is obtained. This k_{cat} is approximately 600-800 s^{-1}, with very little indication of variability in the extensive enzymological literature (Scopes, 1973). Consequently, the overall rate of reaction (Eq. 4) is 10^3-fold or greater than the rate of dissociation of DPG from PGK.

In a set of steady-state experiments designed to probe the molecular formalism (Eq. 1) we have demonstrated that the pathway of transfer of DPG between PGK and GPDH is entirely consistent with the direct transfer mechanism, and inconsistent with the dissociation-diffusion mechanism (Weber and Bernhard, 1982). These steady-state kinetic experiments involve the utilization of high concentrations of PGK-DPG complex as a potential substrate for the GPDH-catalyzed reaction, under conditions of very low concentrations of GPDH. A notable feature of the coupled reaction process, when both enzymes are isolated from halibut muscle sarcoplasmic fluid, is the saturation of overall reaction velocity at higher concentrations of PGK-DPG complex (Weber and Bernhard, 1982). The dependence of reaction velocity on concentration of the PGK-DPG complex follows a Michaelian rate law (Fig. 1). From this velocity dependence a K_M for the enzyme substrate-enzyme complex, as well as a V_{max} for turnover, can be calculated. Note in Figure 1, that the extrapolated V_{max} yields a k_{cat} of the same order of magnitude as the calculated maximal dissociation "off" rate constant calculated on the basis of the equilibrium constant and an aqueous diffusion rate constant for the "on" rate. At first hand this might appear to be a confirmation of the diffusion random-association mechanism. However, note the conditions for the experiment. Unliganded PGK is present in enormous excess over unliganded GPDH. Hence, any dissociation of DPG into aqueous solution would soon be followed by a one-sided competition between PGK ($\sim$10^{-5} M) and GPDH ($\sim$10^{-9} M). Hence, the transfer of the common metabolite cannot possibly proceed via the aqueous environment. Rather, it must proceed via some direct transfer routes where the large excess of kinase sites does not make itself evident. Nevertheless, the fact that the two rate parameters (k_{cat} and k_{off}) are comparable is interesting. It implies to us, that there is nothing unusual about the dissociation of DPG from PGK in the direct transfer process versus the dissociation into the aqueous environment, but instead, that the fate of dissociated ligand within the GPDH-PGK complex is restricted to partition between the two enzymes.

One detail regarding the plot of Figure 1 which was bothersome to us is the relatively low value for k_{cat} under saturation conditions. A wealth of experiments utilizing ATP and 3-phosphoglycerate as a source for diphosphoglycerate has shown that under conditions of PGK-limiting turnover, the reduction of DPG by NADH catalyzed by GPDH proceeds with a turnover number nearly 400-fold faster than this estimated k_{cat} (Huskins, 1979). One potential explanation for this seeming paradox is that one of the substrates in the PGK reaction, lacking in our direct-transfer kinetic experiment, is an effector of the coupled reaction rate. Indeed, both ATP and 3-phosphoglycerate (3-PGA) are activators of the coupled reaction pathway (Weber and Bernhard, 1982). In this regard the 3-PGA is a far more effective activator than is ATP. The dependence of reaction velocity on 3-PGA in the coupled process is shown in Figure 2. Note the saturation phenomenon in the dependence of reaction velocity on 3-PGA concentration. At optimal ATP concentration, and saturating 3-PGA concentration, the k_{cat} calculable from the turnover gives the expected value, assuming synergistic effects for ATP and 3-PGA. In this regard, it is interesting to note that at these concentrations of 3-PGA, sufficient to nearly saturate the transfer reaction velocity, the metabolite has no significant effect on either the rate of GPDH-catalyzed reduction of DPG by aqueous NADH or on the displacement of DPG from the PGK binding site. Thus 3-PGA is not an effector of either the kinase or the dehydrogenase reaction studied singly but is an effector of the direct transfer process. A complete knowledge of the rate and equilibrium association parameters for the substrate of each of the two

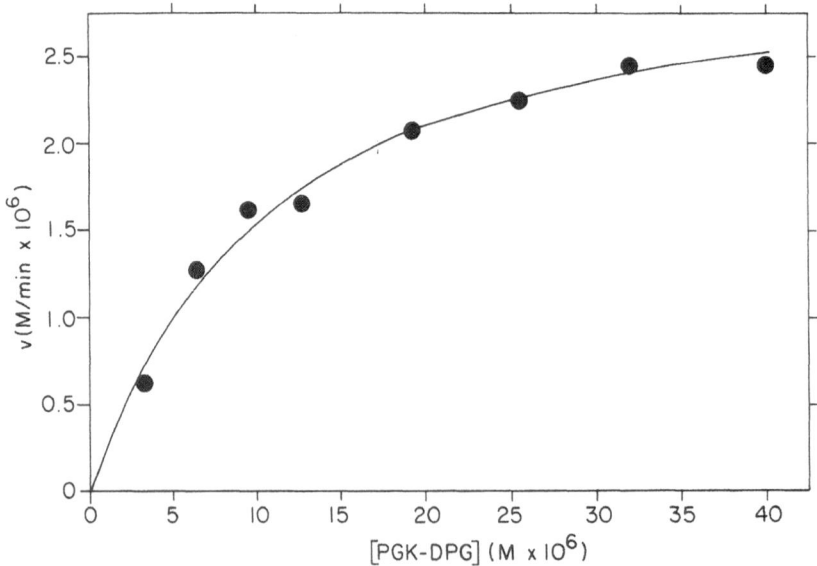

Fig. 1.   Initial rates of GPDH-catalyzed NADH oxidation as a function of PGK-DPG concentration, in 50 mM 2-methyl imidazole buffer, pH 7.4, at 25°C. The solid line is calculated on the basis of Michaelian behavior for PGK-DPG as a competent substrate: K_m = 10.8 μM, k_{cat} = 1.6 sec^{-1}. [GPDH-site] = 0.34 nM.

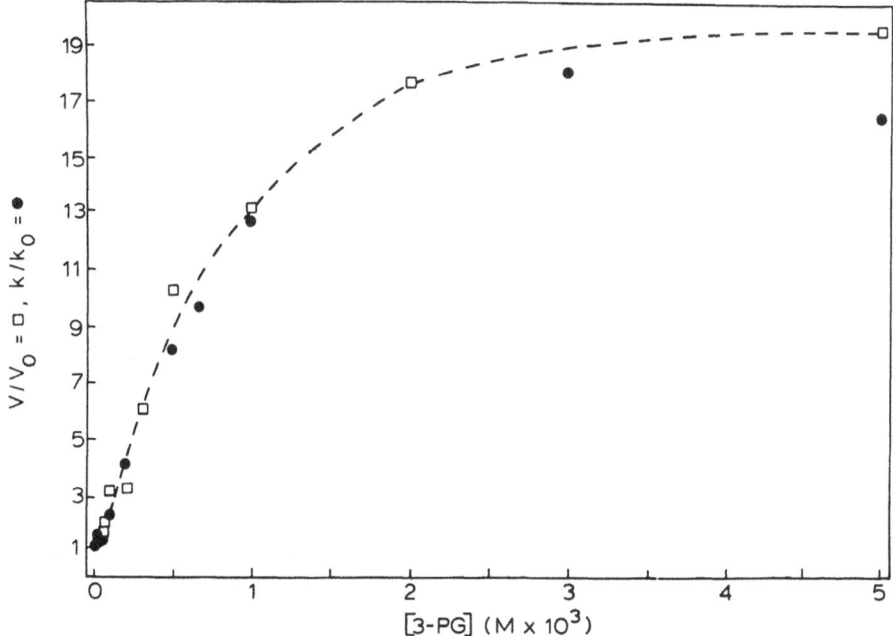

Fig. 2. The dependence of the PGK-DPG-saturated reaction velocity on 3-phosphoglycerate concentration. v/v_0 and k/k_0 are the ratio of velocities (under high PGK-DPG concentrations (■)) and rate constants (under low PGK-DPG concentrations (●)) in the presence and in the absence of 3-PGA respectively. Adapted from Weber and Bernhard (1982) by permission.

enzymes is insufficient to predict the properties of the coupled reaction system.

The rate constant for the dissociation of DPG from PGK is unusual. In general, the "off" rate constant (k_{off}) is comparable to or larger than the enzyme-catalyzed turnover numbers (k_{cat}). In order to probe for the direct transfer mechanism with such enzymes it becomes necessary to substantially "buffer" the aqueous metabolite concentration to a very low value. This can be achieved if high concentrations of an affine enzyme site are realizable. Under such conditions, an aqueous concentration of the common metabolite can be reduced to values far below that of K_M for an E_2-catalyzed reaction (Srivastava and Bernhard, 1984). In this way we have distinguished the E_1-metabolite complex from the aqueous metabolite as a competent substrate for the E_2-catalyzed reaction (Srivastava and Bernhard, 1984, 1985). For example, consider a dehydrogenase (E_2) catalyzed oxidation of NADH. In the presence of a large excess of another dehydrogenase (E_1) the aqueous concentration of NADH can be severely lowered. Under such conditions ([NADH]$_{aqueous}$ << K_M-NADH), the NADH-dependent reaction with S_2 catalyzed by E_2 can be reduced to a predictably very slow rate if aqueous NADH is the only competent substrate (Srivastava and Bernhard, 1984, 1985). The predictions of catalytic reaction velocity (based on the assumption that aqueous NADH is the only competent reductant) are compared with experiments in Table 1. These predictions are derived from the known dissociation constants of E_1-NADH and from the known Michaelian parameters (K_M and V_{max}) for the E_2-catalyzed reaction velocity. Note that sometimes experiment and calculated predictions agree quite precisely, whereas in other cases the

Table 1. Comparison of the Observed Rate of S_2 Reduction in the Presence of E_1, with Those Predicted on the Basis of Aqueous NADH as the Only Competent Coenzyme[a]

[E_1 Site] ($M \times 10^4$)	[E_2] ($M \times 10^{10}$)	Fixed concentrations	Predicted rate[b] ($M/min \times 10^7$)	Observed rate ($M/min \times 10^7$)
E_1=GPDH				
2.20			1.68	10.45
2.48	LDH=0.29	[NADH]$_t$=1.34x10^{-4}M	1.31	8.84
2.76			1.06	6.03
2.87		[Pyruvate]=2x10^{-3}M	1.00	5.47
2.98			0.93	4.82
E_1=LDH				
0.71			1.10	8.04
0.85	GPDH=1.08	[NADH]$_t$=4.95x10^{-5}M	0.69	7.88
1.13			0.39	6.91
1.41		[DPG] =9.0x10^{-5}M	0.30	6.03
1.70			0.21	5.15
E_1=LDH				
0.42			2.45	6.83
0.57	α-GDH=1.38	[NADH]$_t$=3.26x10^{-5}M	1.19	6.29
0.85			0.58	4.34
1.13		[DHAP]=5.0x10^{-4}M	0.39	3.85
1.41			0.30	3.22
E_1=GPDH				
0.16			15.38	14.47
0.32	α-GDH=2.11	[NADH]$_t$=2.25x10^{-5}M	5.71	6.59
0.40			3.43	2.73
0.47		[DHAP]=5.0x10^{-4}M	2.44	2.41
0.63			1.52	1.21
E_1=αGDH				
0.11			21.92	16.88
0.22	GPDH=1.08	[NADH]$_t$=4.95x10^{-5}M	13.55	8.44
0.33			6.70	6.43
0.44		[DPG]=7.0x10^{-5}M	4.17	4.50
0.56			2.98	3.21
0.67			2.34	1.61

[a]Adapted from Srivastava and Bernhard (1985) by permission.
[b]Calculated from the thermodynamic and kinetic parameters of appropriate enzymatic reactions (Srivastava and Bernhard, 1985).

experimental velocity is very much greater than that predicted according to the dissociation-diffusion mechanism. In the latter cases, the complex E_1-NADH must itself be a competent substrate for the E_2-catalyzed reaction and, hence, the coupled reaction process must involve metabolite transfer via an intermediate E_1-E_2 complex. Indeed, this inference is strongly substantiated by the saturation phenomenon observable in the dependence of E_2-catalyzed reaction velocity on E_1-M concentration (Srivastava and Bernhard, 1984, 1985, 1986a). In all cases where E_1-M is a competent substrate for the E_2-catalyzed reaction, we have demonstrated a Michaelian relationship between reaction velocity and E_1-M concentration. A summary of such reactions involving enzyme–enzyme complex formation is given in Table 2.

Among the enzyme–enzyme interactions summarized in Table 2, the interactions between dehydrogenases are particularly noteworthy for two reasons. (1) Due to the extensive utilization of NADH (and NADPH) as a specific coenzyme substrate in many enzyme catalyzed reactions, the possibility of NADH transfer can be experimentally investigated over a wide variety of dehydrogenases. (2) The three-dimensional structures of several dehydrogenases are known to atomic resolution (Buehner et al., 1974; Moras et al., 1975; Banaszak and Bradshaw, 1975; Holbrook et al., 1975; Branden et al., 1975; Branden and Eklund, 1980). Therefore, the potential interactions between pairs can be investigated by computer graphic analysis.

THE MOLECULAR MECHANISM OF DIRECT TRANSFER OF METABOLITES

On the basis of many steady-state kinetic experiments of the type we describe above, involving a diversity of dehydrogenases, we have discerned the general rule for the applicability of the direct transfer mechanism (Srivastava and Bernhard, 1985, 1986a). Direct transfer will occur whenever the two dehydrogenases exert opposite chiral specificity for the transfer of hydrogen between the C_4 of the nicotinamide ring and the substrate. The two classes of dehydrogenases have been designated as "A" and "B". Direct transfer of coenzyme will occur between any A–B pair of dehydrogenases but will not occur between two A or two B dehydrogenases (Srivastava and Bernhard, 1985, 1986a). The experimental results in support of the rule is summarized in Table 3.

From the molecular graphics of three dehydrogenases of known structure, we have proposed a molecular mechanism for the direct transfer of coenzyme between A and B dehydrogenases based on two specific structural details. (1) The different conformations of coenzyme in A

Table 2. K_m Values for E_1-M in the E_2 Catalyzed Reaction[a]

E_1-M	E_2	Constant Substrate	$K_m^{E_1}$-M (Mx10^6)[b]
(H.M.) PGK-DPG	(H.M.) GPDH	NADH	12 ± 2
(H.M.) GPDH-NADH	(H.L.) ADH	Benzaldehyde	13.3
(H.M.) GPDH-NADH	(H.M.) LDH	Pyruvate	5.88 ± 1.21
(H.M.) LDH-NADH	(H.M.) GPDH	DPG	2.6 ± 0.45
(R.M.) αGDH-NADH	(H.M.) LDH	Pyruvate	11.36
(H.M.) LDH-NADH	(R.M.) αGDH	DHAP	7.69
(P.H.) LDH-NADH	(R.M.) αGDH	DHAP	2.86
(H.M.) GPDH-NAD+	(H.M.) LDH	Lactate	10.0
(R.M.) Aldolase-DHAP	(R.M.) αGDH	NADH	8 ± 4

[a]Where the direct transfer of metabolite (M) is demonstrable via E_1-M-E_2 complex formation.
[b]From Weber and Bernhard (1982); Srivastava and Bernhard (1984, 1985, 1986 a,b).

Abbreviations: H.M., Halibut muscle; H.L., Horse liver; R.M., Rabbit muscle; P.H., Pig heart; B.H., Bovine heart.

Table 3. Direct Transfer of Nicotinamide Coenzyme or the Lack Thereof Between Pairs of Dehydrogenase of Known Chiral Specificity[a]

E_1[b]	E_2[c]	Coenzyme	Result of Transfer Experiment	Stereochemistry of E_1, E_2 Pair
GPDH	LADH	NADH	Transfer	B, A
GPDH	MDH	NADH	Transfer	B, A
GPDH	ADH	NADH	Transfer	B, A
GPDH	SDH	NADH	Transfer	B, A
GPDH	LDH	NADH	Transfer	B, A
GPDH	α-GDH	NADH	No Transfer	B, B
α-GDH	LADH	NADH	Transfer	B, A
α-GDH	LDH	NADH	Transfer	B, A
α-GDH	MDH	NADH	Transfer	B, A
α-GDH	GPDH	NADH	No Transfer	B, B
LDH	GPDH	NADH	Transfer	A, B
LDH	α-GDH	NADH	Transfer	A, B
LADH	LDH	NADH	No Transfer	A, A
LADH	MDH	NADH	No Transfer	A, A
LADH	ADH	NADH	No Transfer	A, A
LADH	SDH	NADH	No Transfer	A, A
LADH	α-GDH	NADH	Transfer	A, B
GPDH	LDH	NAD^+	Transfer	B, A
GPDH	G6PD	NAD^+	No Transfer	B, B
G6PD	GDH	NADPH	No Transfer	B, B
G6PD	AlDH	NADPH	Transfer	B, A

[a]Taken from Srivastava and Bernhard, 1984, 1985, 1986a,b.

[b]E_1 is utilized as a coenzyme carrier (E_1 > coenzyme > $K_d^{E_1-coenzyme}$).

[c]E_2 is utilized in catalytic concentration (10^{-9}-10^{-10}M) for reducing a saturating concentration of specific substrate (S_2).

versus B dehydrogenases lead to a non-mirror image relationship between chiral coenzyme molecules when two enzymes with opposite chiral specificity are juxtaposed cleft-to-cleft (Srivastava et al., 1985). For this reason, it is possible to transfer the nicotinamide ring from site A to site B (or vice-versa) without permitting internal molecular rotation (a process which occurs readily and rapidly in the aqueous solvent). (2) In the limited instances where we have had the opportunity to examine the three-dimensional structural coordinates, the molecular surface surrounding the active-site cleft of A dehydrogenases is virtually entirely negatively charged, whereas the complementary molecular surface in a B dehydrogenase (glyceraldehyde-3-PO_4 dehydrogenase) is entirely positively charged (Srivastava et al., 1985). These surrounding surfaces are each sufficiently charged such that A-A or B-B interactions would lead to strong repulsions. Some details of the molecular mechanism for the transfer of coenzyme or the lack thereof between a pair of dehydrogenases are summarized in Figures 3 and 4. The mechanism has been presented in detail elsewhere (Srivastava et al., 1985).

Further details regarding the process of transfer of NADH can be obtained from transient experiments. The direct transfer of NADH from an A dehydrogenase site to a B dehydrogenase site has a built-in signal of specific residents of the coenzyme, due to the spectral differences which

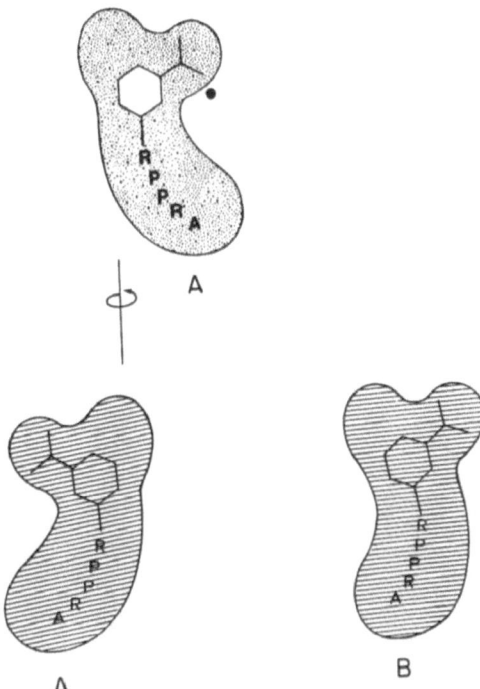

Fig. 3.   Schematic representation of transfer of NAD$^+$ between a pair of
dehydrogenases oriented with the cleft entries juxtaposed. The
coenzyme conformations when bound to A and B dehydrogenases are
indicated. Coenzyme is indicated as occupying all three
dehydrogenase sites so as to illustrate the problem of coenzyme
transfer between two A dehydrogenases (lower left to upper) and
two dehydrogenases of opposite chiral specificity (B to A;
lower right to upper). Note the impossibility of coenzyme
transfer from A to A due to the spatial restrictions on
molecular rotation within the complex (see text). The
transport of coenzyme from B to A can occur partially by
translation alone as is elaborated in Figure 4. Adapted from
Srivastava et al. (1985) by permission.

obtain when the coenzyme is bound in its _anti_ conformation (in the A
dehydrogenases) versus when it is bound in its _syn_ conformation (when the
coenzyme is bound in B dehydrogenases). The spectra of NADH bound to an
A dehydrogenase (LDH), a B dehydrogenase (αGDH), and in bulk water solvent
are shown in Figure 5. Via such signals, the kinetics of transfer of
NADH from an A to a B site can be monitored utilizing stopped-flow
spectrophotometry or spectrofluorometry (Srivastava and Bernhard, 1986b).
An experiment of this kind is illustrated in Figure 6. For comparison,
we show the transient dissociation of coenzyme from a B dehydrogenase
(αGDH) into the aqueous environment. This latter process can be
monitored by the difference in NADH fluorescence emission when the
coenzyme is bound to protein versus when it is free in aqueous solution.
The dissociation process can be made quasi-irreversible by mixing αGDH-
NADH with an excess of NAD$^+$ (Curve 2, Figure 6). Note, perhaps
surprisingly, that the specific rate constant for transfer of NADH from
αGDH to lactate dehydrogenase (LDH) is considerably faster (by
approximately one order of magnitude) than the specific rate constant for
NADH dissociation into the aqueous solution. Under conditions of the

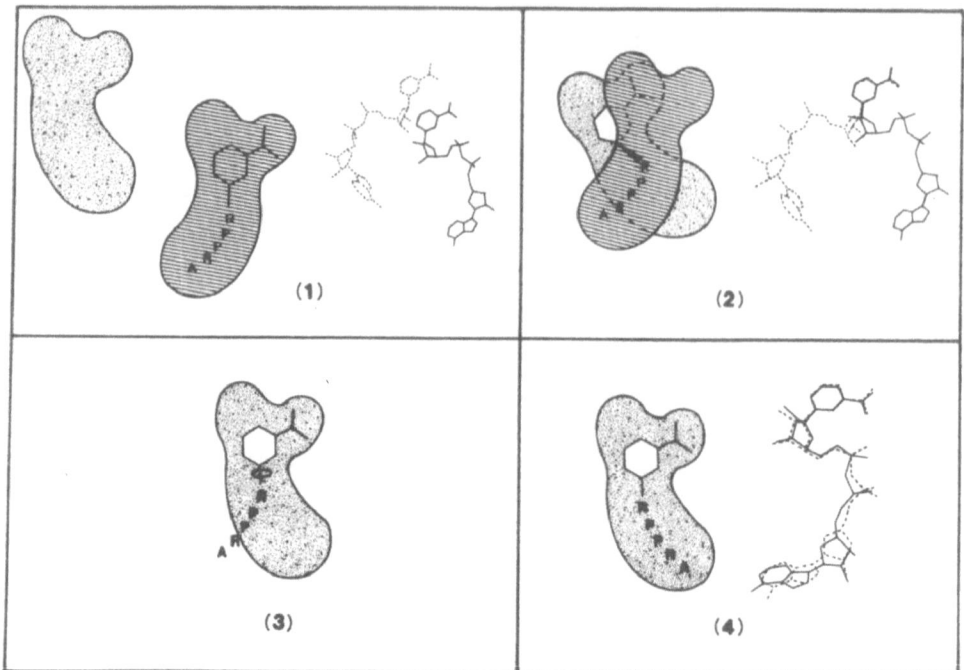

Fig. 4.   The mechanism of direct transfer of NAD$^+$ between a pair of
dehydrogenases of opposite chirality.   Illustrated is the
transfer of NAD$^+$ from a B dehydrogenase site (lower) to an apo
A dehydrogenase (upper).   The sequence of events is
schematized as follows.   1) The two enzyme sites are juxtaposed
at closest contact.   2) The nicotinamide ring is transferred
from site B to site A.   Note the impossibility of transfer of
the dinucleotide structure by translation alone.   3) The front
(B) dehydrogenase dissociates from the rest of the coenzyme
allowing free rotation about the nicotinamide N_1-C_1' glycosidic
bond.   4) Following rotation of $\sim 180^\circ$ about the N_1-glycosidic
bond, the rest of the dinucleotide structure can bind to the A
dehydrogenase site by translation into the second cavity.   The
precise spatial orientation for coenzyme transfer, taken from
computer graphics, is shown alongside the schematic orientation
of the two proteins in the various stages of transfer.   Adapted
from Srivastava et al. (1985) by permission.

experiment, there is virtually no free aqueous NADH.   Virtually all of
the coenzyme is bound either to αGDH or to LDH.   This experiment
demonstrates that the interaction of the two dehydrogenases facilitates
the transfer of coenzyme presumably due to allosteric interactions
between LDH and αGDH-NADH.

   We have tentatively interpreted this and other direct transfer
experiments in terms of an allosteric transition model.   In the case of
NADH transfer among dehydrogenases the model can be summarized by Eq. 5.

$$
\begin{array}{ccc}
\text{NADH (B)} & & \text{(B)} \\
\text{E(holo)} & \longrightarrow & \text{E(apo)} \\
| \text{(A)} & \longleftarrow & | \text{NADH (A)} \\
\text{E(apo)} & & \text{E(holo)}
\end{array}
\qquad (5)
$$

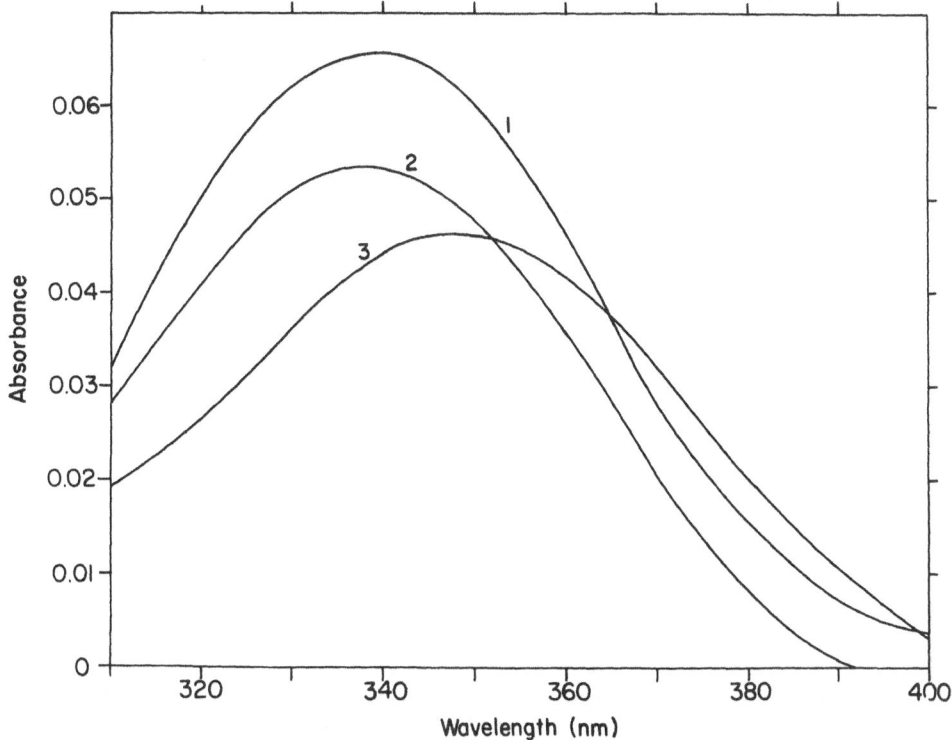

Fig. 5.   Absorption spectra of (1) NADH, (2) LDH-NADH and (3) αGDH-NADH,
in 50 mM Tris-HCl buffer containing 0.35 mM β-mercaptoethanol
and 1 mM EDTA, pH 7.5 at 25°C. Note the shift of NADH spectrum
upon ligation with either LDH or αGDH. From the known
dissociation constants of αGDH-NADH and LDH-NADH we calculate
that NADH is more than 95% saturated in each case.

The "holo" subscript is to denote the protein conformation of the
dehydrogenase which contains the bound NADH. The interacting unliganded
cognate enzyme is assumed to be in the "apo" protein conformation. A
characteristic of all the dehydrogenases of known structure is the
existence of two conformational states, an apo and a holo conformation
(Grau et al., 1981; Grau, 1982; Murthy et al., 1980; Eklund et al., 1981,
1982; Cedergren-Zeppezauer et al., 1982).

In order that this transfer mechanism be effective, the allosteric
equilibrium for the reaction (Eq. 5) must be near unity. Note that the
dissociation rate constants for coenzyme into aqueous solution are
variable, depending on the particular enzyme species (Table 4). In
contrast, the specific rates of transfer of NADH among enzymes are nearly
equal when coenzymes are transferred directly. These results (Table 4)
are achieved both by the acceleration and the deceleration of transfer
rates relative to dissociation rates. This suggests a further feature of
the direct transfer model. The rate of direct metabolite transfer can be
modulated via ligand-dependent enzyme-enzyme interaction (Srivastava and
Bernhard, 1986b).

All of the enzymes we have described are characterized by the
existence of at least two distinctly different protein conformations,

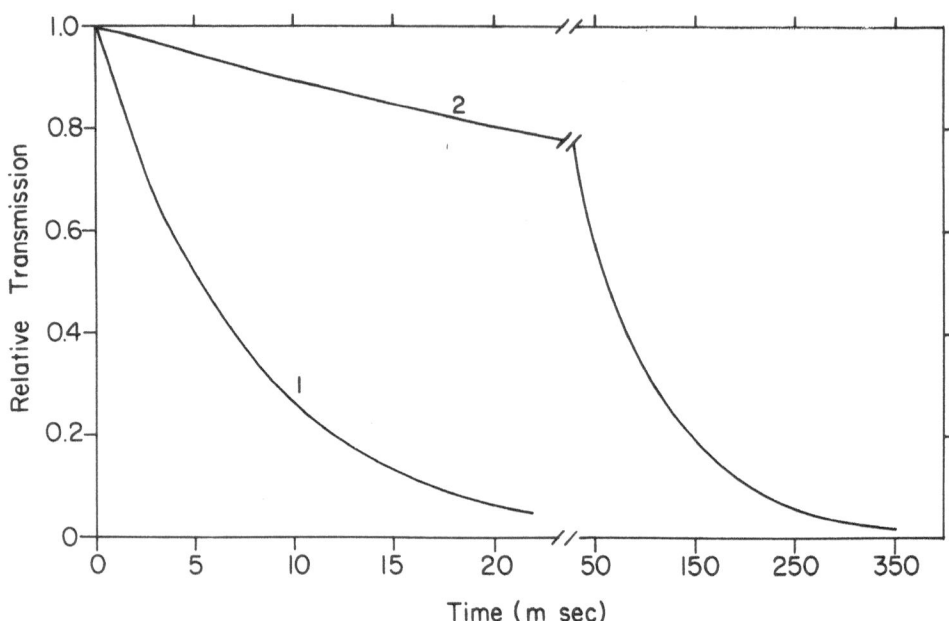

Fig. 6.   Transient rates of transfer of NADH from αGDH. The rate of
          NADH transfer from αGDH to LDH (Curve 1) is compared with the
          rate of dissociation of NADH into the aqueous solvent (Curve
          2).   These stopped-flow kinetic experiments are carried out by
          monitoring the fluorescence emission properties of NADH in
          different resident environments, viz. LDH, αGDH and bulk water
          solvent. Excitation wavelengths for Curve (1) and Curve (2)
          are 360 and 340 nm respectively. Fluorescence emission was
          filtered through 408 nm glass filter. All experiments were
          carried out in 50 mM Tris-HCl buffer, pH 7.5, containing 0.35
          mM β-mercaptoethanol and 1 mM EDTA. Configurations and
          concentration of reactants are shown for each curve.

Curve (1):   αGDH (24.56 μN)
                                  + LDH (50.18 μN)
              NADH (16.24 μM)

Curve (2):   αGDH (18.64 μN)
                                  + NAD$^+$ (15 mM)
              NADH (12.6 μM)

Note that the specific rate constant for transfer of NADH from
αGDH to LDH (k_{trans} = 132.5 sec^{-1}) (Curve 1) is considerably
faster than the specific "off rate" of NADH from αGDH (k_{off} =
10.8 sec^{-1}) (Curve 2).

dependent on the state and nature of metabolite ligations. It is
tempting to speculate that, as in the case of A and B dehydrogenases,
only one of the potential enzyme conformational states is capable of
recognition of the cognate enzyme.   In the case of the dehydrogenases we
have postulated that in the complex each member of the cognate pair of
enzymes is in a particular protein conformation (Srivastava and
Bernhard, 1986b).   This appears to be the case as well with the cognate
pair GPDH-PGK. In this pair it is well known that both PGK and GPDH can
exist in two alternate conformations (Pickover et al., 1979; Banks et
al., 1979; Watson et al., 1982; Murthy et al., 1980).   The PGK

Table 4. Comparison of the Specific Off-Rate Constant (k_{off}) of E_1-NADH Complex and the Transient Rate Constant for Transfer of NADH (k_{trans}) from E_1-NADH to E_2

E_1	E_2	k_{off}^a (sec^{-1})	k_{trans}^b (sec^{-1})
R.M. αGDH	H.M. LDH	9.4 ± 2.8	142 ± 21
P.H. LDH	H.M. GPDH	40.8 ± 8.6	160 ± 18
H.M. LDH	R.M. αGDH	245	172 ± 22
	H.M. GPDH		180 ± 25
H.M. GPDH	P.H. LDH	> 300	152 ± 12
	H.M. LDH		132 ± 8

[a] Determined by monitoring the decrease in fluorescence emission of E_1-NADH (excitation at 340 nm) due to displacement of NADH from E_1 site by excessive concentration of NAD^+.

[b] Determined by monitoring the difference in fluorescence properties of NADH at different enzyme sites. In αGDH-LDH and GPDH-LDH pairs excitation maximum is 360 nm, and 340 nm respectively. Fluorescence emission in each case is filtered through 408 nm glass filters. $E_1 > NADH << E_2$.

conformation is effected by specific ligation. Thus DPG is known to change the PGK conformation from that which predominates in its absence, or in the presence of bound nucleotides (Watson et al., 1982). Similarly, two different conformations have been detected for acyl-GPDH dependent on specific ligation (NAD^+ versus NADH or ATP) (Malhotra and Bernhard, 1968, 1973, 1981).

Previously we have argued that the transfer of NADH from 3-phosphoglyceroyl-GPDH-NADH to LDH-pyruvate occurs with concomitant oxidation of NADH (Scheme 1). We presume that the apo-holo dehydrogenase pair is operative in this case, and that the net result of such

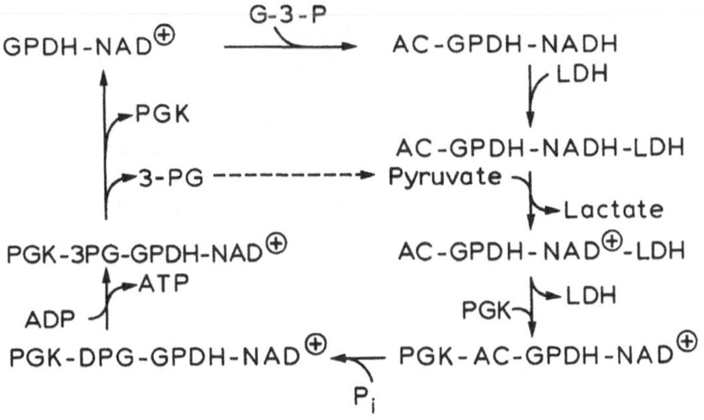

Scheme 1

interaction and reaction is to convert 3-phosphoglyceroyl GPDH-NADH to 3-phosphoglyceroyl-GPDH-NAD$^+$ (Srivastava and Bernhard, 1985). From our previous research, it would appear that acyl-GPDH-NAD$^+$ is the only ternary complex species which interacts with phosphoglycerate kinase (Weber and Bernhard, 1982). In the presence of low concentrations of inorganic phosphate, acyl group of GPDH is transferred as DPG to PGK (Srivastava and Bernhard, unpublished results). It is noteworthy that in the direct transfer of diphosphoglycerate (DPG) from PGK to GPDH, neither the transfer process nor the subsequent reduction of DPG by NADH at the GPDH site is appreciably inhibited by comparable or higher concentrations of apo PGK (Weber and Bernhard, 1982). This implies that the unliganded kinase is not specifically recognized by GPDH, in contrast to the DPG-liganded kinase. Thus the present information suggests that only one protein conformation is recognized by cognate enzyme in both the GPDH-LDH and GPDH-PGK specific interactions. In each case the specific recognition of enzyme pairs according to conformation is "induced" by the interaction of enzyme with specific substrate ligands. This sequence of specific interactions and direct transfers suggests at least one mechanism by which substrates and enzymes together organize the metabolic sequence. If enzyme-enzyme interactions are directly relevant to the mechanism of flow through the metabolic pathway, then ligand-induced interactions leading to altered protein conformations may be relevant to the recognition process among enzymes.

THE FUNCTIONAL SIGNIFICANCE OF THE DIRECT-TRANSFER MECHANISM

We have presented some evidence that for specific pairs of coupled reactions, the direct-transfer mechanism is appropriate under physiological conditions. In the case of the glycolytic pathway, we have demonstrated that a substantial fraction of the reactions proceed via the direct transfer of metabolite (Srivastava and Bernhard, 1986a). A plausible and physiologically attractive hypothesis is that "whole metabolic pathways, or substantial segments of pathways, proceed via the direct-transfer mechanism". In the following paragraphs we present background information which emphasizes the physiological relevance and the significance of our hypothesis. In this regard two facts, often overlooked, are directly relevant.

(1) The intracellular concentration of intermediary metabolites is in general lower than the concentration of affine enzyme sites (Ottaway and Mowbray, 1977; Sols and Marco, 1970; Srivastava and Bernhard, 1986a). By intermediary metabolite we mean metabolites which are synthesized and subsequently utilized in the intermediate course of a metabolic pathway. These intermediates are to be differentiated from precursor molecules, which initiate the pathway, and final products of the pathway. The precursors and final products may be present in concentrations in excess of that of their affine sites (Srivastava and Bernhard, 1986a). Good examples of the latter group of metabolites are glucose, glucose-6-PO$_4$ and fructose-6-PO$_4$, creatine phosphate (in muscle), lactate, and particular amino acid pools.

(2) The concentration of proteins in the cellular fluid is exceedingly high (Sols and Marco, 1970; Ottaway and Mowbray, 1977; Srere, 1981, 1982, 1984). Although there is some difficulty in defining the cellular fluid volume exclusive of cytoskeletal structural material, the concentration of protein can be roughly estimated in the range of 200-400 mg/ml (Srivastava and Bernhard, 1986a). This concentration is almost as high as the concentration of protein in protein crystals such as those

156

utilized for X-ray diffraction studies (Matthews, 1968; Srere, 1981). A good part of the water of the cell must be organized in interaction with the hydrophilic exterior residues on protein surfaces, and in interaction with accompanying ions. A substantial part of the remaining water is utilized for the solvation of precursor and final product pools of a very limited diversity of metabolites (Atkinson, 1969, 1977). The residual intermediary metabolites are in all probability distributed among the active site clefts of the proteins for which they are affine. Substrates, so sequestered from the aqueous environment, are stabilized against covalent reaction with water and other aqueous components. Thus unstable intermediary metabolites (for example, 1,3-diphosphoglycerate, phosphoenolpyruvate and glyceraldehyde-3-phosphate) are preserved for specific enzyme catalyzed reactions, as long as tight binding and an excess of sites over substrate persists. In the extrema, where direct transfer and virtually total ligation of metabolites to proteins are exclusively involved, the "aqueous" solvent may play a nearly inconsequential role in the metabolite transfer. Nevertheless, predictions regarding the bioenergetics and the kinetics of metabolic pathways have been largely derived from individual enzyme-metabolite parameters obtained in aqueous solution (Lehninger, 1975; Mahler and Cordes, 1971; Metzler, 1977).

In addition to the stabilization of molecules which are unstable in aqueous solution, the excessive enzyme site can preserve specific molecular conformations of metabolites (Srivastava and Bernhard, 1986a). If different enzymes are specific for individual metabolite conformations, only the appropriate enzyme bound conformer can be directly transferred to a particular metabolic pathway (Masters, 1977). It is interesting to note that fructose-1,6-diphosphate is produced as the β anomer in the reaction catalyzed by phosphofructokinase, whereas it is utilized as the α anomer in fructose-1,6-diphosphatase (Clarke et al., 1973).

As discussed above for the transfer of NADH between αGDH and LDH, the rate of transfer of metabolite can exceed the unimolecular desorption rate. The major advantage of the direct-transfer mechanism for most intermediary metabolites lies in the relatively high concentration of enzyme-bound species as compared to the aqueous metabolite. To the extent that our results can be generalized to entire metabolic pathways, there appears to be little chance for kinetic competition via the random diffusion mechanism regardless of the viscosity of the cellular fluid (Srivastava and Bernhard, 1986a). The high concentration of enzymes, which determine the extent of bound metabolites, does not allow for rapid diffusion of metabolite even at the low viscosity of in vitro enzyme experiments. Such diffusion controlled velocity would be much slower in the highly viscous cellular fluid.

THE EVOLUTION OF METABOLIC PATHWAY EFFICIENCY

One important question regarding enzymes of a metabolic pathway is the evolution of enzyme function. In the recent past, this question has been attacked by consideration of the evolution of efficient catalysis of individual enzyme-catalyzed reactions (Albery and Knowles, 1976, 1977; Knowles, 1980; Benner, 1982; Nambiar et al., 1983). Notably, Albery and Knowles have considered the evolution of enzyme efficiency in terms of a single enzyme and the unimolecular structural transitions and bimolecular association reactions which constitute the microscopic contributions to the overall catalytic reaction velocity. Albery and Knowles note the

rates of association at physiological concentrations for the substrates
of the triose phosphate isomerase reaction. These bimolecular rates of
association (assumed to proceed via the free diffusion of substrates
through the aqueous environment) lead to velocities comparable to the
overall reaction velocity through the unimolecular pathway of transformations at the enzyme site (Albery and Knowles, 1976). This argument is
experimentally correct if the free aqueous concentrations of reaction
components are nearly the same as the experimentally determined total
concentration of such components in muscle fluid. Under conditions where
the overall reaction velocity approaches the diffusion limit, further
catalytic efficiency can evolve if the diffusion rate of substrate "on"
to the site is the same as the diffusion rate of product "on" to the site
(Albery and Knowles, 1976). Any other situation would give rise to an
imbalance between substrate "on" and product "off", so as to lower the
catalytic efficiency. Basically for this reason, Albery and Knowles
propose that enzymes which have achieved their optimal evolution maintain
an equal balance in partition between enzyme-substrate and enzyme-product
complexes (Albery and Knowles, 1976, 1977; Knowles, 1980). In Table 5
is contained a variety of enzyme catalyzed reactions of metabolic
significance, whose internal equilibria between bound products and bound
reactants have been experimentally determined. Note the universal
tendency towards an internal equilibrium constant near unity. The
experimental data (Table 5) have been cited as convincing arguments for
the evolution of catalytic efficiency to the metabolite diffusion limit
(Knowles, 1980).

Table 5. Equilibrium Constants for Aqueous Versus Enzyme-Bound
Metabolites in Different Enzyme-Catalyzed Reactions

Enzyme	K_{eq} (aqueous)	K_{eq} (enzyme-bound)	References
Triosephosphate isomerase	2.2×10^1	0.6	Albery & Knowles, 1976
Hexokinase	2×10^3	~1.0	Wilkinson & Rose, 1979
Pyruvate kinase	3×10^{-4}	1.0-2.0 $(10.0-15.0)$[a]	
Pyruvate kinase (glycolate reaction)	$> 5 \times 10^1$	1.0-3.0	Nageswara Rao et al., 1978, 1979
Arginine kinase	1×10^{-1}	1.2	
Creatine kinase	1×10^{-1}	~1.0	
Adenylate kinase	4×10^{-1}	1.6	
Phosphoglycerate kinase (yeast)	3×10^{-4}	0.8	
Phosphoglycerate kinase (muscle)	8×10^{-4}	0.5-1.5	Huskins et al., 1982
Phosphoglucomutase	1.7×10^1	0.4	Ray & Long, 1976a,b
Lactate dehydrogenase	1×10^4	1.0-2.0	
Yeast alcohol dehydrogenase	1.4×10^4	0.15-0.25	Nambiar et al., 1983
Horse liver alcohol dehydrogenase	2×10^4	0.10-0.20	
Myosin ATPase	2×10^5	10	Gutfreund & Trentham, 1975

[a]Stackhouse et al. (1985).

158

In a metabolic pathway operating via direct transfer, efficient operation of the pathway must involve the facile transfer of metabolite from enzyme to enzyme everywhere within the path. Optimal transfer through the pathway will occur when the common metabolite is equally affine for the two enzymes involved in the direct transfer (Fig. 7). Thus it would be anticipated that there would be nearly equal partitioning between E_1-S_1 and E_2-S_1. In order that the different transfer processes proceed with nearly equal facility, it must follow that the species E_2-S_2 and E_2-S_3 be partitioned comparably in the internal reaction process. In this way the direct-transfer pathway provides an explanation for the experimental observations, without the necessity of assuming that the free metabolite concentrations in situ happen to be those which balance the bimolecular diffusion rates to the unimolecular rate processes.

A casual investigation of the concentrations of glycolytic enzymes and their intermediary metabolites (Tables 6 and 7) reveals that many of these metabolites are present in insufficient concentration to maintain the diffusion-limit reaction velocity; it is highly likely that a major fraction of the total metabolite concentration is enzyme-bound rather than free in aqueous solution.

Our alternative hypothesis is that evolution of metabolic pathways proceeds towards optimization of the efficiency of metabolite transfer among the enzymes of the pathway. This entails an equalization of thermodynamic stabilization between E_1-S_1 and E_2-S_1. In contrast to the hypothesis of Albery and Knowles, we assume that evolution proceeds towards optimization of the pathway as a whole rather than towards the optimization of individual enzyme catalysts.

SUMMARY

Our model for enzyme catalysis within a metabolic pathway involves the following:

(1) Metabolic products of one reaction are transferred to their sequential site of utilization via direct enzyme-enzyme interactions without exposure of the metabolite to the aqueous environment.

Table 6. Concentration of Individual Glycolytic Enzymes[a]

Enzyme	Site Concentration (μM)
Phosphoglucomutase	31.9
Aldolase	809.3
α-glycerol-P dehydrogenase	61.4
Triose-P-isomerase	223.8
Glyceraldehyde-3-phosphate dehydrogenase	1398.6
Phosphoglycerate kinase	133.6
Phosphoglycerate mutase	235.9
Enolase	540.7
Pyruvate kinase	172.9
Lactate dehydrogenase	296.0

[a]In rabbit muscle sarcoplasm (from Srivastava and Bernhard, 1986a).

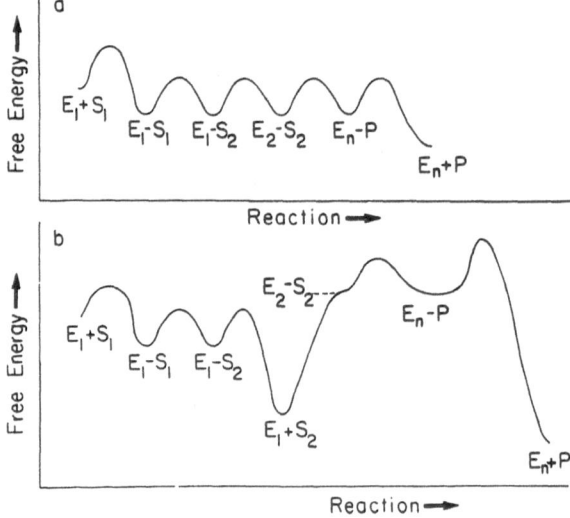

Fig. 7.   Energetic consequences of the direct-transfer mechanism.   The
standard free-energy changes during metabolic interconversion
are illustrated in the equi-energy model (a).   "Realistic" free
energy changes for metabolic interconversions out of, and into,
aqueous solution are indicated in (b).

Table 7.   Concentration of Precursors, Coenzymes and Glycolytic
Intermediates[a]

Metabolite	Concentration (µM)
Glucose-1-PO$_4$	240
Glucose-6-PO$_4$	3,900
Fructose-6-PO$_4$	1,500
Fructose-1,6-di-PO$_4$	80
Dihydroxyacetone-PO$_4$	160
Glyceraldehyde-3-PO$_4$	80
3-phosphoglycerol-GPDH	800
1,3-diphosphoglycerate	50
3-phosphoglycerate	200
2-phosphoglycerate	20
Phosphenol pyruvate	65
Pyruvate	380
Lactate	3,700
Creatine-PO$_4$	26,600
ATP	8,050
Pi	8,000
NAD$^+$	541
NADH	50

[a]In resting rat muscle, except for the concentrations of coenzymes.
The coenzyme concentrations are from rat liver (from Srivastava and
Bernhard, 1986a).

(2) The directly transferred metabolite is partitioned nearly equally between E_1 and E_2.

(3) The equilibrium free-energy change between E-S and E-P is nearly unity when all components are bound to the enzyme site under physiologically relevant conditions.

(4) The metabolic system of enzymes and intermediary metabolites is poised for efficient transfer and interconversion of metabolites.

(5) A substantial fraction of the free-energy change, which would have occurred upon conversion of precursor to final product in dilute aqueous medium, is partitioned into the various enzyme-metabolite complexes formed during the degradation of precursor metabolite.

(6) Metabolic flow is effected by the input of aqueous precursor and/or by the export of final product, either into the aqueous environment or via utilization in other pathways.

REFERENCES

Albery, W.J. and Knowles, J.R., 1976, Biochemistry, 15:5631.
Albery, W.J. and Knowles, J.R., 1977, Angew. Chem., Int. Ed. Engl., 16: 285.
Atkinson, D.E., 1969, Curr. Top. Cell. Reg., 1:29.
Atkinson, D.E., 1977, "Cellular Energy Metabolism and Its Regulation", Academic Press, New York.
Banaszak, L.J. and Bradshaw, R.A., 1975, Enzymes (3rd Ed.), 11:269.
Banks, R.D., Blake, C.C.F., Evans, P.R., Haser, R., Rice, D.W., Hardy, G.W., Merrett, M. and Phillips, A.W., 1979, Nature (London), 279: 773.
Batke, J., Asboth, G., Lakatos, S., Schmitt, B. and Cohen, R., 1980, Eur. J. Biochem., 107:389.
Beeckmans, S. and Kanarek, L., 1981, Eur. J. Biochem.,117:527.
Benner, S.A., 1982, Experentia, 38:633.
Branden, C.-I. and Eklund, H., 1980, in: "Dehydrogenases Requiring Nicotinamide Coenzymes", J. Jeffrey, ed., Birkhauser, Basel.
Branden, C.-I. Jornvall, H., Eklund, H. and Furugren, B., 1975, Enzymes (3rd ed.), 11:103.
Buehner, M., Ford, G.C., Moras, D., Olsen, K.W. and Rossmann, M.G., 1974, J. Mol. Biol., 90:25.
Cedergren-Zeppezauer, E., Samama, J.-P. and Eklund, H., 1982, Biochemistry, 21:4895.
Clarke, M.G., Bloxham, D.P., Holland, P.C. and Lardy, H.A., 1973, Biochem. J., 134:589.
Eklund, H., Samama, J.-P. and Wallen, L., 1982, Biochemistry, 21:4858.
Eklund, H., Samama, J.-P., Wallen, L. and Branden, C.I., 1981, J. Mol. Biol., 146:561.
Fahien, L. and Kmiotek, E., 1978, in: "Microenvironments and Metabolic Compartmentation", P.A. Srere and R.W. Estabrook, eds., Academic Press, New York.
Fahien, L.A. and Kmiotek, E., 1979, J. Biol. Chem., 254:5983.
Fahien, L.A. and Kmiotek, E., 1983, Arch. Biochem. Biophys., 220:386.
Gorringe, D.M. and Moses, V., 1980, Int. J. Biol. Macromol., 2:161.
Grau, U.M., 1982, in: "The Pyridine Nucleotide Coenzymes", J. Everse, Anderson, and K.-S. You, eds., Academic Press, New York.
Grau, U.M., Trommer, W.E. and Rossmann, M.G., 1981, J. Mol. Biol., 151: 289.

Halper, L.A. and Srere, P.A., 1977, Arch. Biochem. Biophys., 184:529.
Holbrook, J.J., Liljas, A., Steindel, S.J. and Rossmann, M.G., 1975, Enzymes (3rd Ed.), 11:191.
Huskins, K.R, 1979, Ph.D. Dissertation, University of Oregon, Dissertation Abstracts Order No. 8005772.
Huskins, K.R, Bernhard, S.A. and Dahlquist, F.W., 1982, Biochemistry, 21:4180.
Knowles, J.R., 1980, Ann. Rev. Biochem., 49:877.
Lehninger, A.L., 1975, Biochemistry, 2nd Ed., Worth Publishers, New York.
MacGregor, J.S, Singh, V.N, Davoust, S, Melloni, E, Pontremoli, S and Horecker, B.L., 1980, Proc. Natl. Acad. Sci. USA, 77:3889.
Mahler, H.R. and Cordes, E.H., 1971, "Biological Chemistry", 2nd Ed., Harper and Row, New York.
Malhotra, O.P. and Bernhard, S.A., 1968, J. Biol. Chem., 243:1243.
Malhotra, O.P. and Bernhard, S.A., 1981, Biochemistry, 20:5529.
Masters, C.J., 1977, Curr. Top. Cell. Reg., 12:75.
Matthews, B.W., 1968, J. Mol. Biol., 33:491.
Metzler, D.E., 1977, "Biochemistry, The Chemical Reactions of Living Cells", Academic Press, New York.
Moras, D., Olsen, K.W., Sabesan, M.N., Buehner, M., Ford, G.C. and Rossmann, M.G., 1975, J. Biol. Chem., 250:9137.
Murthy, M.R.N., Garavito, R.M., Johnson, J.E. and Rossmann, M.G., 1980, J. Mol. Biol., 138:859.
Nageswara Rao, B.D., Cohn, M. and Scopes, R.K., 1978, J. Biol. Chem., 253:8056.
Nageswara Rao, B.D., Kayne, F.J. and Cohn, M., 1979, J. Biol. Chem., 254:2689.
Nambiar, K.P., Stauffer, D.M., Kolodziej, P.A. and Benner, S.A., 1983, J. Am. Chem. Soc., 105:5886.
Ottaway, J.H. and Mowbray, J., 1977, Curr. Top. Cell. Reg., 12:107.
Ovádi, J., Salerno, C., Keleti, T. and Fasella, P., 1978, Eur. J. Biochem., 90:499.
Pickover, C.A., McKay, D.B., Engelman, D.M. and Steitz, T.A., 1979, J. Biol. Chem., 254:11323.
Ray, W.J. and Long, J.W., 1976a, Biochemistry, 15:3993.
Ray, W.J. and Long, J.W., 1976b, Biochemistry, 15:4018.
Salerno, C. and Ovádi, J., 1982, FEBS Letts., 138:270.
Scopes, R.K., 1973, Enzymes, (3rd Ed.), 8:335.
Sols, A. and Marco, R., 1970, Curr. Top. Cell. Reg., 2:227.
Srere, P.A., 1981, Trends Biochem. Sci., 6:4.
Srere, P.A., 1982, Trends Biochem. Sci., 7:375.
Srere, P.A., 1984, Trends Biochem. Sci., 9:387.
Srere, P.A., Halper, L.A. and Finkelstein, M.B., 1978, in: "Microenvironments and Metabolic Compartmentation", P.A. Srere and R.W. Estabrook, eds., Academic Press, New York.
Srivastava, D.K. and Bernhard, S.A., 1984, Biochemistry, 23:4538.
Srivastava, D.K. and Bernhard, S.A., 1985, Biochemistry, 24:623.
Srivastava, D.K. and Bernhard, S.A., 1986a, Curr. Top. Cell. Reg., Vol. 28, in press.
Srivastava, D.K. and Bernhard, S.A., 1986b, Biochemistry, in press.
Srivastava, D.K., Bernhard, S.A., Langridge, R. and McClarin, J.A., 1985, Biochemistry, 24:629.
Stackhouse, J., Nambiar, K.P., Burbaum, J.J., Stauffer, D.M. and Benner, S.A., 1985, J. Am. Chem. Soc., 107:2757.
Sumegi, B. and Alkonyi, I., 1983, Biochim. Biophys. Acta, 749:163.
Sumegi, B., Gilbert, H.F. and Srere, P.A., 1985, J. Biol. Chem., 260:188.
Watson, H.C., Walker, N.P.C., Shaw, P.J., Bryant, T.N., Wendell, P.L., Pothergill, L.A., Perkins, R.E., Conroy, S.C., Dobson, M.J., Tuite,

M.F., Kingsman, A.J. and Kingsman, S.M., 1982, EMBO Journal, 1:1635.
Weber, J.P. and Bernhard, S.A., 1982, Biochemistry, 21:4189.
Welch, G.R., 1977, Prog. Biophys. Mol. Biol., 32:103.
Wilkinson, K.D. and Rose, I.A., 1979, J. Biol. Chem., 254:12567.

ENERGETIC CONSEQUENCE OF DYNAMIC ENZYME-ENZYME INTERACTION

T. Keleti

Institute of Enzymology
Biological Research Center
Hungarian Academy of Sciences
Budapest H-1502, Hungary

INTRODUCTION

In contrast to the widely accepted concept that macromolecular
components within the cytoplasm are freely dissolved, more and more
evidence suggests that proteins associate with each other or with various
membranous and particulate components of the cell (e.g., Keleti et al.,
1977; Welch, 1977; Masters, 1981; Friedrich, 1984, 1985; Keleti et al.,
1985; Srivastava and Bernhard 1985a; Ureta, 1985; and references cited
therein). Moreover, the intracellular distribution of enzymes may not be
a time-invariant property but may vary with the metabolic status of the
cell (Keleti and Welch, 1984).

Enzyme-enzyme interactions occur in many cases between "soluble"
enzymes. These interactions are non-covalent, endowing the complexes
with a dynamic character. Such interactions are assumed to function in
vivo in compartmentation, or channelling, of metabolites. These
phenomena might play a role in metabolic control (Friedrich, 1974, 1985;
Keleti, 1984) and its evolution (Welch and Keleti, 1981).

Enzymes of glycolysis--one of the most ubiquitous metabolic pathways
of energy metabolism--are commonly regarded as "soluble". The possible
interaction between purified aldolase and glyceraldehyde-3-phosphate
dehydrogenase was studied in vitro in considerable detail by kinetic
methods (Ovádi and Keleti, 1978; Grazi and Trombetta, 1980), with the
aid of a "suicide" reaction (Patthy and Vas, 1978), by fluorescence
polarization (Ovádi et al., 1978), by affinity chromatography (Kalman and
Boross, 1982), and by microcalorimetry (Keleti, 1978). The channelling
effect, found originally by rapid kinetic methods (Ovádi and Keleti,
1978), was substantiated by isotope-dilution experiments using HPLC-
isolated intermediates and end-product in a reconstituted three-enzyme
system (Orosz et al., submitted). The third enzyme studied is
triosephosphate isomerase, which was found to form a complex with
aldolase (Gavilanes et al., 1981; Salerno and Ovádi, 1982).

Similarly, direct transfer of the coenzyme NAD(H) between various
dehydrogenases in the cytosol has been postulated on the basis of kinetic
experiments (Srivastava and Bernhard 1984, 1985a, 1985b; Srivastava et
al., 1985).

The first indication of interaction between aldolase and glycerol-3-phosphate dehydrogenase (the enzyme functioning at the branch-point of glycolysis and the reducing equivalent-transporting shuttle mechanism or lipid synthesis) was seen by a rapid kinetic technique and by active-band centrifugation (Batke et al., 1980), by fluorescence polarization (Ovádi et al., 1983), as well as by kinetic methods and mathematical modeling (Ovádi et al., 1985). The coupled reaction between these enzymes can be monitored even within a single co-crystal, by microspectrophotometry (Keleti et al., submitted). The formation of such a co-crystal (myogen A — see Baranowski, 1939; Baranowski and Niederland, 1949) may be an indication of their interaction.

Although search for interactions between glyceraldehyde-3-phosphate dehydrogenase and phosphoglycerokinase gave negative results (Vas and Batke, 1981; Ashmarina et al, 1984), the acylated kinase forms a complex with the dehydrogenase in which direct transfer of the intermediate could be demonstrated (Weber and Bernhard, 1982; Ashmarina et al., 1984). In this complex, even the cooperative interaction between subunits of the dehydrogenase is altered (Ashmarina et al., 1985). This is due to change in the quaternary structure of the dehydrogenase, which is known to be related to both the activity and complex-forming ability of glyceraldehyde-3-phosphate dehydrogenase (Ovádi et al., 1979; Kalman et al., 1980; Vas et al., 1981; Minton and Wilf, 1981; Batke, 1982; Ovádi et al., 1982) and also glycerol-3-phosphate dehydrogenase (Batke et al., 1980; Ovádi et al., 1983).

The major benefit of enzyme-enzyme complex formation is metabolite compartmentation (Friedrich, 1974; Welch, 1977). An important consequence is the decrease (or elimination) of the transient time.

In a consecutive reaction of two enzymes the transient time, τ, is defined generally as (Bartha and Keleti, 1979):

$$\tau = ([E_2]_0 + K_M)/k_2[E_2]_0 \qquad (1)$$

or if $[E_2]_0 \ll K_M$

$$\tau = K_M/k_2[E_2]_0 \qquad (2)$$

(Hess and Wurster, 1970), where $[E_2]_0$ is the total concentration of the second enzyme in the sequence, K_M is its Michaelis constant, and k_2 is the first-order rate constant of the reaction catalyzed by the second enzyme.

If a dynamic complex is formed between the two enzymes catalyzing a consecutive reaction, then $\tau_{overall} < \tau$, where τ is the transient time of the coupled reaction when no complex formation occurs, while $\tau_{overall}$ is the measured transient time when a kinetically significant complex exists.

We have

$$a/\tau' + b/\tau = (c/\tau)_{overall} \qquad (3)$$

where a, b and c are weighting factors which depend on $[E_1E_2]$, on $[E_2]$, and on both $[E_1E_2]$ and $[E_2]$, respectively; τ' is the transient time which can be assigned to the complex between E_1 and E_2, the two enzymes

166

catalyzing the coupled reactions. The existence of τ' yields a non-linear dependence of $1/\tau_{overall}$ on $[E_2]_o$, which may be the diagnostic sign of the enzyme-enzyme complex formation.

Since $1/\tau'$ has the dimension of a first-order rate constant, assuming the validity of the absolute rate theory we write

$$1/\tau' = \kappa(RT/Nh)K^{\#} = \kappa(RT/Nh)\cdot\exp(-\Delta G^{\#}/RT) \tag{4}$$

where R is the gas constant, T the absolute temperature, N the Avogadro constant, h the Planck constant, $K^{\#}$ the equilibrium constant for formation of activated E_1E_2 complex (i.e., $K^{\#} = [E_1E_2]^{\#}/[E_1E_2]$), and κ the transmission coefficient which in this case cannot be considered as always equaling unity. κ equals unity only if the formation of activated complex yields complete channelling; otherwise $0 < \kappa < 1$.

It is to be noted that

$$1/\tau_{overall} = (1/\tau' - 1/\tau)([E_1]_o + [E_2]_o +$$
$$+ K \pm \sqrt{([E_1]_o+[E_2]_o+K)^2-4[E_1]_o[E_2]_o})/2[E_2]_o + 1/\tau \tag{5}$$

where K is the dissociation constant of the E_1E_2 complex and only the positive solution has physical meaning.

From Eqn. (4) it follows (assuming $\kappa = 1$) that if $K^{\#}=1$, $\Delta G^{\#}=0$ and at 298°K, $\tau'_o = 1.6105 \times 10^{-13}$s. If $K^{\#} > 1$, $\Delta G^{\#} < 0$ and $\tau' < \tau'_o$; if $K^{\#} < 1$, $\Delta G^{\#} > 0$ and $\tau' > \tau'_o$. Thus, $\tau' \longrightarrow 0$ if $K^{\#} \longrightarrow \infty$, i.e., all complexes are in activated form and the channel is perfect.

From Eqn. (4), if $\kappa = 1$, we have

$$\Delta G^{\#} = -RT\cdot\ln(Nh/RT\tau') = -2.48\cdot\ln(1.6105\times10^{-13}/\tau') \tag{6}$$

in kJ/mol at 298°K. It follows that the change in transient time - due to the formation or dissociation of enzyme-enzyme complexes - has a direct energetic consequence, which can enhance or lower the energy balance of a coupled reaction and, consequently, that of whole metabolic pathways composed of several coupled reactions.

REFERENCES

Ashmarina, L.I., Muronetz, V.I., and Nagradova, N.K., 1984, Biochem. Internat., 9:511.
Ashmarina, L.I., Muronetz, V.I., and Nagradova, N.K., 1985, Eur. J. Biochem., 149:67.
Baranowski, T., 1939, Hoppe Seyler's Z. Physiol. Chem., 260:43.
Baranowski, T. and Niederland, T.R., 1949, J. Biol. Chem., 180:543.
Bartha, F. and Keleti, T., 1979, Oxid. Commun., 1:75.
Batke, J., 1982, Anal. Biochem., 121:123.
Batke, J., Asboth, G., Lakatos, S., Schmitt, B., and Cohen, R., 1980, Eur. J. Biochem., 107:389.
Friedrich, P., 1974, Acta Biochim. Biophys. Acad. Sci. Hung., 9:159.
Friedrich, P., 1984, "Supramolecular Enzyme Organization. Quaternary Structure and Beyond", Pergamon Press, Oxford, and Akademiai Kiado, Budapest.

Friedrich, P., 1985, in: "Catalytic Facilitation in Organized Multienzyme Systems", G.R. Welch, ed., Academic Press, New York.
Gavilanes, F., Salerno, C., and Fasella, P., 1981, Biochim. Biophys. Acta, 660:154.
Grazi, E. and Trombetta, G., 1980, Eur. J. Biochem., 107:369.
Hess, B. and Wurster, B., 1970, FEBS Lett., 9:73.
Kálmán, M. and Boross, L., 1982, Biochim. Biophys. Acta, 704:272.
Kálmán, M., Nuridsány, M., and Ovádi, J., 1980, Biochim. Biophys. Acta, 614:285.
Keleti, T., 1978, Symp. Biol. Hung., 21:107.
Keleti, T., 1984, in: "Dynamics of Biochemical Systems", J. Ricard and A. Cornish-Bowden, eds., Plenum Press, New York.
Keleti, T. and Welch, G.R., 1984, Biochem. J., 223:299.
Keleti, T., Batke, J., Ovadi, J., Jancsik, V., and Bartha, F., 1977, Adv. Enzyme Regul., 15:233.
Keleti, T., Ovádi, J., and Batke, J., 1987, in: "Towards A Cellular Enzymology", A.A. Klyosov, S.D. Varfolomeev, G.R. Welch, eds., Plenum Press, New York.
Masters, C.J., 1981, CRC Crit. Rev. Biochem., 11:105.
Minton, A.P. and Wilf, J., 1981, Biochemistry, 20:4821.
Ovádi, J. and Keleti, T., 1978, Eur. J. Biochem., 85:157.
Ovádi, J., Salerno, C., Keleti, T., and Fasella, P., 1978, Eur. J. Biochem., 90:499.
Ovádi, J., Batke, J., Bartha, F., and Keleti, T., 1979, Arch. Biochem. Biophys., 193:28.
Ovádi, J., Mohamed-Osman, I.R., and Batke, J., 1982, Biochemistry, 21:6375.
Ovádi, J., Mohamed-Osman, I.R., and Batke, J., 1983, Eur. J. Biochem., 133:433.
Ovádi, J., Batke, J., Mátrai, Gy., and Bartha, F., 1985, Biochem. J., 229:57.
Patthy, L. and Vas, M., 1978, Nature, 276:94.
Salerno, C. and Ovadi, J., 1982, FEBS Lett., 138:270.
Srivastava, D.K. and Bernhard, S.A., 1984, Biochemistry 23:4538.
Srivastava, D.K. and Bernhard, S.A., 1985a, Curr. Topics Cell. Regul., 28, in press.
Srivastava, D.K. and Bernhard, S.A., 1985b, Biochemistry, 24:623.
Srivastava, D.K., Bernhard, S.A., Langridge, R., and McClarin, J.A., 1985, Biochemistry, 24:629.
Ureta, T., 1985, Arch. Biol. Med. Exp., 18:9.
Vas, M. and Batke, J., 1981, Biochim. Biophys. Acta, 660:193.
Vas, M., Lakatos, S., Hajdu, J., and Friedrich, P., 1981, Biochimie, 63:89.
Weber, J.P. and Bernhard, S.A., 1982, Biochemistry, 21:4189.
Welch, G.R., 1977, Prog. Biophys. Mol. Biol., 32:103.
Welch, G.R. and Keleti, T., 1981, J. Theor. Biol., 93:701.

MITOCHONDRIAL HEXOKINASE: INTERACTIONS BETWEEN

GLYCOLYSIS AND OXIDATIVE PHOSPHORYLATION

John E. Wilson

Biochemistry Department
Michigan State University
East Lansing, MI 48824, U.S.A.

In addition to its highly specific requirement for glucose as a substrate, the brain, unlike some other tissues, contains relatively little of the glucose storage form, glycogen. Thus the brain is critically dependent on a continued supply of blood-borne glucose and oxygen to maintain neurological function. This strict reliance on aerobic glucose metabolism is reflected in the organization of relevant enzymes in the brain.

The initial step in metabolism of glucose is phosphorylation, catalyzed by hexokinase. Approximately 80% of the hexokinase activity is associated with the mitochondrial fraction of brain homogenates, in marked contrast to the other glycolytic enzymes which are found primarily in the "soluble" fraction (Johnson, 1960). Work by a number of investigators has indicated that hexokinase is bound to the outer membrane of the mitochondria and, more specifically, to the pore-forming protein in that membrane (Rose and Warms, 1967; Felgner et al., 1979; Linden et al., 1982; Fiek et al ., 1982). Several investigators have presented results indicating that, when bound to the mitochondia, hexokinase has preferential access to intramitochondrially-generated ATP, and that ADP generated by mitochondrially-bound hexokinase is highly effective as a substrate for oxidative phosphorylation (reviewed in Wilson, 1985). Thus mitochondrial binding of hexokinase provides a topological basis for the integration of the initial step of glycolysis and mitochondrial oxidative phosphorylation.

Is this organizational pattern useful in other tissues? A comparison of mitochondrial hexokinase content (expressed as a hexokinase:fumarase ratio) with tissue phosphoglucomutase levels (the latter enzyme taken as an indicator of glycogen metabolism) showed an inverse correlation (Wilson and Felgner, 1977). Such results suggest that association of hexokinase with the mitochondria is metabolically advantageous in tissues that depend heavily on blood-borne glucose, rather than glycogen, as a glycolytic substrate, and that this provides an effective mechanism for coupling mitochondrial ATP formation with glucose phosphorylation.

At least in some tissues, there is good reason to view hexokinase as more akin to a mitochondrial enzyme than to the other enzymes of glycolysis. Thus, during maturation of the brain, increases in hexokinase activity are closely coordinated with increases in the mitochondrial cytochrome oxidase activity, and with increases in adenylate kinase, an enzyme closely linked to energy metabolism as represented by the adenylate pool and which is also at least partly located in the mitochondria, but not with increases in the glycolytic enzymes, lactate dehydrogenase and pyruvate kinase (Wilson, 1972). In muscle, Staudte and Pette (1972) have shown that the activities of a number of glycolytic enzymes, but not hexokinase, were correlated with the levels of glycogen-metabolizing enzymes, while hexokinase was correlated with the activity of the mitochondrial enzyme, citrate synthase. This correlation of hexokinase with mitochondrial enzymes, rather than with other glycolytic enzymes, is also true in a spatial sense, at least in muscle (Pette, 1975) and retina (Lowry et al., 1961).

The binding of hexokinase to mitochondria is rapidly reversible and highly sensitive to physiological levels of glucose 6-phosphate, which solubilizes the enzyme, and P_i, which antagonizes the solubilizing action of glucose 6-phosphate. Several studies (reviewed in Wilson, 1985) have now shown that the soluble/particulate distribution of hexokinase in brain homogenates can be altered by treatments which perturb the concentration of these metabolites in brain. When the enzyme binds to mitochondria, the mitochondrial membrane serves, in essence, as a positive allosteric effector of the enzyme, decreasing its sensitivity to inhibition by glucose 6-phosphate several fold. Thus during conditions such as ischemia, glucose 6-phosphate levels decrease, which in itself should lead to increased activity. In addition, however, a greater proportion of the enzyme is bound to the mitochondria in a manner that decreases its susceptibility to inhibition by glucose 6-phosphate, further amplifying the activity. Thus variations in the degree of interaction between hexokinase and mitochondria may play an important role in regulation of the activity of this enzyme.

REFERENCES

Felgner, P.L., Messer, J.L., and Wilson, J.E., 1979, J. Biol. Chem., 254:4946.
Fiek, C., Benz, R., Roos, N. and Brdiczka, D., 1982, Biochim. Biophys. Acta, 688:429.
Johnson, M.K., 1960, Biochem. J., 77:610.
Linden, M., Gellerfors, P., and Nelson, B.D., 1982, FEBS Lett., 141:189.
Lowry, O.H., Roberts, N.R., Schulz, D.W., Clow, J.E., and Clark, J.R., 1961, J. Biol. Chem., 236:2813.
Pette, D., 1975, Acta Histochem., Supplementary Vol. 14, 47.
Rose, I.A. and Warms, J.V.B., 1967, J. Biol. Chem., 242:1635.
Staudte, H.W., and Pette, D., 1972, Comp. Biochem. Physiol., 41B:533.
Wilson, J.E., 1972, J. Neurochem., 19:223.
Wilson, J.E., 1985, in: "Regulation of Carbohydrate Metabolism", R. Beitner, ed., CRC Press, Boca Raton, FL.
Wilson, J.E. and Felgner, P.L., 1977, Mol. Cell. Biochem., 18:39.

DIFFUSION AND PERFUSION IN THE LIVING CELL:

IMPLICATIONS FOR METABOLIC REGULATION AND ORGANIZATION

Denys N. Wheatley and P. Colm Malone*

Department of Pathology
University of Aberdeen
Foresterhill, Aberdeen AB9 2ZD
Scotland, U.K.

*Sub-Department of Investigative Pathology
University of Birmingham Medical School
Birmingham B15 2TJ
England, U.K.

INTRODUCTION

A tacit assumption throughout much of biology, based on the smallness of size of most metazoan cells, is that diffusion of molecules within the cytoplasm suffices for metabolic purposes. This long-standing premise is challenged. Because cells are rich in protein (20% or more on average), it follows that, irrespective of whether this macromolecular material exists mostly in true solution (which is most improbable) or out of solution in the form of an elaborate cytomatrix and cytoskeleton, the random movements of molecules (especially large and/or reactive ones) will be impeded. However, since molecules of all sizes can be distributed throughout the cytoplasm within a matter of a second or two, the presence of a system which facilitates dispersion is indicated.

While it is rationalized that clearly visible cytoplasmic streaming achieves rapid translocation of solutes in huge cells such as phloem elements, Chara, and Amoeba, a similar mechanism probably exists in all cells, although it would be far more difficult to detect in small cells. It is suggested that the internal matrix of cells, upon which enzyme activity is (mainly) located, requires the constant perfusion of a fluid phase over it in order to support metabolic activity, which would proceed at a negligible rate if carried out in a completely soluble phase (see Welch, this Volume). This principle greatly increases the delivery rate of reactants to the metabolic machinery and is likened to the operations designed by chemical engineers, rather than by test-tube chemists. When the perfusate is enriched with nutrients cellular metabolism may be stimulated, which could in turn make more energy available to accelerate perfusion, thereby providing a self-regulating mechanism underlying and to some extent governing cellular metabolic activity.

Progress in understanding the organization of cellular metabolism depends on a clear conception of the physical nature of the cytoplasm (and nucleus), about which there remains some controversy. We should bear in mind the following:

(i) Cells are too rich in protein for more than a small (?) part to be in true solution.

(ii) Organized structures (such as mitochondria) are detectable in living cells by microscopy, and therefore there is no logical reason why smaller organized structures should not exist at a submicroscopic level.

(iii) In large cells, movement of the more fluid phase can occur -- as indirectly perceived by the movement of visible particles in the process called cytoplasmic streaming. It follows once again, that there is no logical reason why a similar process of streaming should not occur in all cells at a level below microscopic detection.

On the above premises, it is argued that an "endocirculation" (or some active mixing process) may be mandatory in all eukaryotic cells to sustain active metabolism. The alternative is to assume that incoming nutrients, ions, etc., as well as intermediates and products of cellular metabolism, are translocated throughout the cytoplasm solely by their random kinetic motion (diffusion). A major issue dealt within these proceedings concerns the topographical arrangement of enzymes, and the possibility that substrates and products of reaction pathways are transferred directly from one site of action to another ('channeling'). At some stages, e.g., amino-acid access to ribosomes or protein-product targeting to intracellular sites of utilization, there is unlikely to be precise channeling; and we fall back on diffusion as the main mechanism of dispersion and translocation involved. The argument to which attention is here drawn is that in the 'unresting' cell (Gerard, 1940), molecules will be actively dispersed by a process of convected diffusion, which may reach a level in some cells where an obvious cyclosis actually occurs through a series of labile channels as a form of endocirculation (see Wheatley, 1985, for a fuller discussion).

Two rather extreme impressions of the cell have been discussed, one in which organelles waft around rather haphazardly in a soupy cytoplasm (see Crick and Hughes, 1950). The other is one in which most of the cytoplasm is elaborately organized as surfaces, membranes, cytoskeleton, and cytomatrix. Irrespective of which is adopted, the protein content will inevitably lead to much of the aqueous phase being organized (Clegg, 1982). Both the insoluble phase and the structured aqueous phase will reduce diffusion; and the more organized the cytomatrix, the more restricted dispersion of solutes will be in the cytoplasm. From the data of Horowitz et al. (1979), small ions and molecules may still be able to traverse the radius or diameter of a small cell within a second or so; but large and reactive molecules will not diffuse rapidly. It has been estimated that proteins may take about 20 min to traverse the diameter of a HeLa cell (see Wheatley, 1985). Experimental data show that large proteins can be dispersed throughout the cytoplasm of a cell within a matter of 1-2 seconds (Stacey and Allfrey, 1977; see Kohen, these proceedings). It follows that a mechanism exists in cells which can distribute large as well as small molecules throughout the cytoplasm in a similar short time scale (1-2 seconds). This is inconsistent with a model of diffusion operating within cytoplasm. Just as one waits seemingly an eternity for a spoonful of sugar to disperse throughout a cup of tea, the process can be speeded up enormously by stirring it briefly. Similarly, Malone (1981) points out that life forms would not exist at the bottom of Lake Constance if simple diffusion were the only mechanism by which oxygen reaches the depths, since it would take an inordinately long time to reach there. We propose, therefore, that the actively metabolizing cell is dependent on the (continual) movement of the bulk aqueous phase (the sap) for translocation of many molecules within the cytoplasm, and this will allow reactions to proceed often at orders of magnitude faster than would be accomplished by diffusion alone (see Wheatley, 1985). A cautionary note is sounded, therefore, in applying simple diffusion laws in kinetic analyses of cellular metabolism.

Since life may well be a continuation of a surface-based operation from primeval times (see Welch, these proceedings), the mechanism by which cells are capable of sustaining high rates of enzymatic activity along complicated metabolic pathways is perhaps by appropriate topographical arrangement of enzyme complexes on cell surfaces which are perfused by a stream of sap, bringing necessary precursors, ions, co-factors, etc., and removing products and waste. The system is likened to that which would be designed for maximum efficiency by a chemical engineer, rather than be the small scale chemist using a homogeneous reaction mixture in a test-tube. The principle of perfusion greatly increases the delivery rate of required reactants, and can assist in capturing those which are in unduly low concentration (see Wheatley and Inglis, 1980; the topic of delivery rate is discussed more fully in Coulson et al., 1977). In conclusion, the modus operandi suggested for the metabolic organization of the cell is probably far closer to that of perfused surfaces than that of a free solution in which the laws of mass action apply, despite the fact that we continue to use the latter in most of our biochemical analyses.

REFERENCES

Clegg, J.S., 1982, "Biophysics of Water", F. Franks, ed., John Wiley and
    Sons Ltd., Chichester.
Coulson, R.A., Hernandez, T., and Herbert, J.D., 1977, Comp. Biochem.
    Physiol., 56A:251.
Gerard, R.W., 1940, "The Unresting Cell", Harper, New York.
Horowitz, S.B., Paine, P.L., Tluczek, L., and Reynhout, J.K., 1979,
    Biophys. J., 25:33.
Malone, P.C., 1981, Med. Hypotheses, 7:1477.
Stacey, D.W. and Allfrey, V.G., 1977, J. Cell Biol., 75:807.
Wheatley, D.N., 1985, Life Sciences, 36:299.
Wheatley, D.N. and Inglis, M.S., 1980, J. Theor. Biol., 83:437.

ORGANIZATION OF ENERGY METABOLISM:

ENZYMOLOGICAL APPROACHES

INTERACTIONS OF MITOCHONDRIAL MATRIX ENZYMES WITH

MITOCHONDRIAL INNER MEMBRANES

Reginald L. Tyiska, James S. Williams,
Lynn G. Brent, Alan P. Hudson,
Barbara J. Clark, Jack B. Robinson, Jr.,
and Paul A. Srere

Pre-Clinical Science Unit
Veterans Administration Medical Center
and
Biochemistry Department
University of Texas Health Science Center
4500 S. Lancaster Rd.
Dallas, TX 75216, U.S.A.

INTRODUCTION

We have shown previously that citrate synthase (EC 4.1.3.7), mito-chondrial malate dehydrogenase (EC 1.1.1.37), and fumarase (EC 4.2.1.2) bind to the matrix surface of the inner mitochondrial membrane (D'Souza and Srere, 1983). These experiments, along with others, were taken as evidence to support a hypothesis which proposes that there is an organization of Krebs TCA cycle enzymes with their metabolically sequential neighbors and with protein components of the inner membrane (Srere, 1985). The three enzymes studied were selected because of their availability and their stability. Since succinate dehydrogenase (EC 1.3.99.1) and α-ketoglutarate dehydrogenase complex (EC 2.3.1.61, 1.6.4.3, 1.2.4.1) are already known to be tightly bound to the inner membrane, the only remaining Krebs TCA cycle enzymes to be studied were aconitase, (NAD^+)isocitrate dehydrogenase (EC 1.1.1.41) and succinyl CoA synthetase (EC 6.2.1.4). An inner-membrane associated (NAD^+)isocitrate dehydrogenase has been isolated from potato mitochondria (Teszuka and Laties, 1983). We have also shown that certain mitochondrial dehydrogenases (malate, α-ketoglutarate, pyruvate and β-hydroxyacyl CoA) bind to Complex I, an inner membrane protein (Sumegi and Srere, 1984). We could not detect the binding of (NAD^+)isocitrate dehydrogenase to Complex I.

The second consideration of the hypothesis, that there exists an organization of Krebs tricarboxylic acid cycle enzymes, concerns the interaction of sequential Krebs TCA cycle enzymes. Thus far, of the eight possible interactions, only three have been demonstrated: fumarase with malate dehydrogenase (Beeckmans and Kanarek, 1981), malate dehydrogenase with citrate synthase (Halper and Srere, 1979), and succinyl CoA synthetase with α-ketoglutarate dehydrogenase complex (Porpaczy et al., 1983). In this chapter we present evidence for the interaction of mitochondrial aconitase with citrate synthase.

The last consideration we wish to address here is an elaboration of the breadth of this effect in systems other than rat liver. If the Krebs tricarboxylic acid cycle is present in an organized form, and functionality to some degree is dependent on such organization, then these experiments of enzyme interaction with mitochondrial inner-membrane preparations should be present in all eucaryotic systems. We have previously studied fresh tissue from rat, dog, pig, rabbit, cow, and lemon fruit (Moore et al., 1984; Robinson and Srere, 1985a), and here we present data on enzyme binding to mitochondrial inner-membrane preparations from normal and transformed cells in culture and from yeast.

MATERIALS AND METHODS

Materials

The following chemicals were obtained from the indicated sources: $NADP^+$, NAD^+, ADP, DL-isocitrate, bovine serum albumin, cis-aconitate, HEPES, Tris, imidazole, PIPES, oxaloacetic acid, 5.5'(dithio)bis-2-nitrobenzoic acid, pig heart mitochondrial and cytosolic malate dehydrogenases, pig heart citrate synthase, polyethylene glycol (Mr=8,000), and triethanolamine-HCl from Sigma; succinyl-CoA synthetase from Boehringer; pig heart aconitase (both mitochondrial and cytosolic forms) from H. Beinert, Univ. of Wisconsin (Madison, WI); pig heart (NAD^+)isocitrate dehydrogenase and ($NADP^+$)isocitrate dehydrogenase from R. Colman, Univ. of Delaware (Wilmington, DE); (NAD^+)isocitrate dehydrogenase from K. Gabriel and G.W.E. Plaut, Temple Univ. (Philadelphia, PA). Acetyl-CoA is made from CoA (P-L Biochemicals, Milwaukee, WI) (Simon and Shemin, 1953).

Methods

Enzymes were assayed spectrophotometrically according to the following established procedures: aconitase (Fansler and Lowenstein, 1969), (NAD^+)isocitrate dehydrogenase (Plant, 1969), ($NADP^+$)isocitrate dehydrogenase (Cleland et al., 1969), citrate synthase (Srere et al., 1963), succinyl-CoA synthetase (Cha, 1969). Enzymes were dialyzed by the spin dialysis method (Penefsky, 1977), using Biogel P-30. The following buffers were used with the binding studies of the various enzymes: aconitase in 2 mM triethanolamine-HCl, pH 7.0; (NAD^+)isocitrate dehydrogenase in 2 mM Tris-acetate, pH 7.0, 1 mM NAD^+, 1 mM $MnCl_2$, and 20% glycerol; ($NADP^+$)isocitrate dehydrogenase in 2 mM imidazole-acetate, pH 7.0, 1 mM $MnCl_2$, 1 mM $NADP^+$, and 10% glycerol; and citrate synthase and malate dehydrogenase in 2 mM HEPES/KOH, pH 7.0. These conditions were chosen so that each enzyme was stable for the course of the binding assay.

Enzymes were assayed in 1 ml cuvettes (10 mm light path) in a Gilford spectrophotometer, fitted with a Gilford recording assembly. Units are μ mole of product per minute at 25°.

Preparation of Inner Membrane Vesicles

Rat liver mitochondria were isolated according to the procedure of Schnaitman and Greenawalt (1968), with modification (Matlib and Srere, 1976). Inner membrane vesicles were prepared using the procedure of Hackenbrock and Miller-Hammon (1975), as described earlier (D'Souza and Srere, 1983).

Binding Assay

The binding of enzyme to all inner-membrane preparations here was carried out as described previously (D'Souza and Srere, 1983). A total volume of 60 μl, containing 50-100 μg of inner membrane protein and 0-10 μg of enzyme in the appropriate stabilizing buffer for each enzyme, was incubated for 15 min at 0°. The mixture was centrifuged, and both the supernatant solution and the resuspended pellet were assayed for enzyme activity.

Interaction of Enzymes in Polyethylene Glycol

The conditions for these experiments were the same as described by Halper and Srere (1979).

Protein Determination

Protein concentrations were determined by the method of Lowry et al. (1951), using bovine serum albumin as a standard.

Preparation of Mitochondria from Cells in Culture

MA10 and R2C cell mitochondria were prepared (D.A. Freeman, personal communication) as described above. Baby hamster kidney and Ehrlich ascites tumor cells were treated as liver, except that the initial homogenization using the Dounce homogenizer was 30 strokes rather than 18.

Preparation of Yeast Mitochondria and Mitochondrial Membrane Fractions

Yeast mitochondria were prepared by the method of Ainsley et al. (1984). Mitochondrial membranes were prepared by the method of Daum et al. (1982).

RESULTS

Binding of Rat Liver Enzymes to Inner Membrane Vesicles

When mitochondrial aconitase or (NAD^+)isocitrate dehydrogenase is added to vesicles of rat liver mitochondrial inner membrane, binding occurs as a saturable process (Figs. 1 and 2). On the other hand, added cytosolic aconitase and/or $(NADP^+)$isocitrate dehydrogenase failed to bind in significant amounts to these membrane preparations. In separate experiments it was found that membranes contained small amounts of succinyl-CoA synthetase but did not bind added succinyl-CoA synthetase

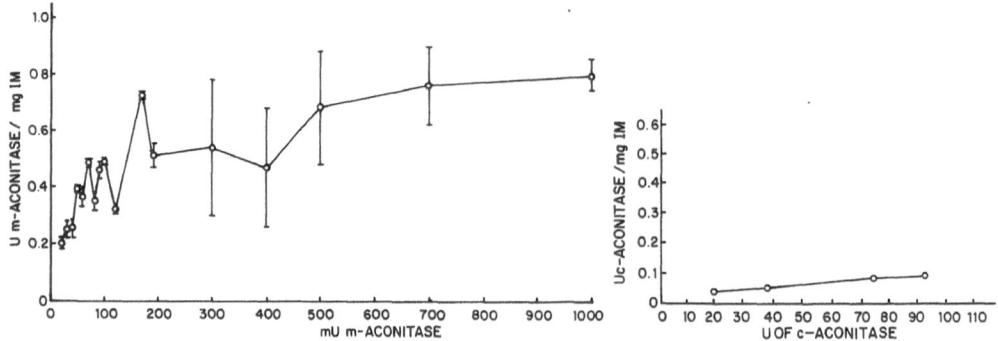

Fig. 1.   Binding Studies of the Mitochondrial (m) and Cytosolic (c)
           Isozymes of Aconitase to Rat Liver Mitochondrial Inner
           Membranes.   Binding was done as described in Materials and
           Methods.   In Part A (left), (m-Aconitase) the error bars
           represent the standard deviation observed in nine experiments
           with the plotted point representing the average values.   In
           part B (right), (c-Aconitase) the data shown are from one
           experiment.

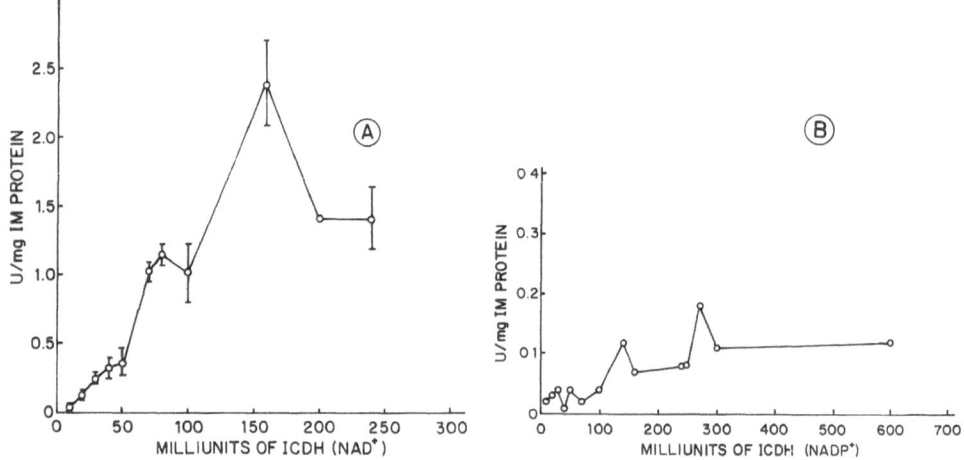

Fig. 2.   Binding Studies of (NAD$^+$)- and (NADP$^+$)-Dependent Isocitrate
           Dehydrogenase to Rat Liver Mitochondrial Inner Membranes.   The
           data shown are the average of six experiments.

(data not shown).   A summary of the binding of these enzymes to inner
membrane is presented (Table 1), along with comparative data for citrate
synthase and matrix proteins.

   The binding of the (NAD$^+$)isocitrate dehydrogenase and aconitase to
inner membranes is sensitive to ionic strength. A concentration of 20 mM
Tris-acetate, pH 7.0, or 20 mM triethanolamine-HCl, pH 7.0, is sufficient
to prevent binding of (NAD$^+$)isocitrate dehydrogenase or mitochondrial
aconitase, respectively (Figs. 3 and 4).   While the binding of
(NAD$^+$)isocitrate dehydrogenase decreased with increasing pH (Figs. 3 and
4), little or no effect of pH was observed with mitochondrial aconitase

Table 1.   Amount of Enzyme Bound to Preparations of Rat Liver Inner
            Membrane

Enzyme	µg bound/mg IM
Citrate Synthase	32.0 ± 14.0
(NAD$^+$)Isocitrate Dehydrogenase	85.0 ± 14.0
m-Aconitase	50.0 ± 14.0
(NADP$^+$)Isocitrate Dehydrogenase	4.0 ± 2.0
c-Aconitase	4.0 ± 1.0
Matrix protein	800.0 ± 400.0

The amount of enzyme bound was calculated from the units of enzyme
bound. The specific activity of each enzyme was determined on the
day of the experiment.   The values are the averages and ranges for at
least three determinations.

binding. Binding is complete within the first five minutes of incubation
(data not shown).

The amount of enzyme precipitated in the absence of membranes was not
greater than 2% of the added enzyme, and these amounts where they
occurred were subtracted from the experimental values.

Additional experiments were carried out to determine if m-aconitase
or (NAD$^+$)isocitrate dehydrogenase could bind to phosphatidyl-choline
liposomes either with or without cardiolipin. No significant binding
(about 2%) occurred to either type of liposome (data not shown).

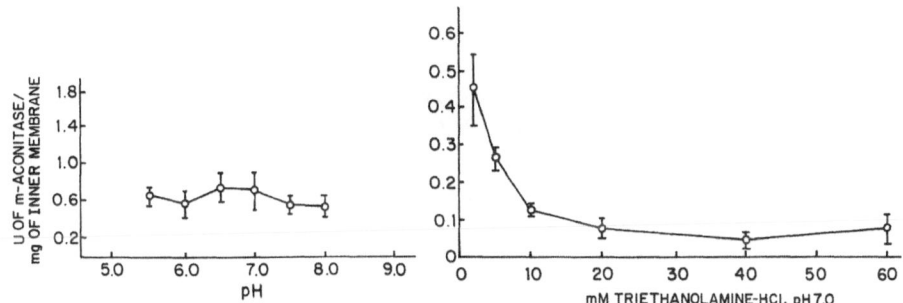

Fig. 3.   Effect of Variations of pH and Ionic Strength on the Binding of
           Mitochondrial Aconitase to Rat Liver Mitochondrial Inner
           Membranes.
    A.   The effect of pH. Binding assays were performed at iso-ionic
           strength with PIPES-KOH, HEPES-KOH, and triethanolamine-HCl
           with overlapping points to estimate buffer effects. The data
           shown are the average ± the range observed in two experiments.
    B.   The effect of ionic strength. Buffer concentration was
           increased over the indicated range and binding measured. The
           data presented are the mean ± the standard deviation seen in
           three experiments.

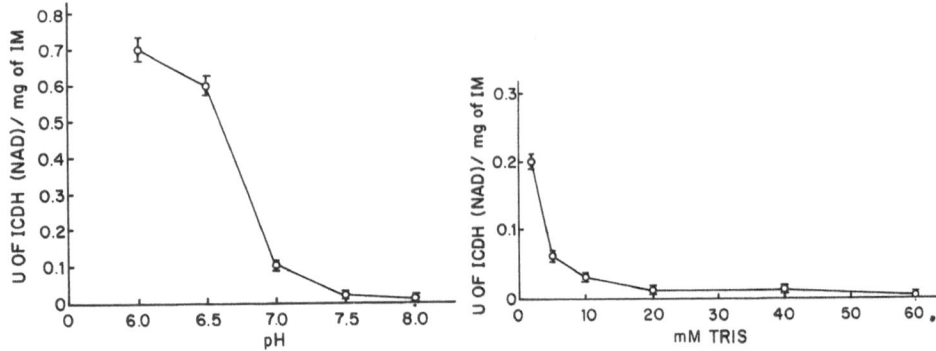

Fig. 4.   Effect of Variation of pH and Ionic Strength on the Binding of
(NAD$^+$)-Specific Isocitrate Dehydrogenase to Rat Liver
Mitochondrial Inner Membranes. See legend to Fig. 3, except
that Tris rather than triethanolamine was used in both parts.

Interaction of Aconitase with Proteins in Polyethylene Glycol

When mitochondrial aconitase is incubated with citrate synthase in
the presence of 14% polyethylene glycol, turbidity was detected (Fig. 5)
that was not observed when c-malate dehydrogenase or serum albumin was
used instead of m-aconitase. The precipitate contains both citrate
synthase and aconitase activity. Cytosolic aconitase did not precipitate

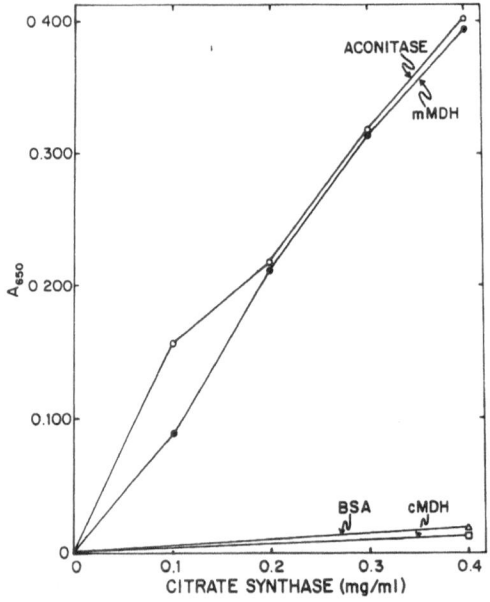

Fig. 5.   Precipitation of Aconitase by Citrate Synthase. The conditions
are 5 mM KPO$_4$, pH 7.0, 14% polyethylene glycol, 80 μg of
protein (mitochondrial aconitase (mAcon), mitochondrial malate
dehydrogenase (mMDH), bovine serum albumin (BSA), or cytosolic
aconitase). A single experiment is shown here, but similar
results were obtained consistently; daily variations in
absolute turbidity were large.

with citrate synthase under these conditions. However, a number of proteins can apparently aggregate with m-aconitase under these conditions (Table 2). The conditions for the interaction between citrate synthase and aconitase are dependent on total protein concentration, ionic strength, and polyethylene glycol concentration. Small variations in these factors caused substantial changes in the absolute results but did not change the specificity that we report.

Interactions with Mitochondrial Preparations from Tumor Cells in Culture

Various neoplastic cells in culture yield mitochondrial inner membranes which bind citrate synthase in a manner similar to that observed in normal tissue (Table 3). These measurements were made on micro-preparations of mitochondria from approximately 100 mg cell protein, but in the instance where comparison is most likely valid (BHK cells vs normal kidney) there was no significant difference.

Interaction of Citrate Synthase with Yeast Mitochondrial Preparations

There is binding of citrate synthase to inner mitochondrial membranes of the lower eucaryotic organism, yeast (Table 4). A protein component of the membrane appears to be implicated in binding, as has been seen in the mammalian system (Robinson and Srere, 1985a). The yeast inner mitochondrial membrane is capable of binding either the pure yeast citrate synthase, or the enzyme purified from pig heart, which implies a conservation of whatever region of citrate synthase is necessary for binding. It is probable from these data that the two enzymes bind in a similar fashion, since the bindings are similarly affected by the treatments of the membrane by trypsin and heat. This conclusion is further substantiated by the experiments (Fig. 6), which show that the two enzymes respond similarly to changes in ionic strength and to the presence of oxaloacetate, although the yeast citrate synthase's interaction with inner membranes is more stable with respect to these parameters.

Table 2. Interaction of Several Enzymes in 14% Polyethylene Glycol

Interaction Studied	Turbidity (A_{650})
Expt I	
Citrate Synthase (160 µg) + m-Aconitase (80 µg)	0.476
Citrate Synthase (160 µg) + c-Aconitase (80 µg)	0.075
Citrate Synthase (160 µg) + Bovine Serum Albumin (80 µg)	0.032
Expt II	
m-Aconitase (80 µg) + Citrate Synthase (80 µg)	0.224
m-Aconitase (80 µg) + Succinyl-CoA Synthetase	0.215
m-Aconitase (80 µg) + ($NADP^+$)Isocitrate Dehydrogenase (80 µg)	0.112
m-Aconitase (80 µg) + m-Malate Dehydrogenase (80 µg)	0.217
m-Acotinase (80 µg) + Bovine Serum Albumin (80 µg)	0.122

See text for details

Table 3. Binding of Citrate Synthase to Mitochondrial Membrane Micro-Preparations from Various Cell Types

Cell Type	n_1*	n_2**	units CS bound per mg membrane protein
Mouse Leydig Cell Tumor (MA10)	3	10	1.0 ± 0.8
Mouse Leydig Cell Tumor (R2C)	2	6	1.3 ± 0.3
Baby Hamster Kidney Cells	2	7	3.0 ± 0.7
Ehrlich Ascites Tumor Cells	2	4	1.7 ± 0.7
Normal Rat Liver	8	24	1.8 ± 0.4
Normal Rat Liver (macro preparation)	10	26	3.2 ± 1.4
Normal Rat Kidney (macro preparation)	1	3	3.0 ± 0.1

* Number of mitochondrial membrane preparations
** Number of individual binding assays

Table 4. Effect of Different Treatments of Yeast Mitochondrial Inner Membranes on Binding of Citrate Synthase

Treatment	Pig Heart Citrate Synthase		Yeast Citrate Synthase	
	U bound/mg	% Inhibition of Binding	U bound/mg	% Inhibition of Binding
none	4.3 ± 0.3	0	3.7 ± 0.2	0
boiling 15 min[a]	0.8	82 ± 3.2	0.3	91 ± 5.4
trypsin[b]	1.4	68 ± 14.6	1.0	74 ± 10.0

[a] 100 µg of yeast mitochondrial IM were boiled for 15 min in 100 µl of 2 mM HEPES, 0.5 mM BME, pH 7.0. The solution was then sonicated and used for the binding assay. The values represent mean ± S.E. of 6 experiments.

[b] 100 µg of yeast mitochondrial IM were incubated at room temperature for 10 min with 2 µg of pancreatic trypsin in 100 µl total volume. The proteolysis was stopped by the addition of 2 µg of egg white trypsin inhibitor. The IM was sedimented at 20,000 rpm for 30 min and then brought up in 100 µl of 2 mM HEPES/NaOH, 0.5 mM mercapto-ethanol, pH 7.0. The values represent mean ± S.E. of 3 experiments.

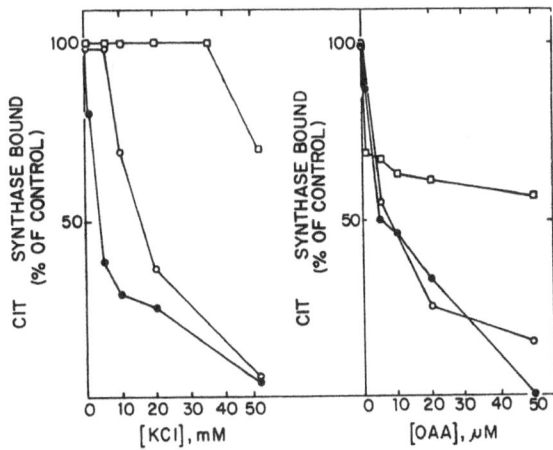

Fig. 6.   Effect of Variation of Ionic Strength and Oxaloacetate
Concentration on Binding of Citrate Synthase to Yeast
Mitochondrial Inner Membrane. **Left Panel:** Effect of ionic
strength.   Binding done in 2 mM HEPES-NaOH, 1 mM dithiothreitol,
pH 7.0 into KCl added as indicated or 0.5 g S-200 purified
protein. 100 μg membrane and 1 unit enzyme in 100 μl total
volume were used. ●-----●, pig heart citrate synthase to
purified protein. O-----O, pig heart citrate synthase to yeast
mitochondrial inner membrane. □ ----- □ , yeast citrate
synthase to yeast mitochondrial inner membrane.
Right Panel. Effect of OAA addition. Symbols and procedures
as described for left panel, with OAA added as indicated.

    We have attempted the purification of the factors necessary for
citrate synthase binding in rat liver (Robinson and Srere, 1985a); and,
while much has been learned about that system, purification of that
protein has proven difficult. We therefore shifted our focus to the
factor present in yeast inner mitochondrial membranes. Initial attempts
at extraction were successful, with the detergent octyl glycoside giving
total solubility of the factor, which, when dialyzed into phosphatidyl
choline liposomes, showed quantitative recovery of all binding. Various
treatments for purification were attempted on the detergent extract, which
are summarized in Table 5.

Table 5.   Purification of Citrate Synthase Binding Factor(s) from Yeast
           Mitochondrial Membranes

Treatment	% Yield*	Specific Activity**
Octyl glucoside solubilization	(100)	2.3
Salt extraction of PEG precipitated OG extract	15	45
Lithium bromide extract of membranes	25	6
Sephacryl S-200 chromatography of PEG precipitated OG extract	39	200

* relative to detergent extract
** numbers shown are units citrate synthase bound per mg protein

Although the gel filtration experiment gives binding activity in high specific activity and good stability (no change after one week at 4°C), there are nonetheless several problems with this preparation. First, the binding activity emerges over a large range of the effective volume of elution of the column, implying that although the binding factor appears to be soluble after detergent treatment, there may not be a discrete molecular species free in solution. Second, efforts to concentrate this activity by standard means led to loss of binding activity. Third, the binding activity after reconstitution in liposomes shows several undesirable properties. These consist of variability in specific activity upon dilution and an inability to bind more than 50% of added citrate synthase no matter what amount of liposomes was added (data not shown).

Since the binding isotherm is not a simple one, we attempted to investigate possible causes of this problem. Isoelectric focusing of either purified yeast citrate synthase or crude extracts of yeast yield at least two peaks (Fig. 7), which implies that there exist multiple forms of the enzyme and which may represent a part of the cause of the inconsistencies observed in the binding studies.

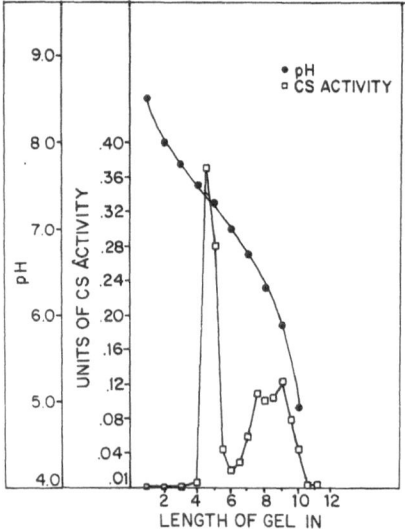

Fig. 7. Separation of Purified Yeast Citrate Synthase by Analytical Isoelectric Focussing. 11.7 units of yeast citrate synthase were applied to a 5% polyacrylamide gel equilibrated with 1.6% ampholytes in the pH range 6-8 and 0.4% in the range of 3-10 and focusing done 1.75 h at 400 volts. The gel was then cut into 5 mm slices and the gel slices extracted into 0.1 ml of 0.1 M potassium phosphate pH 7.4 overnight at 4°, followed by enzyme assay and pH measurement.

DISCUSSION

The enzymes composing the Krebs tricarboxylic acid cycle may exist within mitochondria in an organized form in vivo, which we have called the Krebs tricarboxylic acid cycle "metabolon" (Robinson and Srere, 1985b). We have shown here and elsewhere (D'Souza and Srere, 1983; Sumegi and Srere, 1984; Moore et al., 1984; Robinson and Srere, 1985a) that those enzyme activities previously thought to exist in a soluble form in the mitochondrial matrix possess the ability to bind to the inner surface of the inner membrane, while purified enzymes from other cellular compartments do not possess such binding abilities. We have shown these binding properties exist in yeast, plants, various animal cells, and neoplastic cells in culture. The membrane factors responsible for this binding are extractable from the inner membranes of both animal and yeast systems and can be reconstituted in artificial liposomes. Conventional methods of purification of membrane proteins yield results which imply the factor(s) necessary for binding include a protein (or proteins), as do experiments involving treatment of membranes by trypsin or heat.

There are questions yet to be answered in these studies. Although the specificity and ubiquity of these interactions argue for their physiological role, the sensitivity to increase in ionic strength and presence of metabolites is difficult to rationalize with what is believed to exist inside mitochondria. It is probable that the binding interactions studied here are only a part of the in vivo system of organization of the Krebs tricarboxylic acid cycle. It is obvious that in the physical treatments of the cellular fractions necessary to produce "pure" inner membrane preparation, components not firmly associated with the membrane can be lost, and that the normal intramitochondrial environment of the matrix is not present in these studies. The organization of the Krebs tricarboxylic acid cycle may well be a directed summation of various types of interactions, including (nonexclusively) matrix protein-matrix protein, matrix protein-membrane protein, matrix-protein-lipid, membrane-protein lipid and metabolite regulatory events, such that the interactions studied here may only comprise a small part of the overall organizational structure of the "metabolon". The activity and structure of the cycle organization could be thereby influenced by a number of as yet undefined metabolically significant parameters, to provide appropriate response to changes in cellular energetic and synthetic requirements.

Additionally, the loss of consistency of amount of enzyme bound per mg protein observed in partially purified systems is a matter of concern. It is obvious that a simple protein-protein interaction cannot account for the results observed here, and that other parameters, such as multiple enzyme forms and loss of matrix factors in purification of inner membranes, may well be important in the interactions in vivo. In support of these ideas, we have observed (Robinson and Srere, 1985b) that the Krebs cycle enzymes remain in a bound form when gentle disruptive techniques are used on mitochondria, and that such interactions are not susceptible to reversal over much greater ranges of ionic strength and not susceptible to reversal by metabolites.

ACKNOWLEDGMENTS

The authors wish to acknowledge the technical assistance of Mr. Daniel D. Owens and manuscript preparation by Ms. Penny Perkins. This work was supported by the Research Service of the Veterans

Administration, National Institute of Health, and National Science
Foundation.

REFERENCES

Ainsley, M.W., Hensley, P. and Butow, R.A., 1984, Expression of GC
    clusters in the yeast mitochondrial Var 1 gene, J. Biol. Chem.,
    259:8422.

Beeckmans, S. and Kanarek L., 1981, Demonstrations of physical
    nteractions between consecutive enzymes of the citric acid cycle and
    of the aspartate-malate shuttle, Eur. J. Biochem., 117:527.

Cha, S., 1969, Succinate thiokinase from pig heart, Methods Enzymol.,
    13:62.

Cleland, W.W., Thompson, V.W. and Barden, R.E., 1969, Isocitrate
    dehydrogenase (TPN-Specific) from pig heart, Methods Enzymol., 13:30.

Daum, G., Bohni, P.C. and Schatz, G., 1982, Import of proteins into
    mitochondria, J. Biol. Chem., 257:13028.

D'Souza, S.F. and Srere, P.A., 1983, Binding of citrate synthase to
    mitochondrial inner membranes, J. Biol. Chem., 258:4706.

Fansler, B. and Lowenstein, J.M., 1969, Aconitase from pig heart, Methods
    Enzymol., 13:26.

Hackenbrock, C.R. and Miller-Hammon, K., 1975, Cytochrome c oxidase in
    liver mitochondria, J. Biol. Chem., 250:9185.

Halper, L.A. and Srere, P.A., 1979, Interaction between citrate synthase
    and mitochondrial malate dehydrogenase in the presence of
    polyethylene glycol, Arch. Biochem. Biophys., 184:529.

Lowry, O.H., Rosebrough, N.J., Farr, A.L., and Randall, R.J., 1951,
    Protein measurement with the folin phenol reagent, J. Biol. Chem.,
    193:265.

Matlib, M.A. and Srere, P.A., 1976, Oxidative properties of swollen rat
    liver mitochondria, Arch. Biochem. Biophys., 174:705.

Moore, G.E., Gadol, S.M., Robinson, Jr., J.B. and Srere, P.A., 1984,
    Binding of citrate synthase and malate dehydrogenase to mitochondrial
    inner membranes: tissue distribution and metabolite effects,
    Biochem. Biophys. Res. Commun., 121:612.

Penefsky, H.J., 1977, Reversible Binding of Pi by beef heart mitochon-
    drial adenosine triphosphate, J. Biol. Chem, 252:2891.

Plant, G.W.E., 1969, Isocitrate dehydrogenase (DPN-Specific) from pig
    heart, Methods Enzymol., 13:34.

Porpaczy, Z., Sumegi, B. and Alkonyi, I., 1983, Association between the
    alpha-ketoglutarate dehydrogenase complex and succinate thiokinase,
    Biochem. Biophys. Acta, 749:172.

Robinson,Jr., J.B. and Srere, P.A., 1985a, Organization of Krebs
    tricarboxylic acid cycle enzymes, Biochem. Medicine, 33:149.

Robinson, Jr., J.B. and Srere, P.A., 1985b, Organization of Krebs
    tricarboxylic acid cycle enzymes in mitochondria, J. Biol. Chem.,
    260:10800.

Schnaitman, C. and Greenawalt, J.W., 1968, Enzymatic properties of the
    inner and outer membranes of rat liver mitochondria, J. Cell Biol.,
    38:158.

Simon, E.J. and Shemin, D., 1953, Synthesis of thiol esters, J. Amer.
    Chem. Soc., 75:2520.

Srere, P.A., 1985, in: "Organized Multienzyme Systems," G.R. Welch, ed.,
    Academic Press, New York.

Srere, P.A., Brazil, H. and Gonen, L., 1963, The citrate condensing
    enzyme of pigeon breast muscle and moth flight muscle, Acta Chem.
    Scadn., 17:S129.

Sumegi, B. and Srere, P.A., 1984, Complex I binds several mitochondrial

NAD-coupled dehydrogenases, *J. Biol. Chem.*, 259:15040.
Teszuka, T. and Laties, G.G., 1983, Studies on potato isocitrate
   dehydrogenase, *Plant Physiol.*, 72:959.

SOME ASPECTS OF ENZYME ORGANIZATION IN THE

CITRIC ACID CYCLE

P.D.J. Weitzman and Sarah J. Barnes

Department of Biochemistry
University of Bath
Bath BA2 7AY, England

INTRODUCTION

The citric acid (Krebs) cycle is a central metabolic pathway which occurs almost ubiquitously throughout Nature. Its basic two-fold function of providing energy and biosynthetic intermediates, together with its fundamental chemical unity across widely different organisms, have masked a remarkable diversity in the fine details of the enzymes of the cycle and of the regulatory mechanisms which may operate to control them. That diverse organisms may make different use of, or place different emphasis on, the separate functions of the cycle, and may therefore regulate the pathway in distinct ways, poses the possibility that organisms may differ in their organization of the enzymic apparatus of the cycle.

In this brief contribution to the Workshop we shall consider some aspects of enzyme organization, both intra-enzymic and inter-enzymic, which have emerged in the course of our studies on the citric acid cycle. To serve as a reminder, the cycle is represented simply in Fig. 1, with the associated enzymes indicated.

INTRA-ENZYMIC ORGANIZATION: CITRATE SYNTHASE

The oxidative role of the cycle leads to the generation of the reduced nucleotide, NADH, which may therefore be considered an overall product of the cycle. The reaction forming citrate, catalyzed by citrate synthase, is the only reaction of the cycle in which carbon atoms enter by condensation with a cycle intermediate. Citrate synthase may thus be considered as the "initial" enzyme of the cycle, and it is appropriate to ask whether this initial enzyme is sensitive to feedback inhibition by the end product, NADH.

We first showed that citrate synthase from Escherichia coli is indeed sensitive to specific inhibition by NADH, a property not shared by citrate synthases from eukaryotic sources (Weitzman, 1966). Exploration of the possibility that E. coli typified all bacteria and that this difference in inhibition property separated the citrate synthases of

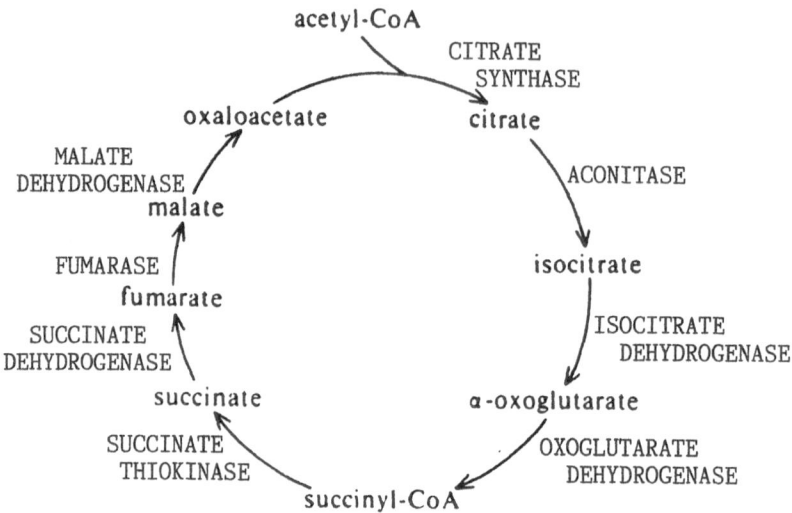

Fig. 1. The citric acid cycle - intermediates and enzymes

prokaryotes from those of eukaryotes revealed a quite unexpected pattern
of enzymic behavior (Weitzman and Jones, 1968; Weitzman, 1981). Only
citrate synthases from Gram-negative bacteria are inhibited by NADH; the
Gram-positive bacterial enzymes are unaffected. Furthermore, AMP (a
"low- energy" metabolic signal) overcomes the NADH inhibition of citrate
synthases from strictly aerobic Gram-negative bacteria (e.g., Pseudomonas
species), whereas the facultatively anaerobic Gram-negative organisms
(e.g., E. coli) do not show this reactivation. Both NADH and AMP act as
allosteric modulators. The obligatory dependence of strict aerobes, but
not of facultative organisms, on the citric acid cycle for energy
production offers a metabolic rationale for this difference in response
to AMP. However, the difference in the response to NADH of citrate
synthases from Gram-negative and Gram-positive bacteria remains to be
explained.

Subsequent studies showed that Gram-negative bacterial citrate
synthases are "large" hexameric enzymes of $M_r \sim 250,000$, whereas the Gram-
positive bacterial and eukaryotic citrate synthases are "small" dimeric
entities of $M_r \sim 100,000$ (see Weitzman, 1981, for review). No NADH-
sensitive citrate synthase has been found which is not of the "large"
type. Moreover, within the hexameric citrate synthase there appears to
be only one type of subunit, leading to the conclusion that the molecular
mechanism effecting inhibition of enzyme activity as a response to the
interaction with NADH may be dependent on the particular organization of
subunits within the hexameric enzyme.

Evidence in support of this was obtained by electron microscopic
examination of citrate synthase purified from a Gram-negative strict
aerobe, Acinetobacter lwoffi (Rowe and Weitzman, 1969). Inhibition of
the enzyme by NADH was shown to be associated with a "swelling" of the
enzyme molecule (movement apart of subunits), which was reversed by the
reactivator, AMP. More recently, we have investigated the effects of
cross-linking this enzyme with the cleavable bifunctional reagent,
dithiobis(succinimidyl propionate), with the aim of "freezing" the
quaternary structure and thereby preventing the NADH-induced molecular

swelling (Mitchell and Weitzman, 1983). Electrophoretic analysis showed the enzyme to be cross-linked; but, contrary to expectation, the cross-linked enzyme still showed inhibition by NADH. However, the inhibition of cross-linked enzyme lacked completely the cooperative, sigmoidal kinetics displayed by the native enzyme, both with respect to NADH inhibition and AMP reactivation. Cleavage of the disulphide bonds in the cross-links by treatment with dithiothreitol restored the sigmoidal characteristics of both inhibition and reactivation. These findings suggest that features of the intra-enzymic organization of the citrate synthase molecule are separately responsible for the sensitivity to regulatory effectors and to the cooperativity of their action.

A further interesting aspect of the molecular organization of citrate synthase is the discovery of two citrate synthases in a mutant of Pseudomonas aeruginosa (Solomon and Weitzman, 1983). Both a "large" form (regulated by NADH and AMP) and a "small" form are present, but the relative proportions of the two forms vary with the stage of growth of the bacterial culture. The "large" form predominates in logarithmic phase, whereas the "small" form is the major component in stationary phase. These observations suggest that the metabolic demands imposed during vigorous logarithmic growth are best accommodated by the organization of citrate synthase into a nucleotide-regulated hexameric structure, whereas the conditions of stationary phase are conducive to the predominance of the "small" nucleotide-insensitive form of the enzyme. Other species of Pseudomonas have also been found to contain these two forms of citrate synthase (Massarini and Cazzulo, 1975; Mitchell and Weitzman, 1986); and it remains to be seen whether, in these cases, the two forms are modifications of a single gene product in response to physiological conditions.

This short summary of some of the properties of citrate synthase should serve to emphasize the disparity between the enzyme from different sources and the role of "organization" in its regulatory behavior.

INTER-ENZYMIC ORGANIZATION: A MULTIENZYME CLUSTER

There has long been interest in the idea that the enzymes of the citric acid cycle may exist in the cell bound together in some form of complex. Many years ago, Green et al. (1948) reported such a complex and called it "cyclophorase"; however, this was soon shown to be the same as the intact mitochondrion (Hogeboom et al., 1948). The possibility of association between citric acid cycle enzymes nevertheless remained an appealing one, and Srere (1972) proposed that "these enzymes are not in random disarray in the (mitochondrial) matrix but are organized in assemblies with a fixed relation to each other and the inner membrane." One of the cycle enzymes, succinate dehydrogenase, is an integral part of the inner membrane, whereas the other enzymes of the cycle are located in the matrix compartment. This locational discontinuity led Srere (1972) to speculate that succinate dehydrogenase may act as an "organizer" for the other citric acid cycle enzymes. He suggested that an organized array of the enzymes could achieve enhanced metabolite throughput at low substrate concentration and might be more efficiently regulated.

The potential advantages of multienzyme complexes, or clusters, have frequently been recognized (e.g., Welch, 1977; Gaertner, 1978; Friedrich, 1984), as has also the likelihood that conditions within the living cell are "crowded" and that many of its constituent proteins/enzymes are located close to each other, perhaps in specific associations, and restricted in their movement (e.g., Srere, 1980, 1981, 1982; Fulton, 1982; Clegg, 1984).

Several reports have appeared on specific interactions between certain combinations of citric acid cycle enzymes: citrate synthase and malate dehydrogenase (Halper and Srere, 1977); citrate synthase and pyruvate dehydrogenase complex (Sumegi et al., 1980; Sumegi and Alkonyi, 1983); fumarase, malate dehydrogenase and citrate synthase (Beeckmans and Kanarek, 1981); and succinate thiokinase and oxoglutarate dehydrogenase complex (Porpaczy et al., 1983). Various methods were employed to detect these interactions, and in all cases the studies were made by mixing together previously isolated (or purified) enzymes. Srere et al. (1978) examined mitochondrial extracts by sucrose density-gradient centrifugation and gel filtration but failed to detect any inter-enzymic complexes.

These investigations were all made on enzymes from animal sources (i.e., of mitochondrial origin). In view of the absence of organellar compartmentation of the citric acid cycle in bacteria and of our previous studies on bacterial citric acid cycle enzymes (Weitzman, 1981), we decided to focus our attention initially on E. coli (Barnes and Weitzman, in preparation). In this we were encouraged by the earlier isolation by Moses and his co-workers (Moses, 1978; Gorringe and Moses, 1980; see also Moses, this Volume) of a multienzyme aggregate from E. coli containing the enzymes of glycolysis. A gentle procedure for cell disruption is required in order to increase the likelihood of preserving delicate enzyme associations, which may not withstand commonly used methods of cell breakage such as ultrasonication. The formation of spheroplasts from E. coli by treatment with lysozyme-EDTA, followed by osmotic lysis of the spheroplasts, offers a route for gentle disruption of the E. coli cells.

The experimental procedure followed that described by Gorringe and Moses (1980), and the lysate was centrifuged to yield a membranous pellet and a supernatant. Gel filtration of the supernatant on Sepharose-4B resulted in two distinct regions of elution of several citric acid cycle enzymes — citrate synthase, aconitase, isocitrate dehydrogenase, succinate thiokinase, fumarase and malate dehydrogenase. The major portions of these enzymes were eluted at distinct positions corresponding to their individual molecular weights. A small proportion, however, was eluted earlier as a high molecular weight cluster along with the multienzyme complexes, pyruvate dehydrogenase and oxoglutarate dehydrogenase. Succinate dehydrogenase, an intrinsic membrane protein, was identified in the centrifuged pellet from the lysate and no activity was released into the supernatant. Passage of the high molecular weight aggregate through a second gel-filtration column resulted in some dissociation, though high molecular weight material was again observed.

When the low molecular weight enzyme fractions were pooled, concentrated by ultrafiltration and re-run on gel filtration, a high molecular weight cluster was again produced, indicating the ability of the individual enzymes to re-associate. However, succinate thiokinase, though present in the pooled fractions, failed to emerge with the other enzymes in the high molecular weight peak. This may be due to the absence of oxoglutarate dehydrogenase in the pooled fractions and is consistent with the reported interaction of succinate thiokinase with oxoglutarate dehydrogenase (Porpaczy et al., 1983).

The association of these enzymes into a cluster is clearly dependent on the solution environment. Glycerol (20%) was routinely present, as very little aggregated material was observed in its absence. The additional presence of polyethylene glycol or high protein (bovine serum albumin) concentration increased the yield of the high molecular weight cluster from the gel-filtration column; these conditions may mimic the intracellular milieu. Conversely, high salt concentration or shifts away from the standard pH 7.5 decreased the yield of high molecular weight material.

The integrity of the high molecular weight cluster of citric acid cycle enzymes was also demonstrated by sucrose density gradient centrifugation. The enzymes in the high molecular weight fraction from gel filtration co-sedimented on sucrose gradient centrifugation, whereas the low molecular weight enzyme fractions sedimented as individual bands.

A sample of high molecular weight material was examined by electron microscopy, after fixing in glutaraldehyde prior to staining with uranyl acetate. Relatively uniform discrete particles were observed with diameters in the range 20-30 nm, consistent with a molecular weight around 2×10^6.

It is significant that when the bacterial cells were disrupted by ultrasonication, no high molecular weight cluster could be found. Furthermore, when the pooled low molecular weight enzyme fractions from gel filtration were sonicated and then concentrated and re-run on gel filtration, no evidence of re-association to an aggregate was found. The effect of sonication may be to destroy structural elements in the enzymes which are necessary for aggregate formation or to damage some other unidentified component(s) (small membrane fragment?) necessary for cluster association.

Essentially similar results have been obtained with lysates prepared from other bacteria, both Gram-negative (Acinetobacter calcoaceticus and Pseudomonas aeruginosa) and Gram-positive (Bacillus subtilis). In all cases, a cluster of enzymes may be isolated containing the activities of fumarase, malate dehydrogenase, citrate synthase, aconitase and isocitrate dehydrogenase. This multienzyme cluster would be capable of catalyzing the sequence of citric acid cycle reactions: fumarate---> malate---> oxaloacetate---> citrate---> isocitrate---> oxoglutarate. We are currently investigating the kinetic behavior of this cluster and comparing it with a mixture of the "free" enzymes to see if the cluster achieves any catalytic enhancement. It will also be interesting to test whether channeling of intermediates occurs and whether the regulatory properties observed in the isolated enzymes (e.g. citrate synthase) are also displayed, or modified, by the cluster.

These findings constitute the first report of the isolation of a citric acid cycle multienzyme cluster from cells. This level of organization may be of particular importance in bacteria, where the absence of mitochondria affords no opportunity for compartmentation of the pathway within those organelles. However, our preliminary examination of gently-disrupted mitochondria from several sources indicates that a similar multienzyme cluster is detectable in such extracts; the cluster may therefore be of general occurrence in all types of cells. It is noteworthy that Beeckmans and Kanarek (see this Volume) have detected association among fumarase, malate dehydrogenase and citrate synthase in extracts of Candida utilis.

Mention was made above of (a) the speculation by Srere (1972) that succinate dehydrogenase may act as an organizer for the other citric acid cycle enzymes, and (b) the report of association between oxoglutarate dehydrogenase and succinate thiokinase (Porpaczy et al., 1983). If the membrane-anchored succinate dehydrogenase were to associate with both the latter 2-enzyme aggregate and the 5-enzyme cluster described in this paper, the resultant supramolecular complex would constitute a citric acid cycle "metabolon" (Srere, 1985). Finally, there is no need to expect that all the citric acid cycle enzyme molecules within a cell must, at any time, be incorporated into such assemblies. Rather, the dynamic association and dissociation ("ambiguity" - Wilson, 1978) of the enzymes of a citric acid cycle complex may confer an additional level of regulation on this multifunctional metabolic pathway.

ACKNOWLEDGEMENTS

We thank the Science and Engineering Research Council for research support and for a studentship to S.J.B.

REFERENCES

Beeckmans, S. and Kanarek, L., 1981, Eur. J. Biochem., 117:527.
Clegg, J.S., 1984, BioEssays, 1:129.
Friedrich, P., 1984, "Supramolecular Enzyme Organization", Pergamon Press, Oxford.
Fulton, A.B., 1982, Cell 30:345.
Gaertner, F.H., 1978, Trends Biochem. Sci., 3:63.
Gorringe, D.M. and Moses, V., 1980, Int. J. Biol. Macromol., 2:161.
Green, D.E., Loomis, W.F., and Auerbach, V.A., 1948, J. Biol. Chem., 172:389.
Halper, L.A. and Srere, P.A., 1977, Arch. Biochem. Biophys., 184:529.
Hogeboom, E.H., Schneider, W.C., and Palade, G.E., 1948, J. Biol. Chem., 172:619.
Massarini, E. and Cazzulo, J.J., 1975, FEBS Lett., 57:134.
Mitchell, C.G. and Weitzman, P.D.J., 1983, FEBS Lett., 151:260.
Mitchell, C.G. and Weitzman, P.D.J., 1986, J. Gen. Microbiol., in press.
Moses, V., 1978, in: "Microenvironments and Metabolic Compartmentation", P.A. Srere and R.W. Estabrook, eds., Academic Press, New York.
Porpaczy, Z., Sumegi, B. and Alkonyi, I.,1983, Biochim. Biophys. Acta, 749:172.
Rowe, A.J. and Weitzman, P.D.J., 1969, J. Mol. Biol., 43:345.
Solomon, M. and Weitzman, P.D.J., 1983, FEBS Lett., 155:157.
Srere, P.A., 1972, in: "Energy Metabolism and the Regulation of Metabolic Processes in Mitochondria", M.A. Mehlmann and R.W. Hanson, eds., Academic Press, New York.
Srere, P.A., 1980, Trends Biochem. Sci., 5:120.
Srere, P.A., 1981, Trends Biochem. Sci., 6:4.
Srere, P.A., 1982, Trends Biochem. Sci., 7:375.
Srere, P.A., 1985, Trends Biochem. Sci., 10:109.
Srere, P.A., Halper, L.A., and Finkelstein, M.B., 1978, in: "Microenvironments and Metabolic Compartmentation", P.A. Srere and R.W. Estabrook, eds., Academic Press, New York.
Sumegi, B. and Alkonyi, I., 1983, Biochim. Biophys. Acta, 749:163.
Sumegi, B., Gyocsi, L., and Alkonyi, I., 1980, Biochim. Biophys. Acta, 616:158.
Weitzman, P.D.J., 1966, Biochim. Biophys. Acta, 128:213.
Weitzman, P.D.J., 1981, Adv. Microb. Physiol., 22:185.

Weitzman, P.D.J. and Jones, D., 1968, Nature (London), 219:270.
Welch, G.R., 1977, Prog. Biophys. Mol. Biol., 32:103.
Wilson, J.E., 1978, Trends Biochem. Sci., 3:124.

ENZYME INTERACTIONS IN THE CITRIC ACID CYCLE

AND THE ASPARTATE-MALATE SHUTTLE

Sonia Beeckmans and Louis Kanarek

Laboratorium voor Chemie der Proteinen
Vrije Universiteit Brussel
Paardenstraat, 65
B-1640 Sint-Genesius-Rode
Belgium

INTRODUCTION

The citric acid cycle is central to energy metabolism in various types of cells, especially that of aerobic organisms. Incoming acetyl-units, which are provided by the degradation of sugars, fatty acids, or amino acids, are completely converted to CO_2 and reducing equivalents. The latter are used for the production of ATP, by a process in which the flow of electrons through the consecutive complexes of the electron transport chain is coupled to the vectorial transport of protons. Since several citric acid cycle intermediates (e.g., oxaloacetate, α-keto-glutarate, succinyl-CoA) are precursors of essential cell components, the cycle also has an important anabolic function. The cycle finds itself in the very peculiar situation whereby certain of its segments have potentially opposite functions, namely bioenergetic and biosynthetic.

In eukaryotes, the citric acid cycle is an exclusively mitochondrial process. It is now generally accepted that all the cycle enzymes are located in the mitochondrial matrix (Ernster and Kuylenstierna, 1970), more or less associated with the inner mitochondrial membrane (Addink et al., 1972; Matlib and O'Brien, 1975; Comte and Gautheron, 1978; Elduque et al., 1982; D'Souza and Srere, 1983). One of the enzymes, succinate dehydrogenase, is entirely embedded in the inner membrane; moreover, this enzyme also belongs to Complex-II of the electron transport chain, thus linking both pathways together.

A general analysis of the regulation of the citric acid cycle is hampered by its complexity and by the multiple interactions with surrounding metabolic pathways. Since certain cycle intermediates are precursors in various biosynthetic reactions, it is a prerequisite that the activity of each segment of the cycle be modulated individually and independently of the other segments. Two main regulatory mechanisms are thought to determine the flux through the citric acid cycle. The first one involves the regulation of individual cycle enzymes, primarily through local metabolite ratios (reviewed by Williamson and Cooper, 1980). The second involves reversible complex formation between various enzymes of the cycle and its surrounding pathways. Such an organization will create a special microenvironment around the cycle: the cell acquires the possibility to maintain a high flux of substrates through the cycle with a moderate number of intermediate molecules.

THE USE OF IMMOBILIZED ENZYMES TO DEMONSTRATE SPECIFIC INTERACTIONS BETWEEN CITRIC ACID CYCLE AND ASPARTATE-MALATE SHUTTLE ENZYMES

Despite several attempts, specific interactions between enzymes of the citric acid cycle have not been demonstrated in aqueous solutions (except for interactions involving the bulky complexes pyruvate dehydrogenase or α-ketoglutarate dehydrogenase [Sumegi et al., 1983; Porpaczy et al., 1983]). Such could, however, be revealed under conditions which are believed to diminish the water concentration, i.e., either by techniques of co-precipitation of enzyme mixtures in polyethylene glycol (Halper and Srere, 1977; Srere et al., 1978; Fahien and Kmiotek, 1983; Fahien et al., 1979; Porpaczy et al., 1983), or, as in our experiments (Beeckmans and Kanarek, 1981; Beeckmans, 1984), by immobilizing one of the enzymes on a solid support (Sepharose beads) and studying adsorption of other enzymes thereto. It should be emphasized that in vivo the water concentration in the mitochondrial matrix is severely reduced. It has been calculated (Hackenbrock, 1968) that the protein concentration in the matrix of actively metabolizing mitochondria (State III according to the definition of Hackenbrock [1966]) should be about 560 mg/ml; this concentration is very close to that in enzyme crystals (Srere, 1982)! The experimental conditions which we chose can thus be considered to mimic the real in vivo situation.

The enzymes involved in our studies are shown in Figure 1. In these experiments we used covalently immobilized fumarase, mitochondrial or cytosolic malate dehydrogenase, or the same enzymes immobilized through specific affinity-purified antibodies, which were bound to Sepharose-Protein A. For direct covalent immobilization, the Sepharose beads were activated by the classical CNBr-method, and the enzymes were stabilized during the coupling step by the addition of 20 mM of substrate (L-malate). Full experimental details on these immobilization methods are given in a previous publication (Beeckmans and Kanarek, 1981).

In Table 1 adsorption of various enzymes to covalently immobilized fumarase and malate dehydrogenase is shown. In all these experiments saturating amounts of enzyme, free in solution (ranging from 0.5 to 3 mg), were applied on a column containing 4 mg of enzyme immobilized on 2 g Sepharose beads (wet weight); 1 mM potassium phosphate + 14 mM 2-mercaptoethanol, pH 7.3, was used as buffer. After 30 min of incubation, the applied solution was passed slowly through the column; and the gel was washed with the same buffer until no more activity came off. The adsorbed enzyme could be eluted with 10 mM potassium phosphate + 14 mM 2-mercaptoethanol, pH 7.3, containing 10 mM L-malate or L-aspartate, with 100 mM potassium phosphate + 14 mM 2-mercaptoethanol, pH 7.3, or with 10 mM potassium phosphate + 150 mM NaCl + 14 mM 2-mercaptoethanol, pH 7.3. The amount of enzyme adsorbed was determined as the activity in the elution fraction; the sum of adsorbed and eluted enzyme activities was always equal to the amount of enzyme applied on the columns.

From the results given in Table 1 we can conclude the following: (1) The observed interactions between the enzymes are specific, since neither of the control enzymes, aldolase and lysozyme, interact with the immobilized enzymes; moreover, the amount of specifically adsorbed enzyme

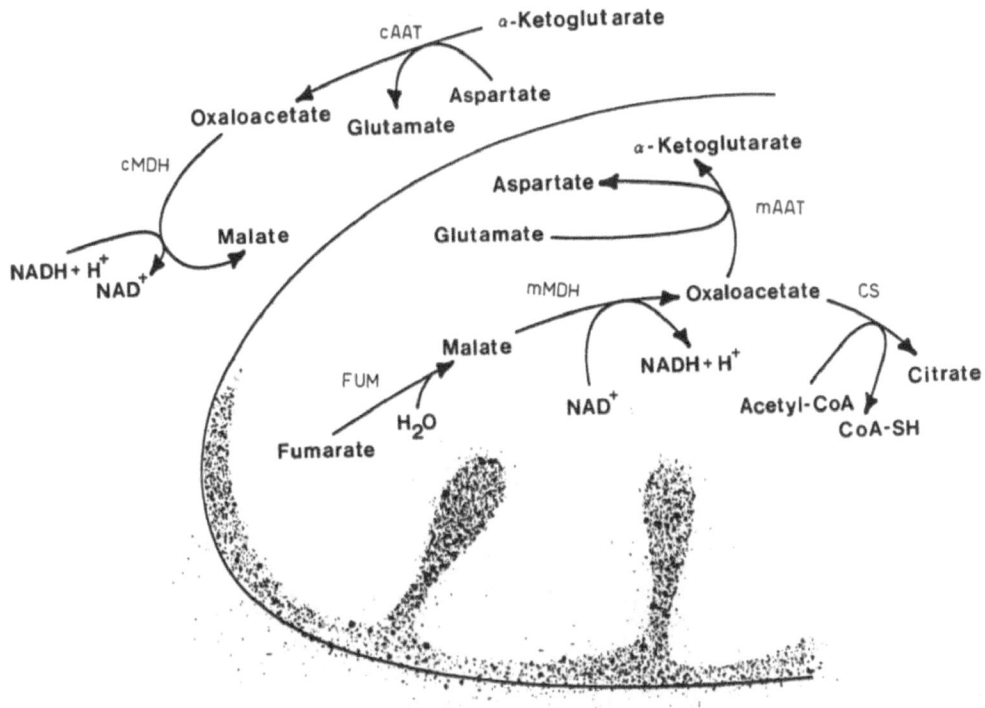

Fig. 1. Enzymes considered in this study.

Table 1. Adsorption of Various Enzymes to Immobilized Fumarase and
Malate Dehydrogenase

| Enzyme applied | Source | Amount of enzyme adsorbed to immobilized enzymes | | | |
| | | FUM | | mMDH | cMDH |
		pig (mg)	chicken (mg)	pig (mg)	(mg)
mMDH	pig	0.25	0.22	-	-
	chicken	0.22	0.27	-	-
cMDH	pig	0	0	-	-
CS	pig	0.27	0.30	0.28	0.23
	chicken	0.23	0.28	0.26	0.18
mAAT	chicken	0	0	0.20	0.10
cAAT	pig	0	0	0.14	0.46
mMDH +	chicken	-	0.18	-	-
mAAT		-	0.13	-	-
aldolase		0	0	0	0
lysozyme		0	0	0	0

was shown to be independent of the amount applied on the column and that the adsorbed enzyme could not be removed by extensive washing with buffer (up to 150 column volumes). (2) Mitochondrial aspartate aminotransferase, which does not specifically interact with immobilized fumarase, can be adsorbed onto the latter when the transaminase is applied together with mitochondrial malate dehydrogenase; this result further stresses the specificity of the observed interactions. (3) The interactions between the different enzymes seem to be electrostatic, since they can be broken by enhancing the ionic strength; metabolites do not seem to play a specific role in the elution of the adsorbed enzymes. (4) Interactions are observed between a chicken and a pig enzyme, whenever there is an interaction between the corresponding two enzymes isolated from the same species; the sites on the surface of the enzymes, involved in the interactions, thus seem to have been conserved throughout the course of evolution.

Further studies of complex formation, involving more than two enzymes, were performed by adding, consecutively, various enzymes to immobilized fumarase (Table 2). The experimental conditions were the same as described for Table 1. After each incubation, the excess of applied enzyme was removed by washing with 1 mM potassium phosphate + 14 mM 2-mercaptoethanol, pH 7.3; but the amount of enzyme specifically adsorbed was not removed in this case. The following conclusions can be drawn from the results shown in Table 2: (1) Part of the mitochondrial malate dehydrogenase, which was adsorbed during the first incubation step, is displaced when the system is incubated with mitochondrial aspartate aminotransferase. (2) The transaminase (which, as was shown in Table 1, has no direct affinity for the immbolized fumarase but binds through malate dehydrogenase) and mitochondrial malate dehydrogenase adsorb in equimolar amounts. (3) Upon incubating the system with citrate synthase (step 3), part of the previously adsorbed transaminase is released from the gel; however, when incubating the system again with transaminase (step 4), part of the citrate synthase is released in favor of the transaminase. The total amount of adsorbed transaminase + citrate synthase remains constant (3.55 nmol). From the latter observations it is tempting to suggest that the enzymes might be organized as one huge

Table 2.   Consecutive Binding of Different Enzymes to Immobilized Chicken Fumarase

| Enzymes applied consecutively | Amount of enzymes adsorbed to immobilized FUM | | | | | |
| | (mg) | | | (nmol)[*] | | |
	mMDH	mAAT	CS	mMDH	mAAT	CS
mMDH	0.290	–	–	4.03	–	–
mAAT	0.200	0.240	–	2.78	2.79	–
CS	0.200	0.110	0.210	2.78	1.28	2.28
mAAT	0.200	0.145	0.170	2.78	1.69	1.85

[*]The amounts of enzymes, expressed in nanomolar quantities, were calculated assuming molecular weights of 72 kd for malate dehydrogenase, 92 kd for citrate synthase, 86 kd for aspartate aminotransferase, and 194 kd for fumarase.

complex, comprising the citric acid cycle (at least the sequence fumarase-malate dehydrogenase-citrate synthase), as well as the aspartate-malate shuttle.

In order to obtain more information on the stoichiometry of the enzyme-enzyme interactions, we immobilized either fumarase or malate dehydrogenase through specific antibodies and studied the adsorption of malate dehydrogenase and fumarase thereon (Table 3). In this approach, most of the surface of the immobilized enzyme remains free and available for interaction (this is in contrast to covalently immobilized enzymes, which are linked to the Sepharose by multi-point attachment). In these experiments, all solutions were buffered with 10 mM Tris-acetate + 14 mM 2-mercaptoethanol, pH 7.3.

The results described above are consistent with a model in which maximally 4 molecules of malate dehydrogenase are bound to 1 fumarase molecule (Figure 2). This complex is able to bind either citrate synthase or aspartate aminotransferase. We propose that these enzymes bind alternatively, in order to allow the cell to perform citric acid cycle or shuttle reactions, depending on metabolic needs (Figure 3).

THE USE OF ELECTROPHORESIS FOR THE DETECTION OF INTERACTIONS BETWEEN CHICKEN HEART FUMARASE AND MITOCHONDRIAL MALATE DEHYDROGENASE

Occasionally, an unexpected comigration of chicken heart fumarase and malate dehydrogenase was observed with polyacrylamide-gel electrophoresis under non-denaturing conditions (Davis, 1964). In such experiments an eluent from pyromellitic-acid-Sepharose affinity chromatography of chicken heart (0.40-0.60 ammonium sulphate fraction) was used as the sample (a detailed description is given elsewhere [Beeckmans and Kanarek, 1982]). The eluent used in this experiment contained almost exclusively fumarase (3.5 nmol/ml) and mitochondrial malate dehydrogenase (16 nmol/ml); it was dialyzed against 10 mM potassium phosphate + 14 mM 2-mercaptoethanol + 10% (v/v) glycerol, pH 7.3. A solution containing 0.6 g Tris + 2.88 g glycine in 1 liter (pH = 8.3) was used as running buffer during electrophoresis, and the final gel concentration was 7% (w/v) in acrylamide. The electrophoresis was performed at 100 volt; the current through each gel (0.8 cm diameter) was less than 1 mA. The enzymes were histochemically stained according to Brewer and Sing (1970), either for fumarase (lane 2) or for malate dehydrogenase (lane 3); in lane 1, the gel was stained for total protein with Coomassie brilliant blue.

Table 3. Adsorption of Mitochondrial Malate Dehydrogenase (Fumarase) to Immobilized Fumarase (Malate Dehydrogenase)

Antibody loaded on protein A-Sepharose (6 nmol)	Antigen immobilized (nmol)		Enzyme adsorbed (nmol)	
1. anti(pig FUM)	pig FUM	6.29	pig mMDH	16.67
	chicken FUM	1.65	chicken mMDH	5.14
2. anti(pig mMDH)	pig mMDH	6.11	pig FUM	2.68
	chicken mMDH	3.33	chicken FUM	1.39

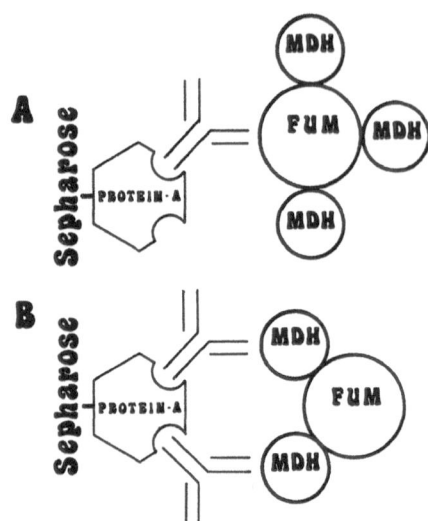

Fig. 2.   Hypothesis on the complex formation between fumarase and
           mitochondrial malate dehydrogenase, when either the former (A)
           or the latter (B) is immobilized via specific antibodies onto
           protein A-Sepharose.

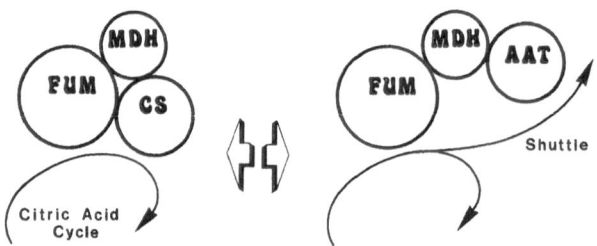

Fig. 3.   Schematical representation of the proposed alternating
           interactions between enzymes of the citric acid cycle and the
           aspartate-malate shuttle.

It is obvious from Figure 4 that both enzymes migrate at the same
rate. This co-migration is unexpected, in view of the large difference
in pI-value of these two enzymes (pI$_{FUM}$ = 8.3, whereas pI$_{MDH}$ = 10;
when run in the absence of fumarase, mitochondrial malate dehydrogenase
migrates out of the gel towards the cathode).

SOME STUDIES WITH YEAST ENZYMES

Although it was not possible to observe interactions between
mammalian citric acid cycle enzymes by simple gel filtration, we found
this technique to be successful, at least to a limited extent, for the
detection of certain associations in extracts of the yeast Candida
utilis. The yeast cells were allowed to autolyze at 37°C in 50 mM
potassium phosphate, containing 1 mM pyromellitic acid, pH 8.5. After 3
hours the slurry was centrifuged, and the enzymes were precipitated with

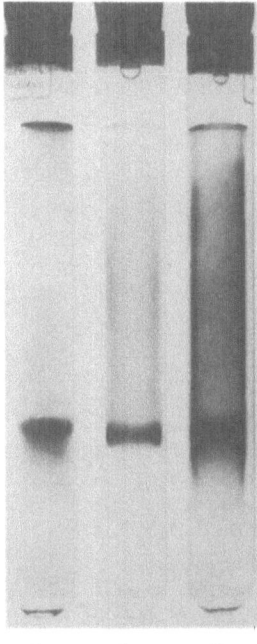

Fig. 4.   Polyacrylamide gel electrophoresis of a mixture of chicken heart
          fumarase and mitochondrial malate dehydrogenase.

65% (saturation) of ammonium sulphate. The precipitate was resuspended
in a small volume of 5 mM potassium phosphate + 1 mM pyromellitic acid,
pH 7.5, dialyzed against the same buffer and afterwards against 10 mM
Tris-acetate, pH 7.5. This solution was applied to a Sepharose-6B column
(1.5 x 90 cm) equilibrated with the same buffer. Results of this gel
filtration are shown in Figure 5. The inset shows the chromatogram in
arbitrary activity units. Obviously, the fumarase activity peak is
highly asymmetric. Moreover, various high molecular weight fractions
were detected (fraction numbers 99 = 355 kd, 101 = 300 kd, 104 = 245 kd,
respectively), whereas the fumarase-activity peak was expected in
fraction number 106 (194 kd, comparable with mammalian fumarases
[Hayman,1961]). When the first part of the chromatogram is enlarged
(Figure 5), part of the malate dehydrogenase and citrate synthase
activity is seen to be associated with these high molecular weight
fumarase fractions. It is concluded that, under these experimental
conditions, associations between fumarase, malate dehydrogenase, and
citrate synthase can be observed in yeast extracts. Since almost all
fumarase activity in our chromatogram is associated in higher molecular
weight species, we suggest that this enzyme plays an important role in
the enzyme-enzyme interactions. It should be emphasized, as mentioned in
Figure 5, that part of the enzyme activity is lost during the course of
this experiment, most probably due to endogenous proteolysis (Pringle,
1975).

FINAL REMARKS

     The above experimental results, and data from the literature

205

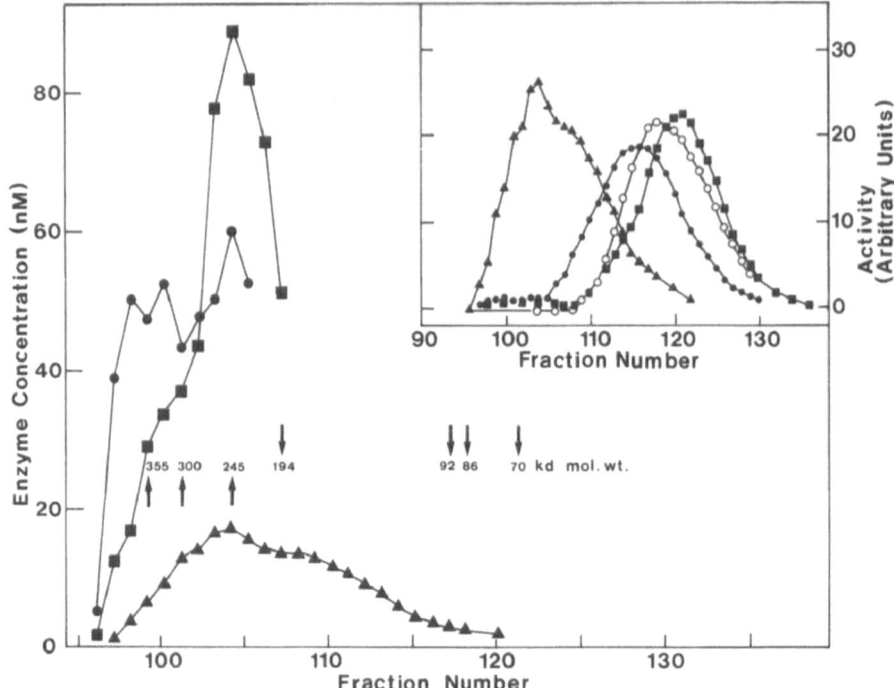

Fig. 5. Gel filtration of a <u>Candida</u> <u>utilis</u> extract on Sepharose-6B in
10 mM Tris-acetate buffer, pH 7.5. Fractions of 1.1 ml were
collected. The different peaks correspond, respectively, to
the following: fumarase (▲), malate dehydrogenase (■),
citrate synthase (●), aspartate aminotransferase (O).
Nanomolar quantities are calculated as in Table 2. Of the
enzymatic activities applied to the column, variable amounts
were recovered in the effluent: fumarase 9%, citrate synthase
40%, aspartate aminotransferase 56%, and malate dehydrogenase
54%.

summarized in Table 4, point to the existence of specific interactions
between various enzymes of the citric acid cycle and its surrounding
pathways. It is suggested that the stoichiometry of the different
enzymes in these complexes might be regulated by local concentrations of
different metabolites, allowing a channeling of intermediates towards
distinct pathways according to the needs of the cell; further experiments
along this line must be performed in order to test the hypothesis.
Nonetheless, present knowledge allows us to speculate that the
regulation of metabolism, and of the citric acid cycle in particular, is
founded on a meticulous interplay between such reversible complex-
formation on the one hand, and the more classical regulation of catalytic
activity of the component enzymes on the other hand.

The fact that most of the interactions are not easily observed in
aqueous solution (Table 4) may be due to important environmental
differences compared to the <u>in vivo</u> situation. Nevertheless, it is also
possible that the apparent lability of the associations might have a
physiological significance; it might allow a cell to interrupt very
precisely any biochemical pathway according to its needs at any time.

Table 4. Interactions of Citric Acid Cycle Enzymes with Enzymes from Surrounding Metabolic Pathways

Observed interaction[a]	Technique used		
	Immobilization studies	PEG coprecipitation	Co-elution upon gel filtration or co-sedimentation
CS - mMDH	A	B,C,D[b]	
CS - FUM	A		interactions could
FUM - mMDH	A		not be demonstrated
mMDH - mAAT	A		using these
CS - mAAT		D[c]	techniques
CS - THIO	E		
PDH - CS			F
KGD - STK		G	G
mMDH - GDH		H	H[b]

Capital letters refer to references: A:Beeckmans and Kanarek, 1981; B: Halper and Srere, 1977; C: Srere et al., 1978; D: Fahien and Kmiotek, 1983; E: Sumegi et al., 1985; F: Sumegi et al., 1980; G: Porpaczy et al., 1983; H: Fahien et al., 1979.

[a] Mammalian enzymes were used.
[b] The interaction was enhanced in the presence of palmitoyl-CoA.
[c] The interaction was only observed in the presence of palmitoyl-CoA. It was shown that the effect of palmitoyl-CoA is specific and not due to detergent-like action, since sodium dodecyl sulphate did not give the same effect.

ABBREVIATIONS

FUM: fumarase; MDH: malate dehydrogenase; CS: citrate synthase; AAT: aspartate aminotransferase; PDH: pyruvate dehydrogenase; KGD: α-ketoglutarate dehydrogenase; STK: succinate thiokinase; GDH: glutamate dehydrogenase; THIO: thiolase; m: mitochondrial; c: cytosolic.

ACKNOWLEDGEMENTS

     This work was supported by grants of the Belgian N.F.W.O. (Nationaal Fons voor Wetenschappelijk Onderzoek); S.B. is "Bevoegdverklaard Navorser" of this Foundation.

REFERENCES

Addink, A.D.F., Boer, P., Wakabayashi, T., and Green, D., 1972, Eur. J. Biochem., 29:47.
Beeckmans, S., 1984, Int. J. Biochem., 16:341.
Beeckmans, S. and Kanarek, L., 1981, Eur. J. Biochem., 117:527.
Beeckmans, S. and Kanarek, L., 1982, Int. J. Biochem., 14:453.
Brewer, G.J. and Sing, C.F., 1970, "An Introduction to Isozyme Techniques," Academic Press, New York.

Comte, J. and Gautheron, D., 1978, *Biochimie*, 60:1299.

Davis, B.J., 1964, *Ann. N.Y. Acad. Sci.*, 121:404.

D'Souza, S.F. and Srere, P.A., 1983, *J. Biol. Chem.*, 258:4706.

Elduque, A., Casado, F., Cortes, A., and Bozal, J., 1982, *Int. J. Biochem.*, 14:221.

Ernster, L. and Kuylenstierna, B., 1970, in: "Membranes of Mitochondria and Chloroplasts", E. Racker, ed., Van Nostrand Reinhold, New York.

Fahien, L.A. and Kmiotek, E., 1983, *Arch. Biochem. Biophys.*, 220:386.

Fahien, L.A., Kmiotek, E., and Smith, L., 1979, *Arch. Biochem. Biophys.*, 192:33.

Hackenbrock, C.R., 1966, *J. Cell Biol.*, 30:269.

Hackenbrock, C.R., 1968, *Proc. Nat. Acad. Sci. U.S.A.*, 61:598.

Halper, L.A. and Srere, P.A., 1977, *Arch. Biochem. Biophys.*, 184:529.

Hayman, S., 1961, Ph.D. Thesis, University of Wisconsin.

Matlib, M.A. and O'Brien, P.J., 1975, *Arch. Biochem. Biophys.*, 167:193.

Porpaczy, Z., Sumegi, B., and Alkonyi, I., 1983, *Biochim. Biophys. Acta*, 749:172.

Pringle, J.R., 1975, *Meth. Cell Biol.*, 12:149.

Srere, P.A., 1982, *Trends Biochem. Sci.*, 7:375.

Srere, P.A., Halper, L.A., and Finkelstein, M.B., 1978, in: "Microenvironments and Metabolic Compartmentation", P.A. Srere and R. Estabrook, eds., Academic Press, New York.

Sumegi, B., Gilbert, H.F., and Srere, P.A., 1985, *J. Biol. Chem.*, 260:188.

Sumegi, B., Guyocsi, L., Alkonyi, I., 1980, *Biochim. Biophys. Acta*, 616:158.

Williamson, J.R. and Cooper, R.H., 1980, *FEBS Lett.*, 117(Suppl.):K73.

EPR MEASUREMENTS OF CONFORMATIONAL CHANGES ON THE E1 COMPONENT

OF THE PYRUVATE DEHYDROGENASE COMPLEX FROM ESCHERICHIA COLI

Klaus Graupe[1] and Hans Bisswanger

Physiologisch-chemisches Institut
Universität Tübingen
Hoppe-Seyler-Strasse 1
D-7400 Tübingen, FRG

INTRODUCTION

One of the most important enzymes of cell metabolism is the PDC[2], which acts as a link between carbohydrate catabolism and, via the tricarboxylic acid cycle, the energy production. Due to this central role, the enzyme is subject to a great number of regulatory influences. Pyruvate and its precursors activate the enzyme complex, while acetyl CoA and metabolites derived from this product, such as tricarboxylic acid cycle intermediates and the high-energy compounds GTP and ATP, are inhibitors (Schwartz et al., 1968; Shen and Atkinson, 1970; Bisswanger, 1972; Bisswanger and Henning, 1971).

How can such an intricate regulatory pattern be managed? The PDC is one of the most highly organized enzyme systems of the cell. Three different enzyme components, each contributing 24 identical polypeptide chains, build up the enzyme aggregate; and six cofactors (TPP, Mg^{2+}, lipoic acid, CoA, FAD, NAD) are necessary to carry out the catalytic reaction sequence (Reed, 1974). Is this complicated structure Nature's answer for accomplishing the regulatory requirements, or is it rather a problem to control such a clumsy crowd of polypeptide chains?

In previous experiments it could be demonstrated that the regulation of the activity of the PDC is essentially limited to the E1 component (Bisswanger, 1974; Bisswanger and Henning, 1971; Saumweber et al., 1981); and a model was postulated, which describes the regulation of enzyme activity on the basis of a slow transition between an inactive and an

[1]Present address: Schering AG, Müllerstr. 170-178, D-1000 Berlin.

[2]Abbreviations: PDC, pyruvate dehydrogenase complex with the three enzyme components: E1, pyruvate dehydrogenase (EC 1.2.4.1); E2, dihydrolipoamide acetyltransferase (EC 2.3.1.12); E3, dihydrolipoamide dehydrogenase (EC 1.6.4.3); TPP, thiamine diphosphate; PCMB-SL, N-4'-(2',2',6',6'-tetramethylpiperidino-1'-oxyl)-p-chloromercuribenzoic acid; MI-SL, N-4'-(2',2',6',6'-tetramethylpiperidino-1'-oxyl)-maleimide.

active state of the enzyme. With this model it was possible to predict all the regulatory effects actually observed with the enzyme complex (Bisswanger, 1984; Horn and Bisswanger, 1983)

A crucial prediction for the model is a slow transition of the enzyme from an inactive to an active state, induced by binding of substrate and cofactors. To establish such conformational changes we modified the enzyme with spin-labels. There is an essential thiol group in the catalytic site of the E1 component, which reacts with p-hydroxymercuribenzoate under concomitant loss of enzymatic activity (Schwartz and Reed, 1970). We used PCMB-SL, a spin-labelled analog of this compound, as a probe for the catalytic site. The dependence of the enzymatic activity on the PCMB-SL concentration is shown in Fig. 1. In the lower concentration range the loss of enzymatic activity proceeds in a linear manner with increasing amounts of PCMB-SL, and extrapolates to one molecule of the spin-label bound per polypeptide chain. Thus, in this range the spin-label reacts with a single thiol group which is essential for catalytic activity. At higher concentrations of PCMB-SL a second, less essential, thiol group becomes accessible.

Fig. 2a/A shows the EPR spectrum of the E1 component modified with PCMB-SL. A broadening of the peaks and a decrease in their relative intensities (especially of the high-field peak) with respect to the unbound spin-label is an indication of a moderate immobilization of the spin-labelled compound due to covalent binding to the enzyme. Addition of TPP, Mg^{2+}, or pyruvate alone to the modified enzyme causes no significant change of the EPR spectrum. However, after simultaneous addition of TPP and Mg^{2+}, a series of additional peaks appear (Fig. 2a/B). This change is not due to direct interactions between the spin-label and the cofactors, since nearly the same spectral alterations are observed when AMP is added to the modified enzyme (Fig. 2b). AMP is an allosteric activator, which binds to a regulatory site distant from the catalytic center (Schrenk and Bisswanger, 1984).

A possible explanation for this spectral change may be the assumption of spin-spin coupling between two adjacent spin-labels. To verify this,

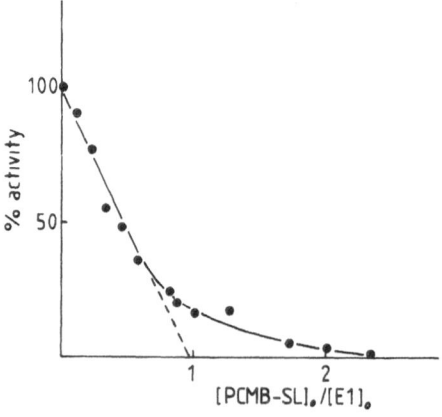

Fig. 1.   Inhibition of the enzymatic activity of the E1 component by PCMB-SL. The spin-label was allowed to react with the E1 component (1.1×10^{-5} M of E1 monomers) at $25°C$ for 10 min. The activity of the enzyme was determined using ferricyanide as electron acceptor (Das et al., 1961).

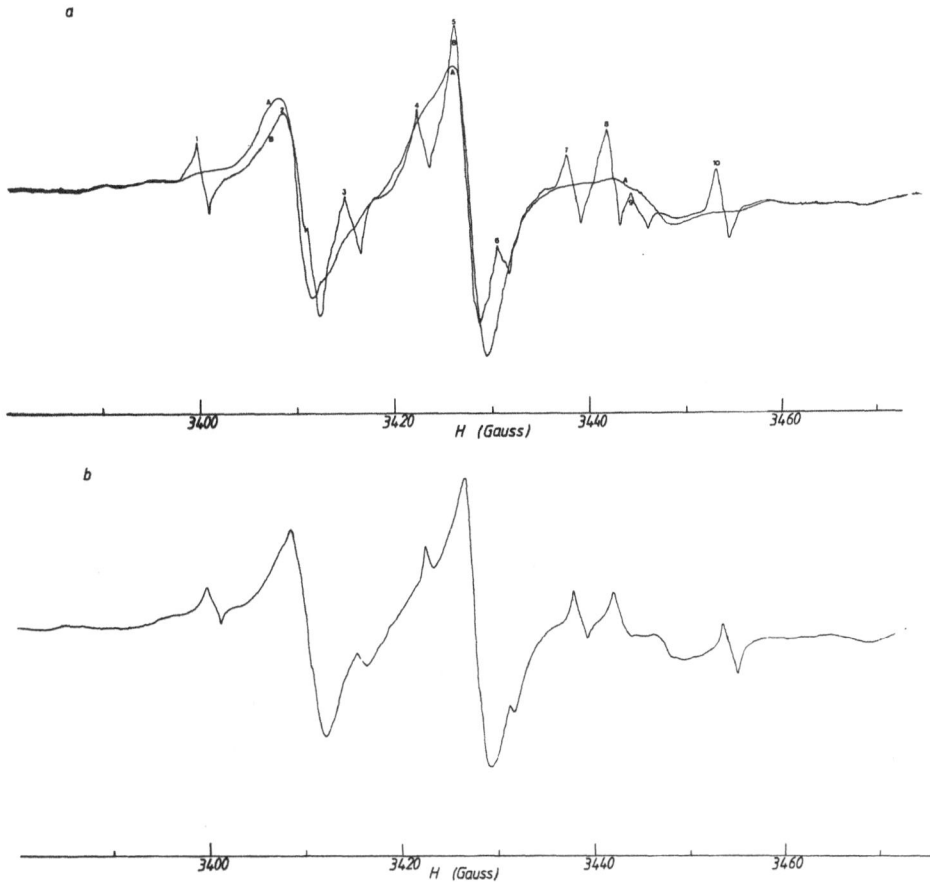

Fig. 2.   EPR spectra of the E1 component labelled with PCMB-SL.
Derivatization of the enzyme with the spin-label was carried
out using a 1:1 molecular ratio.   a) curve A, without
additions; curve B, 12 mM TPP, 160 mM Mg^{2+}; b) 62 mM AMP.
Measurements were done at 25°C in 20 mM TES buffer pH 7.0 with
a 1-G modulation amplitude, a 0.5 sec time constant and 18 mW
microwave power with a Brucker X-band spectrometer B-ER 420.

labelling of the enzyme was done in the presence of TPP and Mg^{2+}.
Schwartz and Reed (1970) reported, that under these conditions p-
hydroxymercuribenzoate reacts preferentially with a peripheral thiol
group of the enzyme without impairing its catalytic activity. Just the
same situation was observed when PCMB-SL was allowed to react with the E1
component in the presence of TPP and Mg^{2+}. A moderate immobilized EPR
spectrum was obtained (Fig. 3a), as was already observed for the E1-PCMB-
SL complex, which was synthesized in the absence of the cofactors (Fig.
2a/A). However, subsequent addition of TPP and Mg^{2+} to the peripheral
spin-labelled enzyme causes no change in the EPR spectrum (Fig. 3b).

Maleimide reacts with the peripheral thiol group without attacking
the central thiol and the enzymatic activity (Papadakis and Hammes,
1977). Spin-labelling of an E1 component modified with N-ethylmaleimide
leads to a moderate immobilized EPR spectrum. Also, in this case
subsequent addition of TPP and Mg^{2+} has no effect upon the shape of the

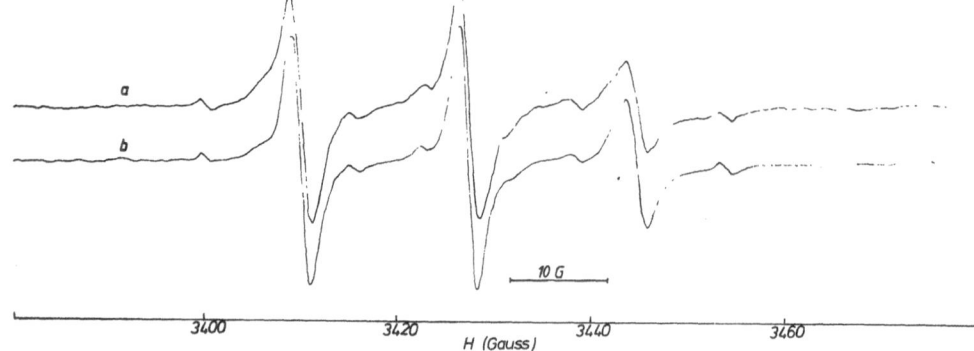

Fig. 3.   EPR spectra of the E1 component labelled with PCMB-SL in the
           presence of 12 mM TPP and 160 mM Mg^{2+}. The labelling was
           carried out with a 1:0.5 molar ratio of enzyme monomer to PCMB-
           SL. After labelling for 10 min at $25^{o}C$ in 20 mM TES buffer pH
           7.0, the cofactors as well as unreacted spin-label were removed
           from the modified enzyme by gel filtration with Sephadex G-10.
           a) without additions; b) with 12 mM TPP, 160 mM Mg^{2+}.

spectrum. Thus, spectral alterations were only observed, when both thiol
groups are accessible for the spin-label. The fact, that the same
spectral changes were brought about by the cofactors TPP and Mg^{2+}, as
well as by the activator AMP, though both types of ligands bind to
different sites on the enzyme, suggests an induction of the same
conformational change leading to the active enzyme state. In this
conformation the two spin-labels are brought into a position which allows
spin-spin coupling.

ACKNOWLEDGEMENTS

     This investigation was performed in collaboration with Dr. W.
Trommer, Kaiserslautern. It was supported by a grant from the Deutsche
Forschungsgemeinschaft, Bonn-Bad Godesberg.

REFERENCES

Bisswanger, H., 1972, Ph.D. Thesis, University of Tubingen.
Bisswanger, H., 1974, Eur. J. Biochem., 48:377.
Bisswanger, H., 1984, J. Biol. Chem., 259:2457.
Bisswanger, H. and Henning, U., 1971, Eur. J. Biochem., 24:376.
Das, M.L., Koike, M. and Reed, L.J., 1961, Proc. Natl. Acad. Sci. USA,
     47:753.
Horn, F. and Bisswanger, H., 1983, J. Biol. Chem., 258:6912.
Papadakis, N. and Hammes, G.G., 1977, Biochemistry, 16:1890.
Reed, L.J., 1974, Acc. Chem. Res., 7:40.
Saumweber, H., Binder, R. and Bisswanger, H., 1981, Eur. J. Biochem.,
     114:407.
Schrenk, D.F. and Bisswanger, H., 1984, Eur. J. Biochem., 143:561.
Schwartz, E.R. and Reed, L.J., 1970, J. Biol. Chem., 245:183.
Schwartz, E.R., Old, L.O. and Reed, L.J., 1968, Biochem. Biophys. Res.
     Commun., 31:495.
Shen, L.C. and Atkinson, D.E., 1970, J. Biol. Chem., 245:5974.

ORGANIZATION OF ENERGY METABOLISM:

IN SITU APPROACHES

ORGANIZATION AND CONTROL OF ENERGY METABOLISM

IN ANAEROBIC MICROORGANISMS

Douglas B. Kell and Robert P. Walter

Department of Botany and Microbiology
University College of Wales
Aberystwyth, Dyfed SY23 3DA, U.K.

INTRODUCTION AND SCOPE

A recurrent question, which dates from the very origins of modern biochemistry itself (see Schlenk, 1985), and which constitutes a major theme of the present conference, concerns the degree of relatedness between the organization and activities of the enzymes of cellular energy metabolism in vivo and their behavior in vitro. At one level, two extreme types of viewpoint, which we may refer to as "holistic" and "reductionist", may be discerned.

The reductionist school would hold that the ability to reconstruct a biochemical pathway, using isolated enzymes in vitro, at a rate, and with a sensitivity to effectors, similar to that obtained in vivo, provides the evidence necessary and sufficient to define the system operating in vivo. Any inability to achieve such a reconstruction may be ascribed to technical difficulties (e.g. denaturation) or to the loss of unidentified enzymes or cofactors required for the pathway.

The holistic view, which is enjoying a renaissance of philosophical interest in the field of quantum physics (e.g. Bohm, 1981; Primas, 1981; Wolf, 1981; Wheeler and Zurek, 1983; Garden, 1984), would hold firstly that any attempt even to measure the properties of a system can have the effect of seriously modifying the "observed" system, and secondly (and especially) that the attempt to extrapolate measurements in vitro to describe a biochemical system in vivo must of necessity fail, for the "disruption" to the real cellular organization occasioned upon enzyme isolation is very severe (Clegg, 1984). In particular, it would be argued, the properties of a system are specified not only by the types and numbers of molecules present but by the way in which the individual molecules of a given type are organized functionally and physically, so that it is the physical organization, as well as the chemical composition, which is different in vitro from that in vivo, and which must be defined if one wishes to understand "the organization of cell metabolism". This is not to mean, of course, that the reductionist view does not permit a rather sophisticated spatiotemporal organization and integration of metabolic pathways; indeed, mathematical networks of fairly simple reaction-diffusion equations incorporating feedback loops

can exhibit highly non-linear behavior (e.g. Stucki, 1978; Hess et al.,
1978; Hess, 1983; Hess et al., 1984), without any need, beyond the simple
diffusion of enzyme and substrate molecules, to invoke collective or
coherent properties (Froehlich and Kremer, 1983) to describe the
organization of the cellular matrix and its associated metabolism.
However, this is not what we have in mind here.

Four our present purposes, we would argue (Welch and Kell, 1986), as
have others (Westerhoff and Chen, 1985), that the _fundamental_
distinctions between the two types of viewpoint lie in the recognition
that, as far-from-equilibrium systems, the types of metabolic pathway
with which we are here concerned do not conform to the ergodic principle,
and that if their organization is of a _microscopic_ nature any _macroscopic_
treatment must be regarded as inappropriate. What do we mean by
"microscopic" in this context?

Consider a generalized metabolic pathway:

$$A \xrightarrow{E_1} B \xrightarrow{E_2} C \xrightarrow{E_3} D \xrightarrow{E_4} E \qquad\qquad \ldots\ldots \text{Scheme 1}$$

Here A, ..., E represent substrate molecules and E_1, ..., E_4
enzymes catalyzing specific reactions. If one makes measurements, as is
usual, of the rate of production under steady-state conditions of E from
A, one will, of course, be measuring the ensemble behavior of numerous
enzyme and substrate molecules of the "same" type. Two extreme
organizational modes are then possible; we will refer to them as bulk (or
delocalized) and microscopic (or localized). In the bulk case, any
molecule of (say) B produced by an individual E_1 molecule may diffuse at
a rate sufficiently rapid to ensure that it is freely available to any
individual molecule of E_2 in the ensemble of interest, such that, apart
from the usual statistical fluctuations (which should be negligible),
molecules of B (i) possess a concentration which has a sharp value, (ii)
exhibit pool behavior and (iii) may adequately be treated macroscopically. In the localized case, no such pool behavior exists; a molecule of B
produced by a given E_1 molecule may act as substrate only for a _specific_
E_2 molecule. The concentration of B seen by an enzyme is no longer equal
to the number of molecules divided by the total volume of the appropriate
compartments within the reaction vessel, and neither the standard
chemical potential nor the chemical activity coefficient of B is
remotedly independent of the state, nature and free energy of the _protein_
matrix (Welch, 1977; Somogyi et al., 1984; Welch and Kell, 1986; Berry et
al., 1985). In other words, the metabolic pathways are in this case
organized, at least functionally, as a "supercomplex", and the system may
be said to exhibit "channelling" behavior:

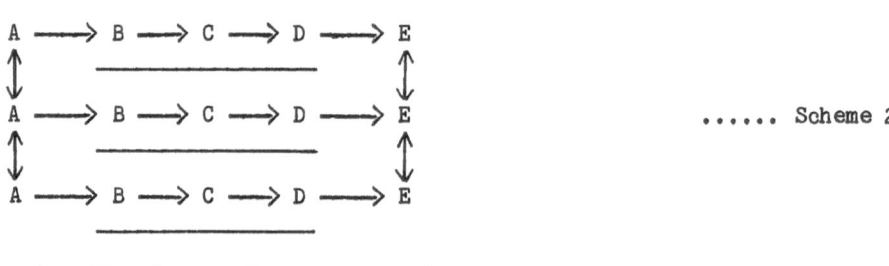

$$\ldots\ldots \text{Scheme 2}$$

pool   direct transfer or   pool
        "channelling"

In the present article, our considerations of this question will be directed mainly to three areas, with reference to some recent experiments which have been carried out in this laboratory: (i) are the free-energy-transducing enzymes of electron transport-linked phosphorylation organized in a "localized" fashion?, (ii) can the so-called double-inhibitor titration method effect a rigorous distinction between localized and pool behavior? and (iii) does localization also occur in "soluble" pathways of energy metabolism such as glycolysis? We will also mention some applied or biotechnological aspects of this question, and will outline certain fundamental difficulties which arise in the description of systems exhibiting localized behavior. However, for reasons of space, we shall limit our primary scope, in so far as it is possible, to systems derived from anaerobic microorganisms.

ORGANIZATION OF ELECTRON TRANSPORT-LINKED PHOSPHORYLATION

As may be gleaned from any biochemistry textbook or relevant monograph (e.g. Nicholls, 1982), we may write the macroscopic process of electron transport-linked phosphorylation (ETP) using the type of shorthand given in Scheme 1, as follows:

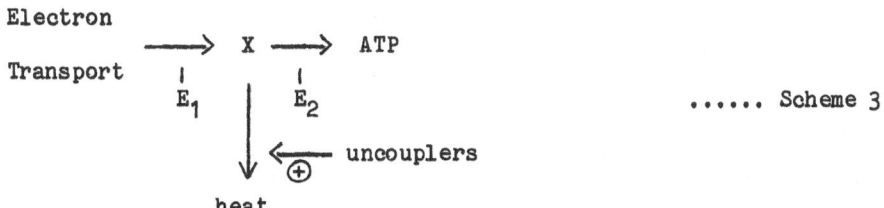

...... Scheme 3

Here E_1 and E_2 represent membranous redox enzymes and ATP synthase enzymes respectively; and drugs called uncouplers can act to inhibit the overall process of ETP, not by inhibiting E_1 or E_2 but by stimulating the decay of "X", a "high-energy intermediate", into heat. Whilst there are many lines of circumstantial evidence which lead one to invoke the presence of additional enzymes (which we have called "protoneural proteins") necessary for ETP (Kell and Morris, 1981; Kell et al., 1981; Kell and Westerhoff, 1985), the above scheme serves as a minimal and widely accepted model. Arguably, the central problem of membrane bio-energetics concerns the nature of this energized intermediate X and the question of whether or not it is constituted, as in the chemiosmotic coupling hypothesis (Mitchell, 1966, 1968; Nicholls, 1982), by the macroscopic proton electrochemical potential difference across a topologically closed vesicle, in the membrane of which are embedded ensembles of E_1 and E_2 molecules (e.g. Kell, 1979; Ferguson and Sorgato, 1982; Kell and Hitchens, 1983; Westerhoff et al., 1984b; Ferguson, 1985; Kell and Westerhoff, 1985; Kell, 1986).

This proton electrochemical potential difference, or protonmotive force (pmf, Δp), is, in the chemiosmotic paradigm, taken to be in equilibrium with that between the bulk aqueous phases that the coupling membrane serves to separate; and thus, within the framework of Scheme 3 above, if X is the protonmotive force then, by definition, such an energy coupling scheme should exhibit classically bulk, delocalized (or pool) behavior. We should like to obtain a rigorous set of criteria by which one might discern whether or not it does so, so as to provide a test of the chemiosmotic coupling hypothesis. However, it is first necessary to mention a somewhat circular argument which has a tendency to creep into discussions of this rather controversial problem.

This argument stems from the well-known finding (e.g. Thayer and Hinkle, 1975; Smith et al., 1976; Hangarter and Good, 1982; Maloney, 1982; Schmidt and Graeber, 1985) that an artificially applied protonmotive force can drive phosphorylation at rates similar to those implicated in native ETP; and, naturally enough, this has widely been taken to constitute powerful support for the veracity of Scheme 3, with the place of X taken by Δp. However, what the data in this type of experiment (op. cit.) actually show in every case is that there is no phosphorylation below a (non-thermodynamically defined) threshold of the applied pmf equal to approximately 150 mV. Thus, in assessing the veracity of a macroscopic energy coupling scheme such as that of Scheme 3, the proper question to ask (Kell, 1986) is whether or not the pmf generated by electron transport exceeds this threshold.

This is a particularly thorny problem, reviewed in extenso elsewhere (e.g. Ferguson and Sorgato, 1982; Ferguson, 1985; Kell and Westerhoff, 1985; Westerhoff et al., 1984; Kell, 1986), but our view is that the most self-consistent explanation of the available data is that the actual pmf generated by electron transport does not in fact exceed this threshold value (Guffanti et al., 1981; Kell and Hitchens, 1982; Hitchens and Kell, 1984; Kell, 1986). The circularity arises from the fact that many signals purporting to measure the pmf can be obtained in a form that apparently corresponds to values of the pmf above the threshold. However, if one can reliably obtain other data which show that energy coupling is localized (in the sense of Scheme 2) then, evidently, Scheme 3 is inappropriate (with Δp as X) and values of the pmf purporting to lie above the threshold must, logically, be in error. The most direct measurements of the membrane potential component of the pmf in mitochondria (Tedeschi, 1980), for instance, do indeed suggest that it is energetically insignificant.

In any event, and whilst much of the present concern (Kell, 1979; Ferguson and Sorgato, 1982; Westerhoff et al., 1984; Ferguson, 1985) about the veracity of the chemiosmotic coupling hypothesis has arisen from the lack of correlation between the reactions of ETP and the apparent pmf, what one is trying to convey is that the avoidance of the circularity referred to above requires that one's arguments and experiments designed to assess the veracity of a macroscopic coupling scheme, such as Scheme 3, should not require the measurement of the concentration of X. One approach, which fulfills this criterion, which can serve in principle to distinguish microscopic from macroscopic coupling schemes (and may therefore be of general interest), and with which we have recently been concerned, is known as the double-inhibitor titration method. Two general types are of interest with reference to Scheme 3.

DOUBLE-INHIBITOR TITRATIONS OF ELECTRON TRANSPORT-LINKED PHOSPHORYLATION

Many species of Rhodospirillaceae may be grown aerobically in the dark or phototrophically under anaerobic (or semi-anaerobic) conditions. Under the latter regime, the bacterial cytoplasmic membrane differentiates to form intracytoplasmic membrane (ICM) invaginations; upon cell disruption these vesiculate to form so-called chromatophores, which possess a cyclic electron transport system and provide a well-coupled and convenient system for the study of ETP. Parenthetically, it may be mentioned that this membrane differentiation is hard to reconcile with the widespread view that membrane protein complexes normally possess lateral diffusion coefficients exceeding 10^{-10} cm^2/s (Kell, 1984), but we will not pursue this issue here (see Harris and Kell, 1985; Kell and Harris, 1985).

In recent work (Hitchens and Kell, 1982a,b, 1983a,b) we have carried out double-inhibitor titrations of photophosphorylation by bacterial chromatophores. With reference to Scheme 3, it is possible to inhibit ATP production using at least three types of inhibitor: electron transport inhibitors (I_1) which inhibit E_1, ATP synthase inhibitors (I_2) which inhibit E_2, or uncouplers. We consider first I_1/I_2-type titrations (Fig. 1). In this case, we first titrate the rate of photophosphoryla-tion, J_p (at saturating light intensity) using a tight-binding and specific I_1-type inhibitor (Fig. 1a). Then, using a fresh batch of chromatophores, we first inhibit photophosphorylation, by say 50%, using a tight-binding and specific I_2-type inhibitor (Fig. 1b, c). The simple view is that, in a delocalized coupling scheme, there is now spare capacity in E_1, so that the residual rate of phosphorylation will be proportionally **less** sensitive to I_1 than in the control chromatophores (Fig. 1b). Conversely, in a fully localized system, the relative titre will be unchanged (Fig. 1c), for one molecule of I_1 will still block one molecule of E_1, whether it is capable of driving phosphorylation in "its" E_2 or not. In practice, the same (or no increase in) titre is obtained in the partially inhibited case (Hitchens and Kell, 1982 a,b; Kell and Hitchens, 1983, and see Venturoli and Melandri, 1982), and similar results are obtained with energy-linked reactions in submitochondrial particles (Baum et al., 1971; Baum, 1978; Westerhoff et al., 1983a,b; Ferguson, 1985) and chloroplast thylakoids (Davenport, 1985), a finding strongly suggestive of a localized coupling system. However, there is a clever and interesting counterargument, which (if true) might still allow this type of finding to be accommodated in a delocalized scheme, and we must needs rehearse it here now.

This argument was first given by Parsonage and Ferguson (1982) (and see Parsonage, 1984; Ferguson, 1985), and is reiterated by Davenport (1985); we may discuss it with reference to Fig. 2, for the case of

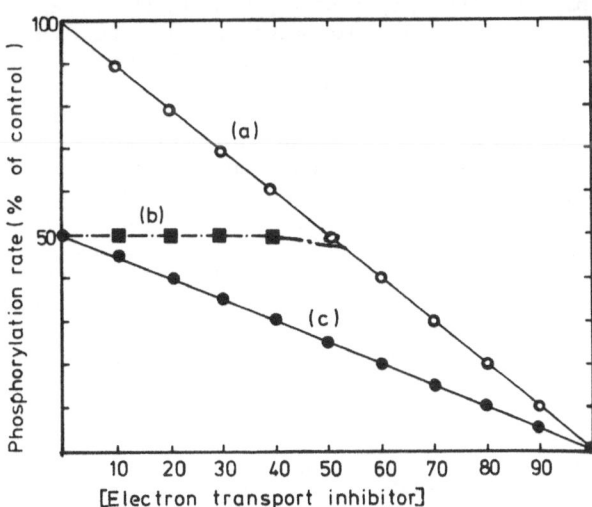

Fig. 1. The principle of an I_1/I_2-type double inhibitor titration of electron transport phosphorylation. For explanation, see text.

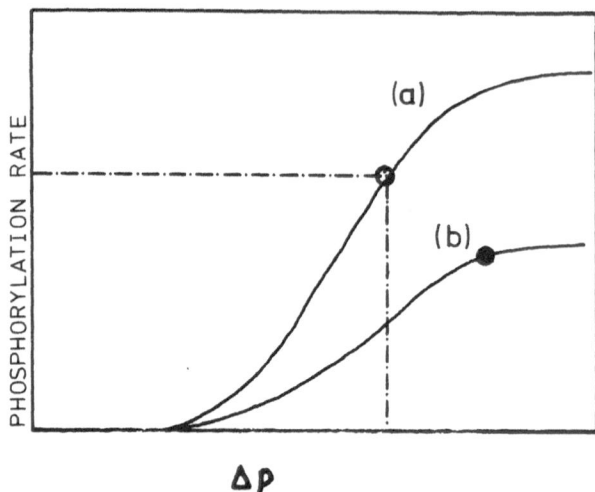

Fig. 2. Possible relationships between the protonmotive force and the
rate of phosphorylation in a double inhibitor titration of
electron transport-linked phosphorylation according to a
delocalized, chemiosmotic coupling model.

chromatophores that have had 0% or 50% of their ATP synthase inhibited.
This illustrates a possible relationship between the pmf and the rate of
phosphorylation (i.e. J_p = [ATP synthases] x f(Δp)). The argument is
that if, in the absence of I_1, the pmf is not saturating for the ATP
synthases, then each titration with I_1 will, simply by decreasing Δp,
take one down the two curves in a similar fashion, so that there will be
no difference in the titre of I_1 whether I_2 is present or not, despite
the real existence of delocalized behavior. Thus the argument requires
(1) that the pmf is <u>not</u> saturating for the ATP synthases and (2) that
there is a decrease in Δp upon inhibition with an I_1-type inhibitor.
Thesis (1) is necessary because if Δp <u>is</u> saturating, and is increased
slightly upon partial inhibition of J_p with I_2 (Parsonage, 1984), then
the initial stages of the titration with I_1 will indeed be less inhibitory when a partial titre of I_2 is present, as expected for a simple
delocalized system (Fig. 2, open circle and closed circle). However, if,
as is found (Venturoli and Melandri, 1982; Berden et al., 1984; Ferguson,
1985), the decrease in J_p occasioned by the partially inhibitory titre of
I_2 is mirrored by a comparable decrease in the number of active (or
potentially active) ATP synthases, then one has to assume, in a delocalized framework, that Δp in the absence of I_1 or I_2 <u>is</u> saturating for
the ATP synthases. In other words (Kell and Hitchens, 1983; Kell,
1986), as is found (Hitchens and Kell, 1982b), these titrations are
symmetrical, in that partial inhibition of J_p with an appropriate titre
of I_1 does not increase the titre of I_2 (Hitchens and Kell, 1982b). Thus
part (1) of the counterargument fails.

Since part (2) of the counterargument should require measurements of
the pmf, it is subject to the circularity described above. However, it
is worth mentioning that in the initial part of a titration curve of J_p
with an I_1-type inhibitor (Kell et al., 1978; Sorgato et al., 1980), or
upon decreasing the flash frequency in photosynthetic systems (Venturoli
and Melandri, 1982), the usual methods claimed to measure the pmf do not
in fact suggest that it decreases from its value in the absence of

inhibitor. More importantly, in well-coupled systems as isolated (e.g. Ferguson et al., 1976), including bacterial chromatophores (Jackson et al., 1981; Venturoli and Melandri, 1982), the P/2e ratio is unchanged upon decreasing the rate of electron transport. Not only does this finding eliminate arguments based upon "energy leaks" (Hitchens and Kell, 1982a), but, we should like to stress, it also eliminates the argument that one can explain these titrations in terms of a decrease in Δp. This is because, with a chemiosmotic scheme, the P/2e ratio cannot be a _less_ sensitive function of Δp than is J_p, for if one attempts to uncouple (or reduce Δp), the respiratory chain should attempt to act faster to _try_ and maintain Δp (and thus J_p) at its initial value. Thus since the P/2e ratio is not decreased by an I_1-type inhibitor, then neither should the pmf be; and this also eliminates part (2) of the counterargument.

It is worth mentioning that one might _artefactually_ obtain a _delocalized_ result of the type shown in Fig. 1 for trivial reasons, such as if the system is heterogeneous, or contains inhibitor-binding sites which are not involved in coupled electron-transport phosphorylation, etc. However, we confine our considerations here to the well-defined systems in which a localized coupling mechanism does seem to be the only _defensible_ explanation of the data obtained.

It is appropriate to mention at this point the metabolic control theory developed and described by Kacser and Burns (1973) and by Heinrich and Rapoport (1974) (and see Kacser, this volume), and which has been reviewed by Groen et al. (1982a) and by Westerhoff et al. (1984a). The theorems contained therein provide a rigorous mathematical treatment with which one may describe the extent to which each enzyme in a system such as that of Schemes 1 to 3 controls the pathway flux, and of the importance of different effectors in this control. In particular, the "flux-control coefficient" of an enzyme is given by the fractional change in the pathway flux divided by the fractional change in the amount of that enzyme (as these changes tend to an infinitesimal amount), and these coefficients may obviously be determined experimentally by the use of specific inhibitors (e.g. Groen et al., 1982a, b). Further, according to the Summation Theorem, the sum of the flux-control coefficients of each of the enzymes in the pathway will equal identically 1, bearing in mind that "leaks", or enzymes removing substrates from the pathway, will tend to have _negative_ flux control coefficients. However, what we wish to concentrate upon is the implicit _assumption_ of pool behavior for each substrate that is built into these metabolic control theories.

Now, it should be obvious that for a strictly localized system, each of the main pathway enzymes will have a flux-control coefficient of 1, since inhibiting x% of the enzymes of a given type will reduce the pathway flux by x%, _whichever enzyme is chosen_, so that the sum of the flux-control coefficients will exceed 1. Thus the violation of the summation theorem also constitutes a powerful criterion for a localized system (Kell and Hitchens, 1983; Kell and Westerhoff, 1985). A full description of the application of this approach to the analysis, in more mathematical terms, of the I_1/I_2 type of double-inhibitor titration described above, is given by Westerhoff and Kell (1985), and is not repeated here. One point is, however, worth making. The flux-control coefficients may be obtained, in principle, _either by increasing or decreasing_ the amount of the enzyme of interest. However, the addition of molecules of a given exogenous enzyme will have no effect (i.e. flux-control coefficient = 0) on the pathway flux if the coupling is "perfectly" localized, whilst inhibiting the enzymes of the same type will cause a proportional inhibition of pathway flux (flux-control coefficient = 1); this difference in behavior also constitutes a useful criterion of "channelling".

The other main type of double-inhibitor titration of ETP (see Kell and Hitchens, 1983; Ferguson, 1985; Westerhoff and Kell, 1985; Kell and Westerhoff, 1985; Kell, 1986) concerns the titration of J_p with an uncoupler, in the presence and absence of a partially inhibitory titre of I_2 (Hitchens and Kell, 1982b, 1983a,b). In terms of the chemiosmotic analysis ("counterargument") given above (see Fig. 2), the ATP synthase inhibitor probably raises, and certainly cannot lower, the pmf in the absence of uncoupler, so that one should expect that the uncoupler is equally or less potent when a partially inhibitory titre of I_2 is present. In practice, and provided that the experiments are not done under non-stationary conditions, when almost any behavior is possible (Cotton and Jackson, 1983), the uncouplers act more potently when a fraction of the ATP synthases is inhibited. Similar data are obtained in thylakoid photophosphorylation (Davenport, 1985) and in ATP-driven reversed electron transport in submitochondrial particles (Westerhoff et al., 1983; Berden et al., 1984), and may be explained, within a localized framework, in terms of the (most) rate-limiting step for uncoupling being not the diffusion of uncoupler molecules to their sites of action but the uncoupling step itself (e.g. Hitchens and Kell, 1983a,b).

A point worth stressing, and apparently not appreciated by some workers (O'Shea and Thelen, 1984; Davenport, 1985), is that the starting points for each of these two types of titration (with I_1 or uncoupler) are the same, so that whatever I_2 does in one case (if one wishes to defend a delocalized model) it must do in the other: if I_1 stays equally potent when a partially inhibitory titre of I_2 is present then so should the uncoupler, if the only means by which I_1 and uncouplers affect J_p is by decreasing the pmf. That I_2 can affect the titration behavior differently constitutes perhaps the simplest argument necessary to illustrate that this type of approach shows that the energy coupling systems of ETP do not interact via a macroscopic, delocalized high-energy intermediate.

In outlining the double-inhibitor titration approach, we have concentrated on bioenergetic systems, for most of the work has been done on them. However, we believe that the approach has a general utility for discerning whether or not a metabolic pathway is organized macroscopically or microscopically. In this regard, a nice example approximating this type of approach, involving the demonstration of a "replitase supercomplex" in DNA synthesis, is given by the work of Pardee and Reddy (1983).

We now wish to turn to another approach to understanding the organization of energy metabolism, in this case of glycolysis in the clostridia.

ORGANIZATION OF GLYCOLYSIS IN CLOSTRIDIUM PASTEURIANUM

Clostridium pasteurianum is a fermentative obligate anaerobe, which derives the free energy necessary for growth by glycolysis (using the EMP pathway) to produce acetate and butyrate in approximately equimolar amounts (Thauer et al., 1977). Glucose is taken up via a phosphotransferase system (PTS) (Booth and Morris, 1982; Mitchell and Booth, 1984); and in intact cells of strain 6013-ES1, which lacks granulose phosphorylase (Mackey and Morris, 1974), continuing glycolytic acid production is strictly dependent upon an exogenous carbon source. As may be seen in Fig. 3, the addition of glucose to a washed cell suspension causes, after a lag of approximately 20-30s, a steady rate of acid production. The lag might be ascribed to the build-up of pools of glycolytic intermediates, but, for reasons which will become apparent, is more likely associated with the optimal poising of the adenine (and perhaps pyridine) nucleotide pools. Permeabilization of the cell membrane by treatment with an appropriate concentration of toluene:ethanol (1:10), a very common method of cell permeabilization (Felix, 1982), releases intracellular cofactors such as ATP (data not shown) and leads to an immediate inhibition of glycolysis (Fig. 3). (Under these conditions, without toluene and with periodic adjustment of the pH, a steady rate of acid production may be maintained for as long as 30 minutes; and the major acidic products may be shown by gas chromatography to be acetate and butyrate, with a small proportion (up to 10%) of D- and L-lactate.

The product of the PEP-dependent PTS system is glucose-6-phosphate (G6P) (which does not penetrate the membrane of intact cells of this organism (Booth and Morris, 1982)), and the question arises concerning the extent to which glycolysis from G6P is inhibited in toluenized cells, according to the following reasoning. If all glycolytic intermediates (together with cofactors such as NAD(H) and ATP) exhibit "pool" behavior, then the ability to restore flux through the glycolytic pathway using G6P, NAD and ATP alone should be severely compromised, since most enzymes, including those of this pathway (e.g. Kotze, 1968; Uyeda and Kurooka, 1970) have $S_{0.5}$ values in the mM range. Thus, in a reaction mixture containing 1.25 mg dry weight cells/ml, with a specific enclosed volume of 2.12 ml/l (Clarke et al., 1982), the dilution of cytoplasmic "pool" constituents upon toluenization will be 450-fold, so that their concentration (if their _in vivo_ concentration corresponds to the $S_{0.5}$

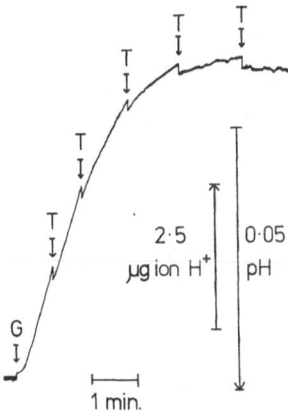

Fig. 3. Glycolytic acid production by intact cells of Clostridium pasteurianum 6013-ES1. Clostridium pasteurianum 6013-ES1 was grown in a glucose minimal medium (Clarke et al., 1982) and washed and resuspended in 25 mM triethanolamine phosphate pH 6.5/50 mM KCl/5 mM $MgSO_4$/0.05% cysteine-HCl. All procedures were carried out anaerobically, and pH changes were recorded as described (Hitchens and Kell, 1982a) at a temperature of 37°C. The reaction volume was 8 ml, and 10 mg dry weight cells were present. At the arrows indicated, 20 mM glucose (G) and 0.02 ml aliquots of toluene:ethanol (1:10) (T) were added.

values of the enzymes for which they are substrates) will be lowered by a factor of at least 100, giving an initial glycolytic rate which would (Fig. 3) be immeasurably small. Alternatively, and even if the lag observed in the initiation of steady-state glycolytic acid production by intact cells (transient time) upon addition of glucose were to be caused by the build-up of pools of glycolytic intermediates, the lag would be increased by a similar amount (Keleti, 1984), and would also escape detection.

In contrast, if glycolysis, say from G6P to pyruvate, were to be organized as a 'supercomplex' (e.g. Mowbray and Moses, 1976; Ottaway and Mowbray, 1977; Gorringe and Moses, 1980; Masters, 1981; Friedrich, 1984; Keleti, 1984), whether physically or just functionally, such that glycolytic intermediates did not exhibit "pool" behavior, then the rapid and extensive restoration of glycolytic flux from G6P might require the addition only of adenine and pyridine nucleotide cofactors. Fig. 4 illustrates an experiment designed to obtain information on this point.

Fig. 4(a) illustrates acid production in toluenized cells of Clostridium pasteurianum 6013-ES1, indicating a contribution to acid production by ATP hydrolysis (cf. Hitchens and Kell, 1982a), whilst Fig. 4(b) shows that a maximal rate of glycolytic acid production is strictly dependent upon the addition of adenine and pyridine nucleotides. There is some variability in the lag phase, and the traces have been chosen to illustrate the range that we have so far experienced (from virtually no lag to ca. 4 min); this variability is ascribed predominantly to the requirement for an optimum poise of the ATP/ADP ratio (since both nucleotides are required at reasonable concentrations for a maximum rate of glycolysis). Since exogenous CoA is not present in this assay, glycolysis to acetate and butyrate is not observed, but acid production continues until the G6P is exhausted and recommences immediately upon injection of another aliquot of 5 mM G6P (not shown).

The steady-state rate of acid production in this case (Fig. 4), when corrected for the ATP hydrolase activity, is approx. 0.08 mol/min/mg dry-weight cells, corresponding to approx. 51% of the rate of acid production of intact cells in this run. Whilst a restoration of glycolytic acid production to 100% of the rate exhibited in intact cells has not so far

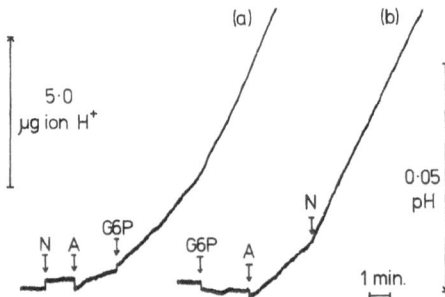

Fig. 4.   Glycolytic acid production by toluenized cells of Clostridium pasteurianum 6013-ES1. Acid production was measured with the system, and in the medium, described in the legend to Fig. 3. 20 mg cells were present, and were toluenized with 0.1 ml toluene:ethanol (1:10), and 1 mM NAD+ (N), 2 mM ATP (A) and 5 mM glucose-6-phosphate (G6P) were added at the arrows indicated.

been obtained, we would mention that it is known that organic solvents _per se_ can noncompetitively inhibit glycolysis in intact cells under conditions in which the cell membrane is not permeabilized (Nogodiwathana et al., 1977; Millat et al., 1982). Also, toluenization does release a certain amount of intracellular protein (not shown). In any event, the relatively small lags and large rates of acid production observed in toluenized cells would seem to argue in favor of an organization of glycolysis in this organism, that does not allow the intermediates to exhibit pool behavior. The dependence of acid production upon catalytic concentrations of both ATP and NAD serves as a control to show (i) that multi-step reactions are being observed and (ii) that cell permeabilization has indeed been effected by the toluene treatment.

Although toluenized cells have often been used to assay the activities of single enzymes (Felix, 1982), including those of glycolysis (Serrano et al., 1973), we are not aware of any report in which a significant segment of the glycolytic pathway of a prokaryote has been shown to function en bloc, without added intermediates, in this way, although Clegg (1984) has obtained and discussed similar data in dextran sulphate-permeabilized mouse L-cells. Further work will be aimed at assessing the ability or otherwise of exogenous glycolytic intermediates to decrease the specific radioactivity of the products derived from radioisotopic G6P, and at assessing the extent to which glycolytic intermediates are released to the extracellular space during G6P dissimilation, together with other comparable approaches discussed by Friedrich (1984) and by Keleti (1984). However, we should like to mention that the type of metabolic organization evidenced by the present type of observation has the interesting and biotechnologically significant corollary, that multistep and complex biotransformations of membrane-impermeant xenobiotics of commercial importance might be effected by this type of permeabilized-cell system.

In our final section, we wish to draw attention to yet another type of system in which we believe that one should also invoke a more microscopic organization of cellular metabolism than is commonly construed, illustrated with reference to two recently discovered anaerobic bacteria.

ENERGY TRANSDUCTION BY THE ATP SYSTEM AND THE "BIOLOGICAL QUANTUM"

When discussing the organization of cellular metabolism from an energetic standpoint, it is usual to consider that the free energy-generating reactions of microbial catabolism (or photosynthetic electron transport) are stoichiometrically coupled to the anabolic, free energy-requiring reactions of biosynthesis by means of the adenine nucleotide pool, as illustrated in Fig. 5 (Westerhoff et al., 1982) for microbial growth on glucose. In this figure, ΔG_p, the phosphorylation potential, is generally poised at a value of approx. -44 kJ/mol (-10.5 kcal/mol) (Thauer et al., 1977), such that it should be more negative than the free energy of anabolism (ΔG_a) is positive and, more importantly, if ATP molecules are formed stoichiometrically, should be less negative than is the free-energy change of catabolism (ΔG_c). In other words, it is considered that the free energy of catabolism is conserved as a "biological quantum" of "packets" corresponding to a (free) energy of ca. 44 kJ/mol, any excess catabolic energy being dissipated as heat, presumably as variable frictional losses in "molecular machines" (Welch and Kell, 1986), with such free-energy conservation being generally

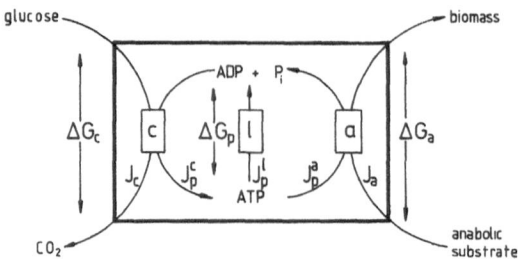

Fig. 5. Energy coupling in microbial growth, in which the catabolic
reactions (c) are stoichiometrically coupled to the anabolic
reactions (a) by means of the adenine nucleotide pool. "l"
represents "leak" reactions which result in a loss of free
energy and an _apparent_ non-stoichiometry between the catabolic
synthesis and the anabolic utilization of ATP. For further
explanation see text, where we discuss in particular the
possibility of a _sub-integral_ stoichiometry between catabolism
and ATP synthesis itself.

effected by substrate-level or electron transport-linked phosphorylation.

However, there are at least two anaerobic bacteria which cannot
conform simply to the above, macroscopic, continuum picture, viz.
Thermoproteus neutrophilus (Fischer et al., 1983) and _Syntrophomonas
wolfei_ (McInerey et al., 1981); for their ΔG_c values are significantly
less negative than -44 kJ/mol. As discussed by Thauer and Morris (1984),
the only mechanistic escape from this paradox is to posit a non-integral
stoichiometric coupling between some reversed electron-transfer reaction
and an ATP hydrolase reaction, so that even macroscopic schemes, such as
the chemiosmotic scheme outlined (Thauer and Morris, 1984), render
meaningless the concept of "the biological quantum". However, the
existence of threshold behavior in membrane energy-coupling systems (see
Kell, 1986), a characteristic of quantal free energy-coupling devices
(Welch and Kell, 1986), indicates that coupling systems of this type
cannot be arbitrarily flexible, and both provides an illustration of a
system in which the concept of a macroscopic "concentration" fails and
which, in this case, from a _macroscopic_ viewpoint, would appear to
violate the Second Law. Further work, to determine exactly whether the
coupling in these systems is microscopic or macroscopic, perhaps using
the double-inhibitor titration method, seems warranted.

ACKNOWLEDGEMENTS

We thank the Biotechnology Directorate of the Science and Engineering
Research Council, U.K., for the award of a studentship to RPW and for
financial support, Professor Gareth Morris for many stimulating and
illuminating discussions, Anthony Pugh for photographic assistance and
Sian Evans for typing the manuscript.

REFERENCES

Baum, H., 1978, Coupling between the charge-separating devices of the
    mitochondrion: intra- or extramembrane? _in_: "The Molecular Biology

of Membranes", S. Fleischer, Y. Hatefi, D.H. MacLennan and A. Tzagoloff, eds., Plenum Press, New York.

Baum, H., Hall, G.S., Nalder, J. and Beechey, R.B., 1971, On the mechanism of oxidative phosphorylation in submitochondrial particles, in: "Energy Transduction in Respiration and Photosynthesis", E. Quaglieriello, S. Papa and C.S. Rossi, eds., Adriatica Editrice, Bari.

Berden, J.A., Herweijer, M.A. and Cornelisson, J.B.J.W., 1984, ATP synthase and energy coupling, in: "H^+-ATPase (ATP synthase): Structure, Function, Biogenesis of the FOF1 complex of coupling membranes", S. Papa, K. Aitendorf, L. Eraster, and L. Packer, eds., Adriatica Editrice, Bari.

Berry, M.N., Grivell, A.R. and Wallace, P.G., 1985, Electrochemical aspects of metabolism, Compr. Treatise of Electrochem., 10:347.

Bohm, D., 1981, "Wholeness and the Implicate Order", Routledge and Kegan Paul, London.

Booth, I.R. and Morris, J.G., 1982, Carbohydrate transport in Clostridium pasteurianum, Biosci. Rep., 2:47.

Clarke, D.J., Morley, C.D., Kell, D.B. and Morris, J.G., 1982, On the mode of action of the bacteriocin butyricin 7423, Eur. J. Biochem., 127:105.

Clegg, J.S., 1984, Properties and metabolism of the aqueous cytoplasm and its boundaries, Amer. J. Physiol., 246:R133.

Cotton, N.P.J. and Jackson, J.B., 1983, Titrations of ATP synthesis with uncoupling agents do not provide evidence of localized high energy intermediates in electron transport phosphorylation in bacterial chromatophores, FEBS Lett., 161:93.

Davenport, J.W., 1985, Double inhibitor titrations of photophosphorylation are consistent with delocalized coupling, Biochim Biophys. Acta, 807:300.

Felix, H., 1982, Permeabilized cells, Anal. Biochem., 120:211.

Ferguson, S.J., 1985, Fully delocalized chemiosmotic or localized proton flow pathways in energy coupling? A scrutiny of experimental evidence, Biochim. Biophys. Acta, 811:47.

Ferguson, S.J. and Sorgato, M.C., 1982, Proton electrochemical gradients and energy transduction processes, Ann. Rev. Biochem., 51:185.

Fischer, F., Zillig, W., Setter, K.O. and Schreiber, G., 1983, Chemiolithotrophic metabolism of anaerobic extremely thermophilic archaebacteria, Nature, 301:511.

Friedrich, P., 1984, "Supramolecular Enzyme Organization," Pergamon Press, Oxford.

Froehlich, H. and Kremer, F., eds., "Coherent Excitations in Biological Systems," Springer-Verlag, Heidelberg.

Garden, R.W., 1984, "Modern Logic and Quantum Mechanics," Adam Hilger, Bristol.

Gorringe, D.M. and Moses, V., 1980, Organization of the glycolytic enzymes in Escherichia coli, Int. J. Biol. Macromol., 2:161.

Groen, A.K., van der Meer, R., Westerhoff, H.V., Wanders, R.J.A., Akerboom, T.P.M. and Tager, J.M., 1982a, Control of metabolic fluxes, in: "Metabolic Compartmentation", H. Sies, ed., Academic Press, New York.

Groen, A.K., Wanders, R.J.A., Westerhoff, H.V., van der Meer, R. and Tager, J.M., 1982b, Quantification of the contribution of various steps to the control of mitochondrial respiration, J. Biol. Chem., 257:2754.

Guffanti, A.A., Bornstein, R.F. and Krulwich, T.A., 1981, Oxidative phosphorylation by membrane vesicles from Bacillus alcalophilus, Biochim. Biophys. Acta, 635:619.

Hangarter, R.P. and Good, N.E., 1982, Energy threshold for ATP synthesis in chloroplasts, _Biochim. Biophys. Acta_, 681:397.

Harris, C.M. and Kell, D.B., 1985, On the dielectrically observable consequences of the diffusional motions of lipids and proteins in membranes. 2. Experiments with microbial cells, protoplasts and membrane vesicles, _Eur. Biophys. J._, in press.

Heinrich, R. and Rapoport, T.A., 1974, A linear steady-state treatment of enzymatic chains. General properties, control and effector strength, _Eur. J. Biochem._, 42:89.

Hess, B., 1983, Non-equilibrium dynamics of biochemical processes, _Hoppe-Seyler's Z. Physiol. Chem._, 364:1.

Hess, B., Goldbeter, A. and Lefever, R., 1978, Temporal, spatial and functional order in regulated biochemical and cellular systems, _Adv. Chem. Phys._, 38:363.

Hess, B., Kuschmitz, D. and Markus, M., 1984, Dynamic coupling and the time-patterns of glycolysis, _in_: "Dynamics of Biochemical Systems", J. Ricard and A. Cornish-Bowden, eds., Plenum Press, New York.

Hitchens, G.D. and Kell, D.B., 1982a, On the extent of localization of the energized membrane state in chromatophores of _Rhodopseudomonas capsulata_ N22, _Biochem. J._, 206:351.

Hitchens, G.D. and Kell, D.B., 1982b, Localized energy coupling during photophosphorylation by chromatophores of _Rhodopseudomonas capsulata_ N22, _Biosci. Rep_, 2:743.

Hitchens, G.D. and Kell, D.B., 1983a, Uncouplers can shuttle between localized energy coupling sites during photophosphorylation by chromatophores of _Rhodopseudomonas capsulata_ N22, _Biochem. J._, 212:25.

Hitchens, G.D. and Kell, D.B., 1983b, On the functional unit of energy coupling in photophosphorylation by bacterial chromatophore, _Biochim. Biophys. Acta_, 723:308.

Hitchens, G.D., and Kell, D.B., 1984, On the effects of thiocyanate and venturicidin on respiration-driven proton translocation in _Paracoccus denitrificans_, _Biochim. Biophys. Acta_, 766:222.

Jackson, J.B., Venturoli, G,, Baccarini-Melandri, A. and Melandri, B.A., 1981, Photosynthetic control and estimation of the optimal ATP:electron stoichiometry during flush activation of chromatophores from _Rhodopseudomonas capsulata_, _Biochim. Biophys. Acta_, 636:1.

Kacser, H. and Burns, J.A., 1973, The control of flux, _Symp. Soc. Exp. Biol._, 27:65.

Keleti, T., 1984, Channelling in enzyme complexes, _in_: "Dynamics of Biochemical Systems", J. Ricard and A. Cornish-Bowden, eds., Plenum Press, New York.

Kell, D.B., 1979, On the functional proton current pathway of electron-transport phosphorylation. An electrodic view, _Biochim. Biophys. Acta_, 549:55.

Kell, D.B., 1984, Diffusion of protein complexes in prokaryotic membranes: fast, free, random or directed?, _Trends Biochem. Sci_, 9:86.

Kell, D.B., 1986, Localized protonic coupling: overview and critical evaluation of experimental techniques, _Meth. Enzymol._, in press.

Kell, D.B., Clarke, D.J. and Morris, J.G., 1981, On proton-coupled information transfer along the surface of biological membranes and the mode of action of certain colicins, _FEMS Microbiol. Lett._, 11:1.

Kell, D.B. and Harris, C.M., 1985, On the dielectrically observable consequences of the diffusional motions of lipids and proteins in membranes. I. Theory and overview, _Eur. Biophys. J._, in press.

Kell, D.B. and Hitchens, G.D., 1982, Proton-coupled energy transduction by biological membranes. Principles, pathways and practice, _Faraday Discuss. Chem. Soc._, 74:377.

Kell, D.B. and Hitchens, G.D., 1983, Coherent properties of the

membranous systems of electron-transport phosphorylation, in:
"Coherent Excitations in Biological Systems", H. Fröhlich and F.
Kremer, eds., Springer-Verlag, Heidelberg.

Kell, D.B., John, P. and Ferguson, S.J., 1978, On the current:voltage
relationships of energy coupling membranes: phosphorylating membrane
vesicles from Paracoccus denitrificans, Biochem. Soc. Trans., 6:1292.

Kell, D.B. and Morris, J.G., 1981, Proton-coupled membrane energy
transduction: pathways, mechanisms and control, in: "Vectorial
Reactions in Electron and Ion Transport in Mitochondria and
Bacteria", F. Palmieri, E. Quaglierello, N. Siliprandi and E.C.
Slater, eds., Elsevier/North Holland, Amsterdam.

Kell, D.B. and Westerhoff, H.V., 1985, Catalytic facilitation and
membrane bioenergetics, in, "Organized Multienzyme Systems: Catalytic
Properties", G.R. Welch, ed., Academic Press, New York.

Kotze, J.P., 1968, 1-phosphofructokinase, a new glycolytic enzyme from
Clostridium pasteurianum W5, S. Afr. J. Sci., 11:349.

Karsinskaya, I.P., Marshansky, V.N., Dragunova, S.F. and Yaguzhinsky,
L.S., 1984, Relationships of respiratory chain and ATP-synthetase in
energized mitochondria, FEBS Lett., 167:176.

Mackey, B.M. and Morris, J.G., 1974, Isolation of a mutant strain of
Clostridium pasteurianum defective in granulose degradation, FEBS
Lett., 48:64.

Maloney, P.C., 1982, Energy coupling to ATP synthesis by the proton-
translocating ATPase, J. Membr. Biol., 67:1.

Masters, C.J., 1981, Interactions between soluble enzymes and subcellular
structure, CRC Critical Rev. Biochem., 11:105.

McInerney, M.J., Bryant, M.P., Hespell, R.B. and Costerton, J.W., 1981,
Syntrophomonas wolfei gen. nov. sp. nov., an anaerobic, syntrophic,
fatty acid-oxidizing bacterium, Appl. Env. Microbiol., 41:1029.

Millat, D.G., Griffiths-Smith, K., Algar, E. and Scopes, R.K., 1982,
Activity and stability of glycolytic enzymes in the presence of
ethanol, Biotechnol. Lett., 4:601.

Mitchell, P., 1966, Chemiosmotic coupling in oxidative and photosynthetic
phosphorylation, Biol. Rev., 41:445.

Mitchell, P., 1968, "Chemiosmotic Coupling and Energy Transduction",
Glynn Research Ltd., Bodmin.

Mitchell, W.J. and Booth, I.R., 1984, Characterization of the Clostridium
pasteurianum phosphotransferase system, J. Gen. Microbiol., 130:2193.

Mowbray, J. and Moses, V., 1976, The tentative identification in
Escherichia coli of a multienzyme complex with glycolytic activity,
Eur. J. Biochem., 66:25.

Nagodawithana, T.W., Whitt, J.T. and Cutaia, A.J., 1977, Study of the
feedback effect of ethanol on selected enzymes of the glycolytic
pathway, J. Am. Soc. Brewing Chem., 35:179.

Nicholls, D.G., 1982, "Bioenergetics", Academic Press, London.

O'Shea, P.S. and Thelen, M., 1984, On the logic of the application of
double-inhibitor titrations for the elucidation of the mechanism of
energy coupling, FEBS Lett., 176:79.

Ottaway, J.H. and Mowbray, J., 1977, The role of compartmentation in the
control of glycolysis, Curr. Top. Cell. Reg., 12:107.

Parsonage, D., 1984, Ph.D. Thesis, University of Birmingham.

Parsonage, D. and Ferguson, S.J., 1982, Titration of ATP synthase
activity with an inhibitor as a function of the rate of generation of
protonmotive force: implications for the mechanism of ATP synthesis,
Biochem. Soc. Trans., 10:257.

Primas, H., 1981, "Chemistry, Quantum Mechanics and Reductionism",
Springer-Verlag, Heidelberg.

Reddy, G.P.V. and Pardee, A.B., 1983, Inhibitor evidence for allosteric
interaction in the replitase multienzyme complex, Nature, 304:86 and
658.

Schlenk, F., 1985, Early research on fermentation - a story of missed opportunities, Trends Biochem. Sci., 10(b):252.

Schmidt, G. and Graeber, P., 1985, The rate of ATP synthesis by reconstituted CF_oF_1 liposomes, Biochim. Biophys. Acta, 808:46.

Serrano, R., Ganceso, J.M. and Ganceso, C., 1973, Assay of yeast enzymes in situ, Eur. J. Biochem., 34:479.

Smith, D.J., Stokes, B.O. and Boyer, P.D., 1976, Probes of initial phosphorylation events in ATP synthesis by chloroplasts, J. Biol. Chem., 251:4165.

Somogyi, B., Welch, G.R. and Damjanovich, S., 1984, The dynamic basis of energy transduction in enzymes, Biochim. Biophys. Acta, 768:81.

Sorgato, M.C., Branca, D. and Ferguson, S.J., 1980, The rate of ATP synthesis by submitochondrial particles can be independent of the protonmotive force, Biochem. J., 188:945.

Stucki, J.W., 1978, Stability analysis of biochemical systems. A practical guide, Progr. Biophys. Mol. Biol., 33:99.

Tedeschi, H., 1980, The mitochondrial membrane potential, Biol. Rev., 55:171.

Thauer, R.K., Jungermann, K. and Decker, K., 1977, Energy conservation in chemotrophic anaerobic bacteria, Bacteriol. Rev., 41:100.

Thauer, R.K., and Morris, J.G., 1984, Metabolism of chemotrophic anaerobes: old views and new aspects, Symp. Soc. Gen. Microbiol., 36:123.

Thayer, W.S. and Hinkle, P.C., 1975, Kinetics of adenosine triphosphate synthesis in bovine heart submitochondrial particles, J. Biol. Chem., 250:5336.

Uyeda, K. and Kurooka, S., 1970, Crystallization and properties of phosphofructokinase from Clostridium pasteurianum, J. Biol. Chem., 245:3315.

Venturoli, G. and Melandri, B.A., 1982, The localized coupling of bacterial photophosphorylation, Biochim. Biophys. Acta, 680:8.

Welch, G.R., 1977, On the role of organized multienzyme systems in cellular metabolism: a general synthesis, Progr. Biophys. Mol. Biol., 32:103.

Welch, G.R. and Kell, D.B., 1986, Not just catalysts. Molecular machines in bioenergetics, in: "The Fluctuating Enzyme," G.R. Welch, ed., Wiley, Chichester.

Westerhoff, H.V. and Chen, Y., 1985, Stochastic free energy transduction, Proc. Natl. Acad. Sci., 82:3222.

Westerhoff, H.V., Colen, A-M., and van Dam, K., 1983a, Metabolic control by pump slippage and proton leakage in 'delocalized' and more localized chemiosmotic energy-coupling schemes, Biochem. Soc. Trans., 11:81.

Westerhoff, H.V., Helgerson, S.L., Theg, S.M., van Kooten, O., Wikstrom, M., Skulachev, V.P. and Dancshazy, Zs., 1983b, The present state of the chemiosmotic coupling theory, Acta Biochim. Biophys. Acad. Sci. Hung., 18:125.

Westerhoff, H.V., Groen, A.K. and Wanders, R.J.A., 1984a, Modern theories of metabolic control and their applications, Biosci. Rep., 4:1.

Westerhoff, H.V., Lolkema, J.S., Otto, R. and Hellingwerf, K.J., 1982, Thermodynamics of growth, Biochim. Biophys. Acta, 683:181.

Westerhoff, H.V. and Kell, D.B., 1985, A control theoretical analysis of inhibitor titration assays of metabolic channelling, Comments Mol. Cell. Biophys., in press.

Westerhoff, H.V., Melandri, B.A., Venturoli, G., Azzone, G.F. and Kell, D.B., 1984b, A minimal hypothesis for membrane-linked free-energy transduction. The role of independent, small coupling units, Biochim. Biophys. Acta, 768:257.

Wheeler, J.A. and Zurek, W.H., 1983, "Quantum Theory and Measurement," Princeton University Press, Princeton.

Wolf, F.A., 1981, "Taking the Quantum Leap," Harper and Row, San Francisco.

ELECTRON AND PROTON FLOW IN HEPATOCYTES DURING FATTY ACID

AND PYRUVATE OXIDATION: IMPLICATIONS FOR ENERGY

TRANSDUCTION AND HEAT PRODUCTION

M.N. Berry, R.B. Gregory, A.R. Grivell,
J.W. Phillips, and P.G. Wallace

Department of Biochemistry
School of Medicine
Flinders University of South Australia
Bedford Park, South Australia 5042
Australia

INTRODUCTION

Since the discovery of the process of oxidative phosphorylation, biochemical textbooks have tacitly assumed that most, if not all, of mammalian mitochondrial respiration is tightly coupled to ATP turnover, so that any increase in cellular oxygen uptake (J_O) must necessarily indicate an augmented demand for ATP. On the other hand, much evidence has now accrued that the increase in J_O observed when hepatocytes are presented with substrates, particularly fatty acids (Berry, 1974a; Williamson et al., 1969; Debeer et al., 1974; Berry et al., 1983a), cannot be wholly or even mainly accounted for by increased utilization of ATP in biosynthetic processes such as gluconeogenesis and urea formation (Berry, 1974a; Berry et al., 1983a; Hems et al., 1966; Krebs et al., 1964). The mechanism by which oxygen consumption is stimulated to a greater extent than predicted from any increased metabolic activity of the cells has not been established unequivocally. Possibilities include uncoupling of the mitochondria (Scholz et al., 1984; Soboll and Stucki, 1985), changes in eficiency of coupling, perhaps by alterations in the H^+/e^- ratio of mitochondrial proton pumping (Nicholls, 1974; Pietrobon et al., 1981), induction of futile cycles of ATP synthesis and hydrolysis (Debeer et al., 1974; Newsholme and Crabtree, 1976; Katz and Rognstad, 1976; Plomp et al., 1985), or stimulation of some other pathway such as reversed electron flow (Berry et al., 1983a). None of these explanations seems entirely satisfactory.

In a previous examination of this problem (Berry et al., 1983a) it was noted that the increase in J_O induced by addition of fatty acid to hepatocytes was due entirely to a stimulation of β-oxidation (J_O^F), Krebs cycle activity (J_O^F) being unchanged. When an extra ATP demand was created by the further addition of a gluconeogenic substrate to the cells, J_O^K increased but J_O^F did not. J_O^F was strongly depressed by the ATP-synthase inhibitor oligomycin, whereas β-oxidation was much less sensitive to this agent. We concluded from these studies that the oxidation of acetyl CoA, derived from palmitate, is directly coupled to ATP synthesis, whereas β-oxidation itself is not, but rather may be involved in other processes, such as heat production. However, we were unable with the techniques then at our disposal to reach a firm conclusion about the relationship between hepatic β-oxidation and energy metabolism. Subsequently, Dr. Martin Brand of the University of Cambridge suggested that we examine the relationship between hepatic fatty acid oxidation and mitochondrial membrane potential ($\Delta\Psi$). The results of those studies are reported in this chapter.

Significance of the Mitochondrial Membrane Potential

According to Mitchell (1961, 1976), electron transport associated with mitochondrial respiration results in the extrusion of protons from the mitochondrial matrix, thereby establishing a proton electrochemical gradient ($\Delta\mu H^+$) across the mitochondrial inner membrane. The concentration component (ΔpH) is small enough in animal mitochondria to be discounted (Cohen et al., 1978). The electrical component ($\Delta\Psi$) is considered to be a true, bulk-phase membrane potential, i.e. a trans-membrane electrical potential difference between the mitochondrial matrix and the aqueous phase outside the inner membrane. It can be detected in various ways. We have employed the organic cation triphenylmethylphos-phonium (TPMP$^+$), which is believed to be taken up by the mitochondria and partitioned between mitochondria and cytoplasm in proportion to the magnitude of $\Delta\Psi$. Corrections must be made for binding — indeed it has been suggested that the uptake of TPMP$^+$ may represent binding to an energized membrane rather than concentration in the mitochondrial matrix (Higuti, 1984).

The distribution of TPMP$^+$ or a similar agent is considered to reflect the proton electrochemical gradient, or protonmotive force (Mitchell, 1961, 1976), which is thought to drive the re-entry of protons into the matrix via the ATP synthase, thereby bringing about the synthesis of ATP by an as yet undefined mechanism. The potential dis-sipation of $\Delta\mu H^+$ during ATP synthesis is believed to be prevented by a compensatory increase in proton extrusion coupled to respiration. Even in the absence of ATP synthesis, some backflow of protons to the matrix would be anticipated; and this also is thought to be compensated by respiration-dependent proton extrusion (Ferguson and Sorgato, 1982). According to chemiosmotic principles, when the membrane potential is at steady state ($\Delta\Psi$ constant) the rate of proton re-entry into the mitochondrial matrix must be exactly matched by the rate of proton extrusion. Moreover, if mitochondria possess a constant resistance to the re-entry of protons to the matrix via passive leaks, it can be anticipated that the steady-state level of $\Delta\Psi$ will be proportional to the rate of respiration (by analogy with Ohm's Law) (Ferguson and Sorgato, 1982). This is occasionally, but by no means regularly, observed (O'Shea and Chappell, 1984). Most studies with TPMP$^+$ or similar agents have been carried out with isolated mitochondria. However, Hoek et al. (1980) and Brand and Felber (1984) have demonstrated that this approach can also be successfully applied to the measurement of $\Delta\Psi$ within the mitochondria of intact animal cells, and we have followed their procedures.

Our findings suggest that Krebs cycle oxidations and β-oxidation fulfill different functions in regard to normal mitochondrial energy transduction, in that they supply separate currents to two distinct but cooperative facets of the energy transducing processes. Accordingly, these studies provide support for proposals advanced by Srere and others that the enzymes of the mitochondrial matrix are not distributed at random, but are associated in organized complexes and may be bound to the inner mitochondrial membrane (Matlib and O'Brien, 1975; Srere et al., 1978; Beeckmans and Kanarek, 1981; Sumegi and Srere, 1984; Sumegi et al., 1985). Our results do not appear compatible with the concept of a single delocalized electrochemical proton gradient across the inner mitochondrial membrane as the sole competent energized "intermediate" in oxidative phosphorylation. Rather, they provide strong support for the existence of an intramembranous energy-transducing process that can transfer energy directly to the integral membrane enzymes of the ATP synthesis and translocating apparatus (Rosing et al., 1977).

METHODS

Carbonylcyanide p-trifluoromethoxyphenylhydrazone (FCCP), rotenone, antimycin, and oligomycin were obtained from Sigma, St. Louis, MO. Myxothiazol was from Boehringer, Sydney, N.S.W., Cab-O-Sil from Cabot Corporation, Boston, Mass., and [^{14}C]triphenylmethylphosphonium iodide (TPMP$^+$) and ACS II scintillation fluid from Amersham, Bucks. Tetradecylglycidic acid (TDGA) was a generous gift from Dr. Gene F. Tutwiler of McNeil Pharmaceutical.

Isolated liver cells from Hooded Wistar rats (250-280 g body wt), starved for 24 h to deplete liver glycogen, were prepared by a modification (Berry, 1974b) of the method of Berry and Friend (1969). The cells (90-120 mg wet wt) were incubated at 37°C in 2 ml of a balanced bicarbonate-saline medium containing albumin, 2.5% (w/v), and 2 μM TPMP$^+$, with a gas phase of 95% O_2, 5% CO_2 (Berry et al., 1983a). Consumption of O_2 was measured in the presence of CO_2 by a manometric method (Krebs et al., 1974). At the end of the incubation period and before protein precipitation, a 0.6 ml portion of the suspension was sampled for the determination of the intracellular mitochondrial membrane potential (Hoek et al., 1980; Brand and Felber, 1984). Preliminary studies showed that TPMP$^+$ enters the cells and reaches a steady-state distribution between cells and medium within 10 min. The cell suspension was centrifuged at 8,000xg for 1 min in an Eppendorf centrifuge and the supernatant rapidly removed. A measured portion (0.5 ml) was mixed with 10 ml of ACS II containing 25 g Cab-O-Sil per liter, and its radioactivity determined by liquid scintillation counting. After determination of the wet weight of cells, the inner mitochondrial membrane potential ($\Delta\Psi$) was calculated from the partitioning of TPMP$^+$ (Hoek et al., 1980; Brand and Felber, 1984). Allowance was made for alterations in distribution relating to the existence of a plasma membrane potential assumed to be 33 mV and for passive adsorption of the cation to the mitochondrial membrane. The possible contribution of ΔpH to $\Delta\mu$ H$^+$ was considered sufficiently small to be disregarded (Cohen et al., 1978; Hoek et al., 1980).

Metabolites were measured by standard enzymic techniques (Berry et al., 1983a) on neutralized perchloric acid extracts of the incubated cells and incubation medium, except in the case of Pi which was measured on cell pellets obtained by rapid centrifugation (Degenaar, 1983). Assays were performed automatically on a Cobas BIO Analyzer (Roche, Sydney, N.S.W.) and the optical extinction data transmitted directly to a PDP 11/44 computer (DEC, Mass.) for processing.

Using calculations which have been validated previously (Berry et al., 1983a), the measured value for J_O was partitioned into J_O^F representing β-oxidation of endogenous or added palmitate and Krebs cycle respiration (J_O^K). In brief, the quantity of O_2 consumed in producing acetyl CoA from palmitate (J_O^F) can be assessed by measuring the amount of acetyl CoA channelled into acetoacetate and 3-hydroxybutyrate or combusted in the Krebs cycle. Krebs cycle respiration represents the difference between J_O and J_O^F ($J_O^K = J_O - J_O^F$). Analogous calculations can be employed to derive the quantity of O_2 consumed in pyruvate oxidation to acetyl CoA (J_O^P).

RESULTS

Relationship Between Fatty Acid Oxidation and Mitochondrial Membrane Potential

The addition of palmitate to liver cells from starved rats stimulated J_O almost 50%, the increase being entirely due to enhancement of O_2-uptake due to β-oxidation (J_O^F), Krebs cycle respiration (J_O^K) being unaltered. The increase in J_O was not accompanied by a proportional increase in $\Delta\Psi$ (Fig. 1). This failure of $\Delta\Psi$ to rise in parallel with

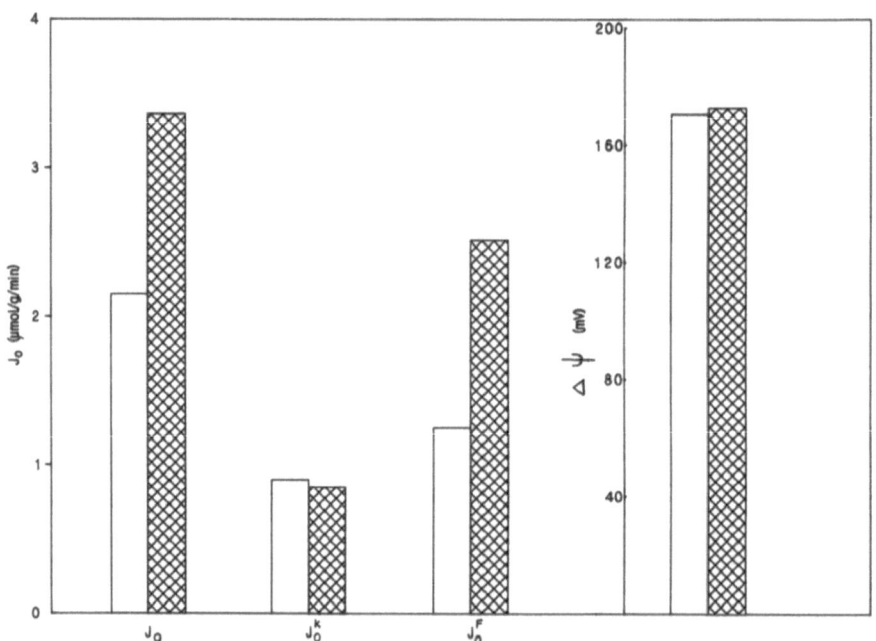

Fig. 1. Respiration of hepatocytes from starved rats incubated in the presence of 2 mM palmitate, showing the contribution to total oxygen uptake (J_O) of the Krebs cycle (J_O^K) and β-oxidation (J_O^F). The mitochondrial membrane potential ($\Delta\Psi$) is also shown.

☐ Endogenous   ⊠ 2 mM Palmitate

the increase in J_O implies that any coupled proton extrusion must have been exactly balanced by proton re-entry to the matrix. Alternatively, the electron transport associated with J_O^F may not have been linked to proton extrusion capable of contributing to $\Delta\Psi$ (Nicholls, 1974; Pietrobon et al., 1981).

To gain further insight into this problem, additional studies were conducted in a manner similar to experiments with isolated mitochondria (Ferguson and Sorgato, 1982), in that cells were exposed to graded concentrations of various inhibitors and J_O and $\Delta\Psi$ measured. A representative plot is presented in Fig. 2, which shows the results of experiments in which the rate of palmitate oxidation was varied by addition of TDGA, a potent inhibitor of palmitoyl-carnitine transferase (Kiorpes et al., 1984), in combination with the respiratory inhibitor rotenone. A value for $\Delta\Psi$ of 120 mV was attained at J_O of about 1 μmol/g/min. Above this respiration rate a three-fold increase in J_O caused an increase in $\Delta\Psi$ of only 40 mV. A similar non-linear relationship between J_O and $\Delta\Psi$ was found when 3-hydroxybutyrate was the added substrate (Fig. 2). However, J_O in the presence of 3-hydroxybutyrate was never as great as that observed in the presence of fatty acid.

The non-linearity of these plots is not strictly in accord with classical chemiosmotic theory, which predicts that J_O should be proportional to $\Delta\Psi$. To examine this discrepancy we compared $\Delta\Psi$ with the

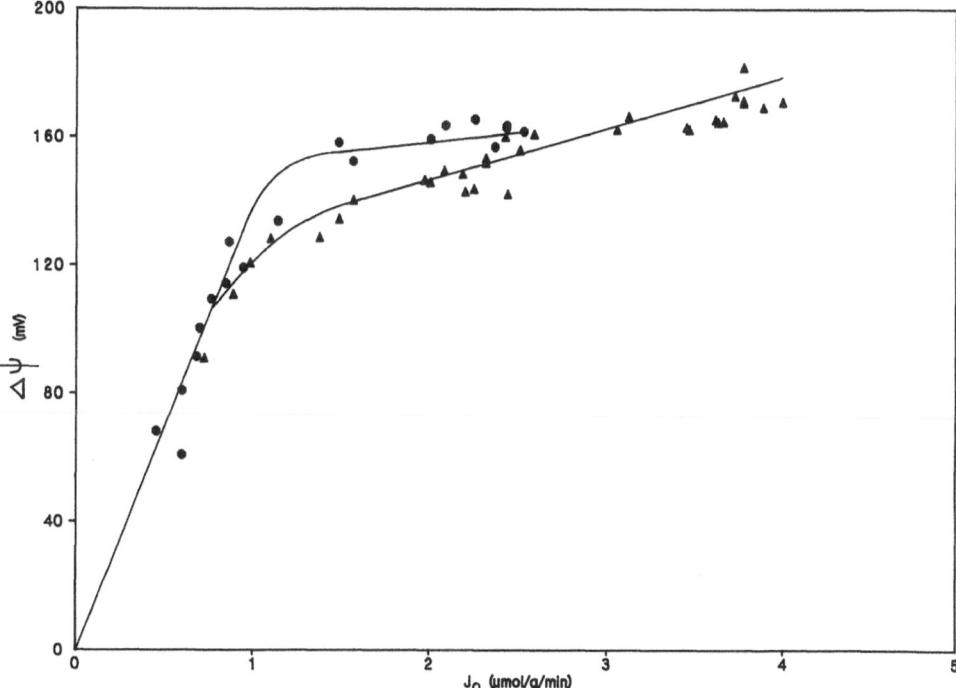

Fig. 2.   Respiration of hepatocytes from starved rats incubated in the presence of 2 mM palmitate and a range of concentrations of TDGA (0-0.25 μM) and rotenone (0-8 μM) [▲————▲]; or 10 mM 3-hydroxybutyrate, 1 μM TDGA and a range of concentrations of antimycin (0-1.2 μM) or myxothiazol (0-1.2 μM) [●————●].

cellular phosphorylation potential (ΔG_p). Fig. 3 illustrates the very
good agreement between $\Delta\Psi$ and ΔG_p, down to values for the latter of circa
390 mV.   The anticipated fall in [ATP] with declining ΔG_p was
exaggerated by a concomitant loss of total adenine nucleotides, presumably by deamination of AMP (Woods et al., 1970).   Nevertheless, [ATP]
did not fall below 50 μM, possibly because of binding of adenine nucleotide to ATP synthase and other enzymes.

In view of the close relationship between $\Delta\Psi$ and ΔG_p (Fig. 3), the
implication can be drawn from Fig. 2 that there is a threshold value for
$\Delta\Psi$ at which ATP synthase activity commences, corresponding to the
inflection point on the plot of J_o vs $\Delta\Psi$ (circa $\Delta\Psi$, 120 mV or ΔG_p, 420
mV).   In agreement with this it was found that high concentrations of
oligomycin or carboxyatractyloside tended to linearize the current-
voltage curve by inhibiting J_o at levels of $\Delta\Psi$ above 100 mV (Fig. 4) while
having little effect on the curve at lower values of $\Delta\Psi$.   We infer from
these studies that the first portion of the curve J_o vs $\Delta\Psi$ (a-b, Fig. 4)
is not directly associated with ATP synthesis, since it is unaffected by
oligomycin or carboxyatractyloside, but rather might reflect an initial
"energization" of the inner mitochondrial membrane as a necesssary
prerequisite for ATP generation.   The subsequent slower but linear rise
in $\Delta\Psi$ with increasing J_o corresponds to measurable increases in cellular
[ATP] and a parallel rise in ΔG_p.   At any given J_o in this range (b-c,
Fig. 4) a particular steady-state level for ΔG_p can be maintained.

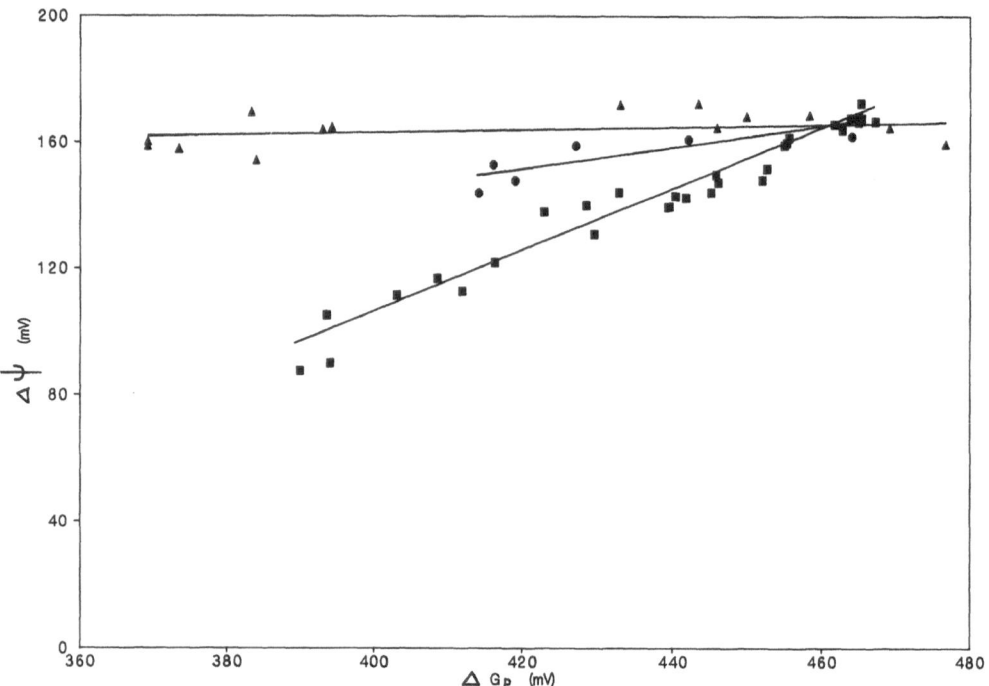

Fig. 3.   Plots of mitochondrial membrane potential ($\Delta\Psi$) vs.
phosphorylation potential (ΔG_p).   Hepatocytes from starved rats
were incubated with 2 mM palmitate, 10 mM lactate.   Other
additions were antimycin (0-1.2 μM) or myxothiazol (0-1.2 μM)
[■——— ■]; oligomycin (0-1.8 μM) [▲———▲];
carboxyatractyloside (0-250 μM) [●——— ●].

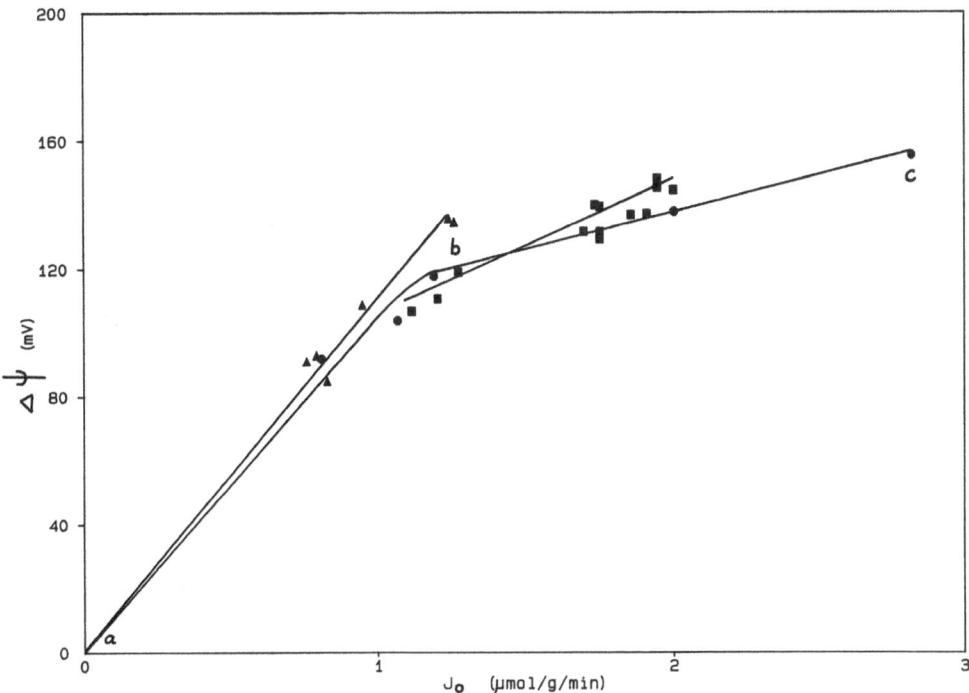

Fig. 4. Respiration of hepatocytes incubated in the presence of 2 mM palmitate, rotenone (0-8 μM) [●———●]; or 2 mM palmitate, rotenone (0-8 μM) and oligomycin (0-1.2 μM) [▲———▲]; or 2 mM palmitate and carboxyatractyloside (0-250 μM) [■———■].

Moreover, this relationship applies both in the absence of added substrate and when cellular respiration is stimulated by fatty acid oxidation. Hence the highest values of ΔG_p are seen in the presence of fatty acid.

It is noteworthy that under most experimental conditions $\Delta\Psi$ and ΔG_p change in parallel. However in the presence of oligomycin or carboxyatractyloside $\Delta\Psi$ does not decline despite substantial depression of J_o and a corresponding drop in ΔG_p (Fig. 3). The fall in ΔG_p in the presence of carboxyatractyloside indicates that it must reflect mainly cytoplasmic [ATP]/[ADP], since mitochondrial [ATP]/[ADP] would be expected to rise on inhibition of the adenine nucleotide translocase. The slope of the plot of $\Delta\Psi$ vs ΔG_p is 1 in the presence of inhibitors of electron transport. This is only half that predicted by the chemiosmotic hypothesis but is appropriate for the stoichiometry of the reaction catalyzed by ATP synthase (Mitchell, 1961, 1976; Campo et al., 1984). However, caution in interpretation of these results is required, in that ΔG_p was measured on the basis of total cellular [ATP] and [ADP], and ΔpH was not determined. The possibility must also be considered that $\Delta\Psi$ is underestimated by measurements made with TPMP⁺. Nevertheless, it would appear that $\Delta\Psi$ reflects the electrical energy stored across (or within) the inner mitochondrial membrane, whereas ΔG_p reflects the driving force for cytoplasmic ATP-linked events.

Regulation of Cellular Respiration in the Presence of Increased Demand for ATP

Addition of a substrate to liver cells frequently initiates synthetic processes which consume ATP. In order to study the effects of ATP synthesis on $\Delta\Psi$, liver cells were incubated with palmitate plus lactate, a circumstance that stimulates glucose formation and ATP turnover. J_o was considerably greater in the presence of this substrate combination than when palmitate alone was added. Moreover, the relationship between $\Delta\Psi$ and J_o was again non-linear, the curve flattening sharply at $\Delta\Psi$ 140 mV (Fig. 5), implying, according to the chemiosmotic hypothesis, an increased rate of proton backflow to the matrix above this level of $\Delta\Psi$. The question arose as to whether this proton current represented a non-specific leak or whether it reflected flow through the proton channel (F_o) within ATP synthase (Fillingame, 1980), associated with the ATP formation required for gluconeogenesis. That the latter was more likely was inferred from the suppression by oligomycin of the extra respiration induced by lactate addition at inhibitor concentrations considerably lower than those required to reduce fatty acid oxidation in the absence of added lactate (Fig. 5) and the observation that glucose synthesis was highly sensitive to the magnitude of $\Delta\Psi$, ceasing below about 130 mV (Fig. 6). Thus, the flattened portion of the J_o vs $\Delta\Psi$ curve (Fig. 5) correlated with circumstances in which glucose was being synthesized.

A more detailed analysis of the data showed that the relationships between J_o and $\Delta\Psi$ reflected to a considerable degree an increase in J_o^K as a consequence of gluconeogenesis (Fig. 7). Indeed a linear relationship between J_o^K and the rate of glucose synthesis could consistently be demonstrated (Fig. 8), supporting the conclusion that in these

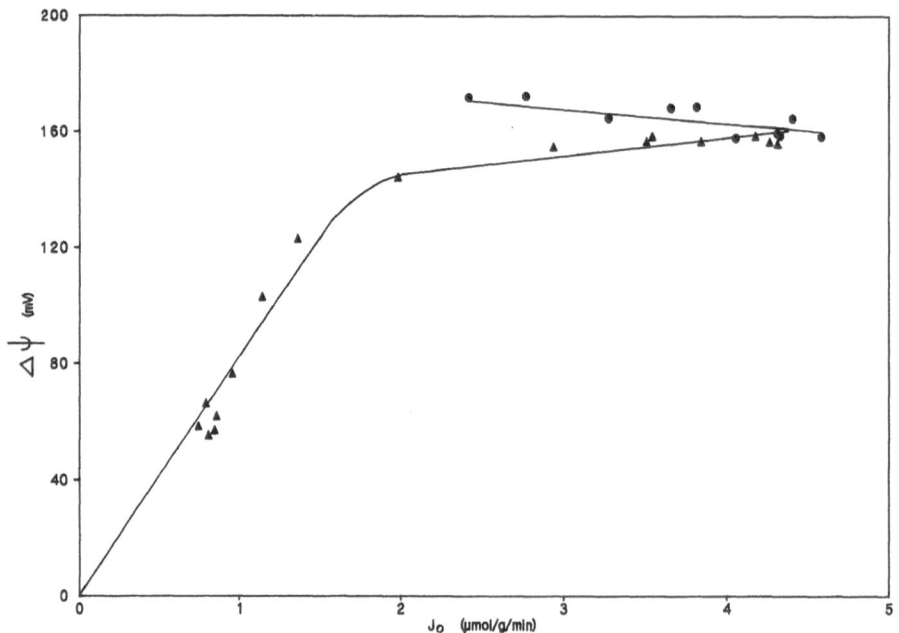

Fig. 5.   Respiration of hepatocytes incubated in the presence of 2 mM palmitate, 10 mM lactate and myxothiazol (0-1.2 µM) [▲——▲]; or 2 mM palmitate, 10 mM lactate, and oligomycin (0-1.2 µM) [●———●].

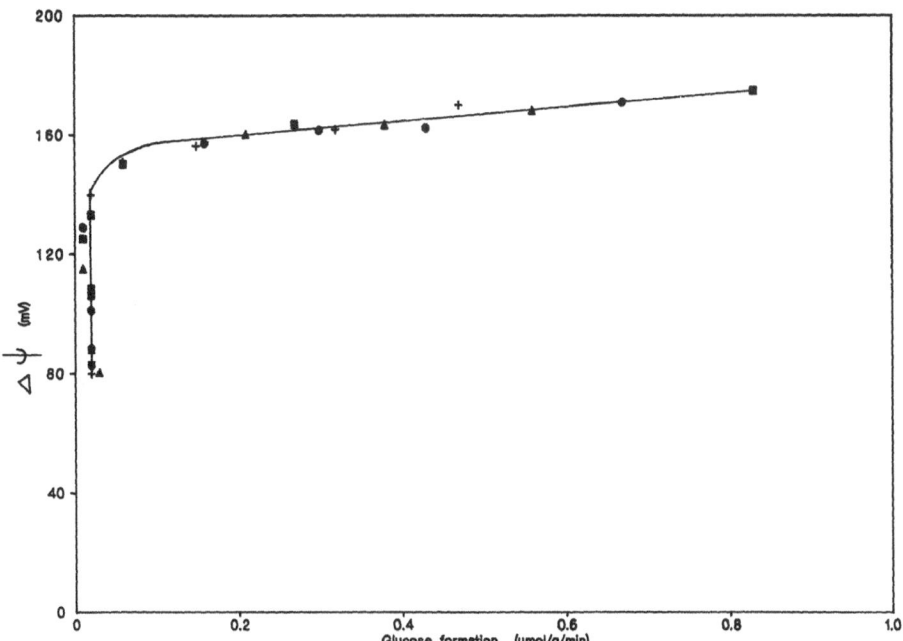

Fig. 6.   Glucose formation by hepatocytes from starved rats incubated in
          the presence of 2 mM palmitate, 10 mM lactate and a range of
          concentrations of TDGA (0-0.25 µM) and rotenone (0-8 µM).

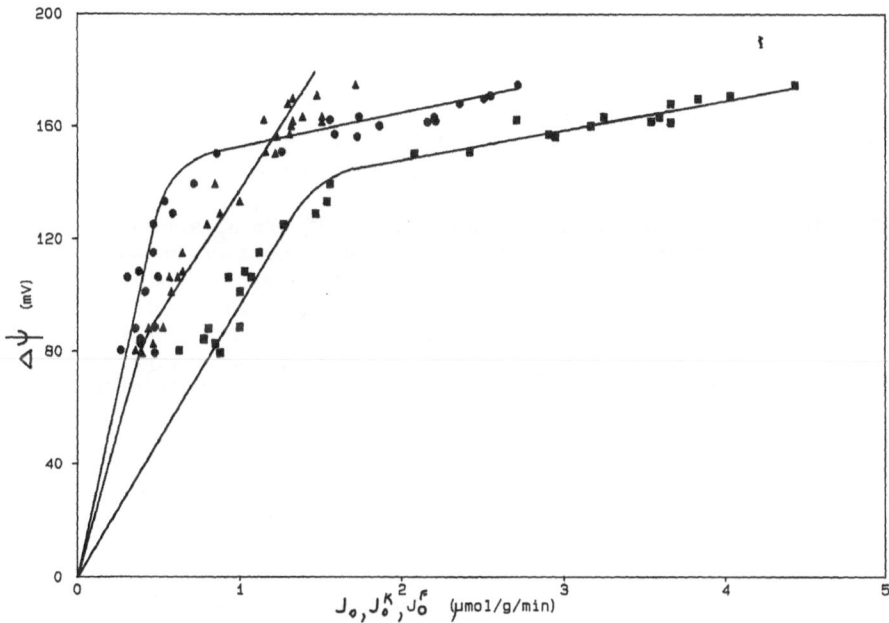

Fig. 7.   Respiration of hepatocytes from starved rats incubated in the
          presence of 2 mM palmitate, 10 mM lactate and a range of
          concentrations of TDGA (0-0.25 µM) and rotenone (0-8 µM).
          ■——■ Total respiration (J_o); ●——● Krebs cycle oxygen
          uptake (J_o^K); ▲——▲ β-oxidation oxygen uptake (J_o^f).

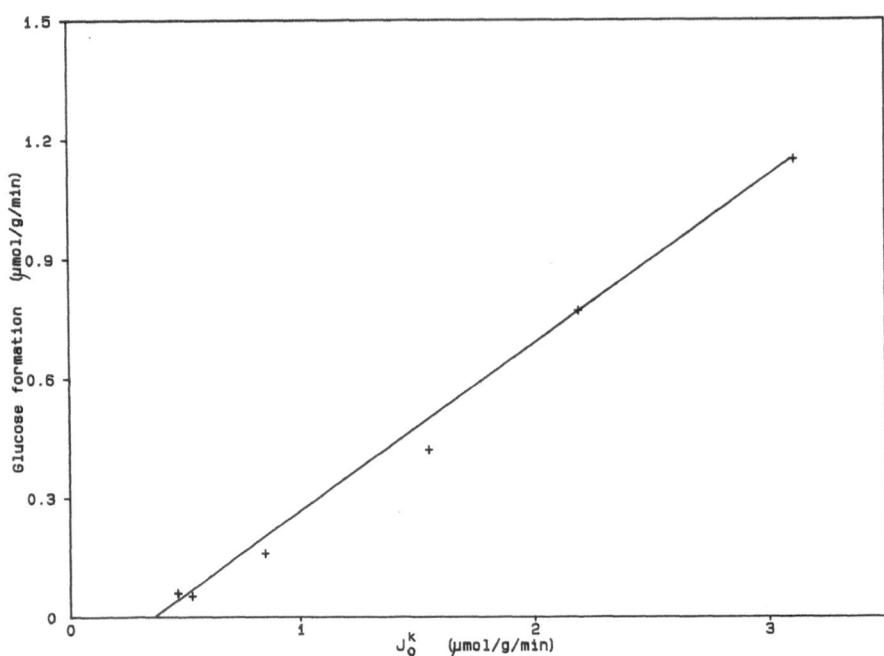

Fig. 8.   Glucose formation and Krebs cycle oxygen uptake (J_O^K) in
          hepatocytes from starved rats incubated in the presence of 2 mM
          palmitate, 10 mM lactate and rotenone (0-8 μM).

experiments there was enhanced proton backflow to the matrix through the
ATP synthase at high $\Delta\Psi$, associated with the increased rate of
regeneration of ATP utilized for gluconeogenesis. In contrast, J_O^F was
unresponsive to ATP demand and was not affected by the rate of glucose
synthesis, but instead increased as a linear function of $\Delta\Psi$ or ΔG_p.

The differences in the characteristics of J_O^K and J_O^F were well
illustrated by the use of oligomycin. Fig. 9a demonstrates the
considerably greater sensitivity of J_O^K to this agent, an initial
inhibition of J_O^K appearing at an oligomycin concentration of 0.8 μM. A
60% higher level of oligomycin was required to initiate depression of J_O^F.
Moreover, a number of markers of cellular energy state changed drama-
tically in the presence of an oligomycin concentration sufficient to
inhibit J_O^F (Fig. 9b).

As a result of these findings, we examined the oxidative metabolism
of other substrates. Pyruvate is both a gluconeogenic precursor and a
major metabolic fuel for cells from starved rats, particularly when
endogenous fatty acid oxidation is blocked with TDGA. As expected, nonproportional relationships were found when J_O and J_O^K were plotted against
$\Delta\Psi$ (Fig. 10). Again a linear correlation between the rates of J_O^K and
glucose synthesis was observed (data not shown). On the other hand, the
O_2-consumption representing oxidation of pyruvate to acetyl CoA (J_O^P) was
directly proportional to $\Delta\Psi$ (and ΔG_p) (Fig. 10), but showed no change in
relationship to the varying demand of gluconeogenesis.

Liver cells incubated with pyruvate and oligomycin showed little
decline in $\Delta\Psi$ over a wide range of inhibitor concentrations, even though
J_O was reduced to less than 50% of normal at the highest concentration of

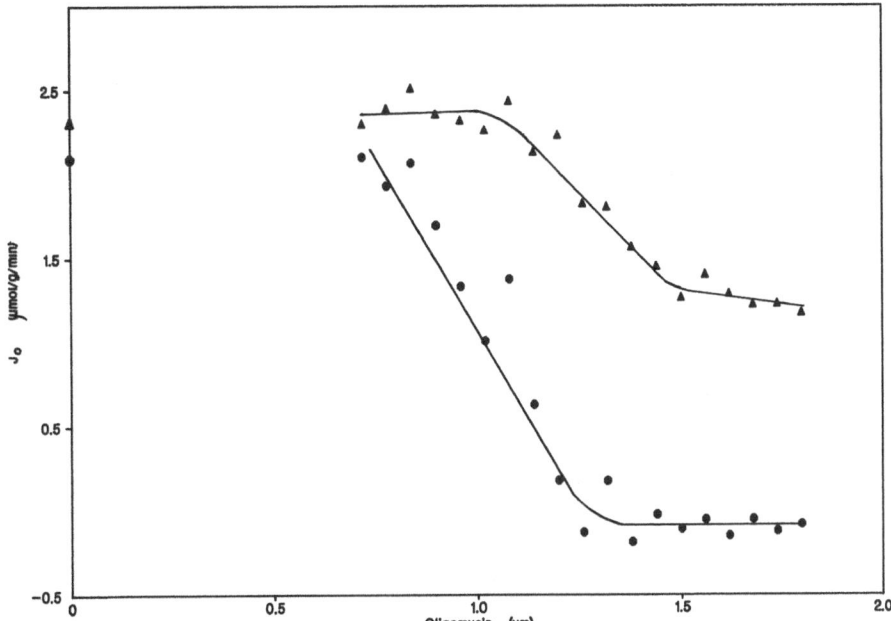

Fig. 9a. Effects of oligomycin on the respiration of hepatocytes
incubated in the presence of 2 mM palmitate, 10 mM lactate.
●——● Krebs cycle oxygen uptake (J_o^K)
▲——▲ β-oxidation oxygen uptake (J_o^F)

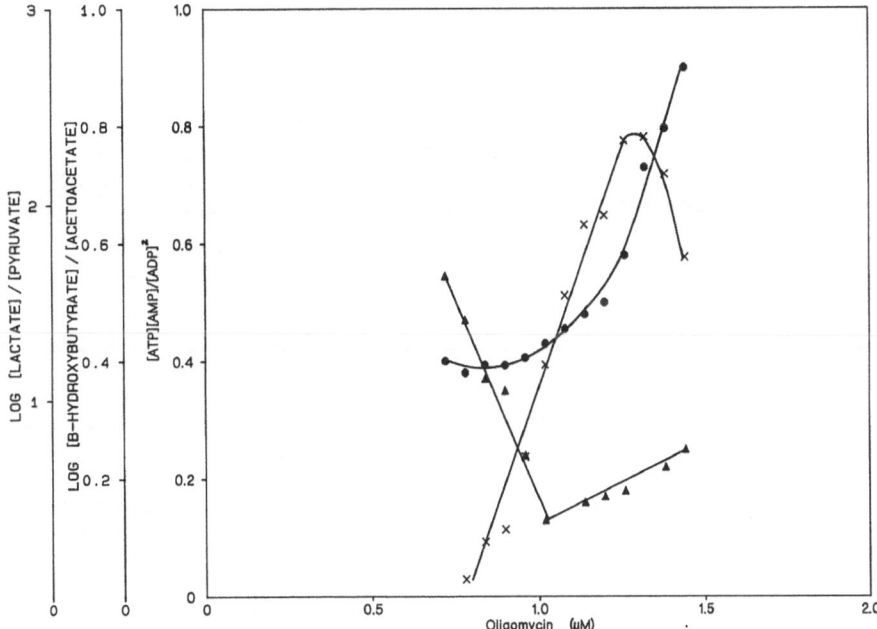

Fig. 9b. Effect of oligomycin on hepatocytes incubated in the presence
of 2 mM palmitate, 10 mM lactate.
●——● log ([lactate]/[pyruvate])
X—— X log ([β-hydroxybutyrate]/[acetoacetate])
▲——▲ $\dfrac{[ATP][AMP]}{[ADP]^2}$

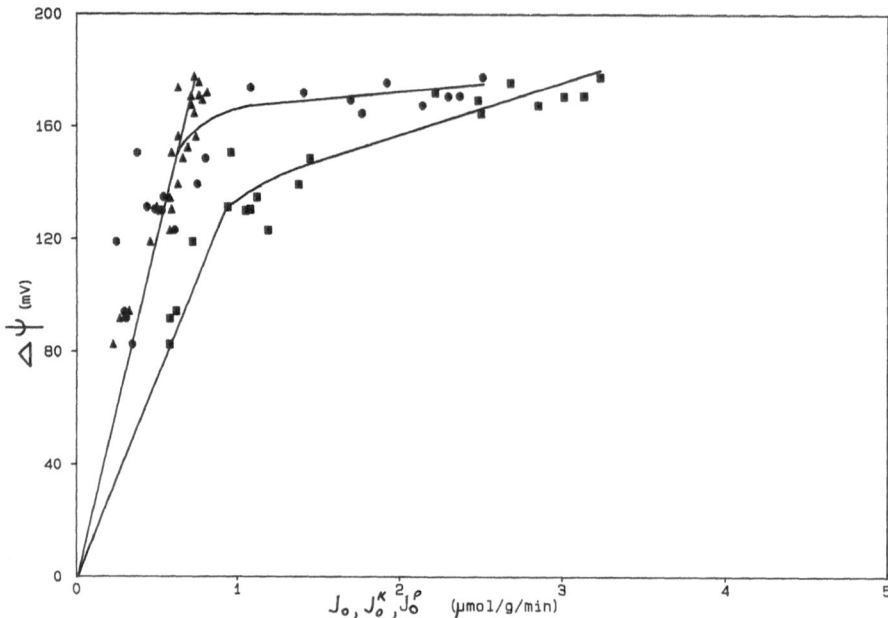

Fig. 10. Respiration of hepatocytes from starved rats incubated in the
presence of 10 mM pyruvate, 1 μM TDGA and a range of
concentrations of rotenone (0-8 μM), antimycin (0-1.4 μM),
myxothiazol (0-1.2 μM) or oligomycin (0-1.0 μM).
■——■ Total respiration (J_o); ●——● Krebs cycle oxygen
uptake (J_o^K); ▲——▲ Oxygen uptake for pyruvate
oxidation to acetyl CoA (J_o^P).

oligomycin tested; ΔG_p fell in parallel with J_o. Analysis of the
respiratory data shows that J_o^P was unaffected by oligomycin levels below
1.2 μM, although J_o^K was markedly reduced (Fig. 11), the rate of J_o^K
declining proportionally with the rate of gluconeogenesis.

DISCUSSION

<u>Is All Mitochondrial Respiration Obligatorily Coupled to Proton Flow and
ATP Synthesis?</u>

    When a gluconeogenic precursor such as lactate or pyruvate is added
to liver cells, respiration is stimulated; and a proportional
relationship is observed between the rate of glucose formation and J_o^K.
We infer from this that the O_2-consumption reflected by Krebs cycle
oxidations is stoichiometrically coupled to generation of the proton
current (proticity) postulated by chemiosmotic theory to be obligatorily
required for mitochondrial ATP synthesis. In keeping with this,
uncoupling agents cause a marked stimulation in J_o^K (Berry et al., 1983a).

    In the absence of a gluconeogenic precursor, however, fatty acids <u>per
se</u> stimulate hepatic cellular O_2-uptake. This stimulation of J_o is
prevented by high concentrations of oligomycin or carboxyatractyloside
(Debeer et al., 1974; Berry et al., 1983a; Plomp et al., 1985),
indicating that the mechanism responsible for the increase in J_o must
involve both ATP synthase and adenine nucleotide translocase.

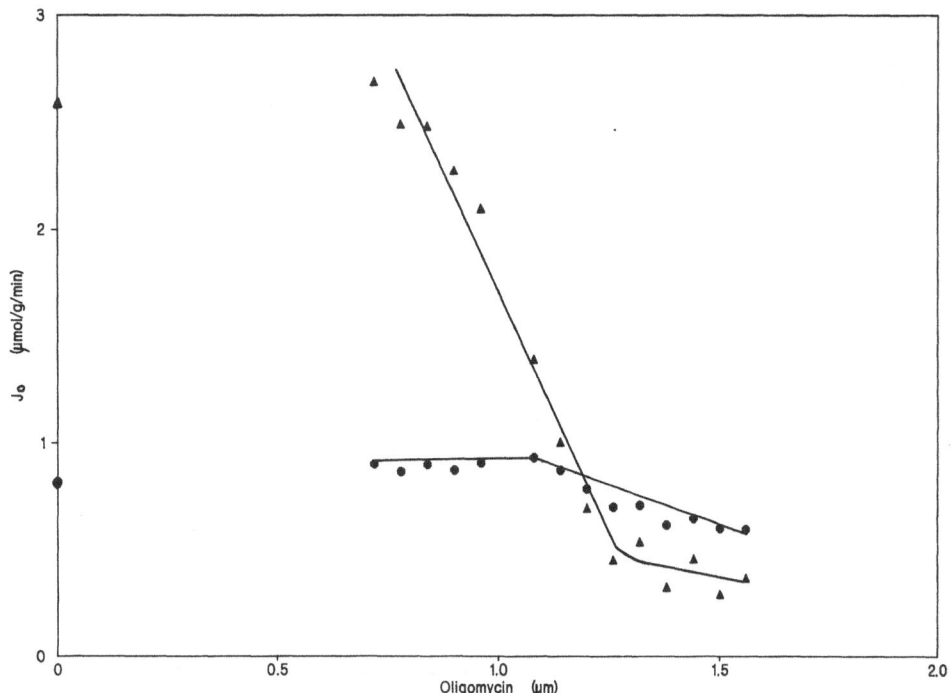

Fig. 11.   Effects of oligomycin on the respiration of hepatocytes
incubated in the presence of 10 mM pyruvate, 1 μM TDGA.
▲——▲ Krebs cycle oxygen uptake (J_O^K)
●——● Oxygen uptake for pyruvate oxidation to acetyl CoA
(J_O^P)

Nevertheless, our results are not readily explained on the basis that
fatty acid addition causes an increased turnover of cytoplasmic adenine
nucleotides due, as has been suggested for example, to an enhanced
operation of the plasma membrane (Na + K)-ATPase (Plomp et al., 1985).
Liver cells in the absence of added fatty acid are capable of maintaining
normal cation gradients across the plasma membrane (Krebs et al., 1974).
Unless it is argued that fatty acids (ranging from C6 to C16) (Berry et
al., 1983b) increase membrane permeability to these cations, a phenomenon
which has not been reported, there is no obvious reason why increased
activity of the (Na + K)-ATPase should occur on addition of fatty acid.
Current theories of respiratory control would predict that the initial
signal of increased cytoplasmic turnover of ATP would be an elevation of
cytoplasmic [ADP], which through well-recognized mechanisms would bring
about a stimulation of respiration. However, in our experiments ΔG_p
increased proportionately with J_O, so that cytoplasmic [ADP] fell as J_O
increased.

It is also difficult to find compelling theoretical reasons why
exposure of hepatocytes to normal concentrations of fatty acid should
induce rapid cytoplasmic ATP turnover in the fasted state. The
initiation of a massive futile cycle of ATP synthesis and hydrolysis by
addition of fatty acid would be expected to lead to a fall in the
cytoplasmic [ATP]/[ADP] ratio rather than the observed rise. In any
case, cycling to an extent sufficient to increase the respiratory rate by
over 50% would seem highly unlikely in view of the reported limited
capacity of such cycles (Newsholme and Crabtree, 1976; Katz and Rognstad,
1976).

Even so, alternative explanations for the stimulation of J_O by fatty acid based on a literal interpretation of chemiosmotic theory are not readily found, since the hypothesis assumes that the sole mechanism of mitochondrial energy transduction depends on the bi-directional flow of protons across the inner membrane, energy being stored in a transmembrane bulk-phase electrochemical proton gradient that is utilized for synthesis of ATP. Hence, if increased ATP turnover is not occurring, to account for our results it is necessary to invoke some form of proton pump "slippage" (Pietrobon et al., 1983), a concept that has not met with universal acceptance (Ferguson and Sorgato, 1982).

Energization of the Mitochondrial Membrane

These considerations have led us to examine the possibility that mitochondria may possess energy-transducing systems alternative to that described by chemiosmotic theory. The majority of those that have been suggested involve as an essential feature of their mechanism direct energization of the inner mitochondrial membrane by means of intramembranous charge separation, independent of the bulk aqueous phase (Azzi, 1969; Nordenbrand and Ernster, 1971; Malpress, 1984). In some instances a "localized" version of the chemiosmotic theory is postulated, in which its general principles are maintained, except that the proton current is thought to be confined to the membrane, flowing within it or as a "protoneural network" on its surface (Williams, 1978; Kell and Morris, 1981; Westerhoff et al., 1984). Others place less emphasis on current flow but argue that the energy transfer between membrane and ATP synthase leads to conformational change in the enzyme, that facilitates ATP synthesis and release (Rosing et al., 1977; Boyer, 1975, 1977; Schlodder et al., 1982).

Support for this latter view comes from experiments in which pulsating high-voltage electric fields of microsecond duration have induced ATP synthesis in submitochondrial particles from beef heart and isolated ATP synthase (Teissie et al., 1981; Knox and Tsong, 1984). Experiments of a similar nature have also been carried out with a chloroplast preparation (Witt et al., 1976). These studies, taken in conjunction with the observation that in chloroplasts a transient electric field is detected prior to the commencement of proton flow (Schlodder et al., 1982), compels serious consideration of the possibility that an electrical energization of the inner mitochondrial membrane is a fundamental feature of the energy-transducing process.

The likelihood that more than one mode of energy transduction exists in mitochondria is supported by the experiments in which we have demonstrated that the nature of the respiratory stimulation, on addition of substrate to hepatocytes, is dependent on whether or not an overt demand for ATP exists. The exposure of liver cells to fatty acid alone causes an increase in J_O^F but is without effect on J_O^K. Conversely, subsequent addition of lactate or other gluconeogenic precursor stimulates J_O^K but has no influence on J_O^F. Similarly, when liver cells are incubated with pyruvate, dramatic differences between the regulation of J_O^F and J_O^K are seen.

These observations lead us to consider the possibility that at least two interacting electrical circuits might be present in the mitochondria of liver cells: one, supplied by J_O^K corresponding to the proton flow associated with ATP synthesis and turnover; the second, powered by J_O^F or J_O^P representing the energy flow required for the conversion of the mitochondrial membrane to an electrically charged energized state necessary for the synthesis of ATP. In fact, only a relatively small fall in J_O^F or J_O^P leads to a decline in ΔG_p sufficient to prevent gluconeogenesis. The proposal that electron transfer to oxygen involves separate circuits, one for provision of proton current, the other for generation of an electrically energized inner membrane is, of course, a marked departure from the standard chemiosmotic model, which regards the inner membrane essentially as an inert partition (Mitchell, 1961, 1976).

It appears from our experiments, and those of others, that the ATP-synthase is inoperative below a certain critical voltage ($\Delta\Psi$ < 110 mV). Evidence for this has come from studies on a variety of experimental preparations. Using bacterial chromatophores, Clark et al. (1983) demonstrated that most of the non-ohmic proton flow passes through ATP synthase and that there is an exponential dependence of conductance on membrane potential. Seren et al. (1985) recently reported that F_O channels in submitochondrial particles have a gated response to membrane potential, being inoperative below approximately 110 mV. Our comparable findings with intact liver cells greatly increase the likelihood that this voltage-sensitive "gating" is of physiological significance and not an experimental artifact. We suggest, therefore, that J_O^F and J_O^P serve to energize the inner mitochondrial membrane, the stored energy thence being transferred to the integral membrane enzymes. The importance of generation of membranous electric charge during energy transduction has been emphasized by others (Higuti, 1984; Malpress, 1984). The magnitude of this charge, generated in our experiments by J_O^F or J_O^P, is reflected in the value of $\Delta\Psi$ which is developed as a consequence of membrane energization.

Some indirect evidence for the importance of membrane energization comes from experiments with uncoupling agents, in which ΔG_p and intracellular [ATP] are reduced to immeasurably low levels. Even in the presence of exogenous substrate, isolated hepatocytes from starved rats generally do not survive this treatment. Their respiration falls below 0.5 µmol/g/min, and they leak enzymes and take up trypan blue. However, in the presence of added pyruvate, respiration is sustained, with the cells generating acetyl CoA and accumulating ketone bodies; trypan blue is excluded and enzyme leakage does not occur. Trypan blue is a negatively charged dye, and hence it is predictable that it would be excluded from mitochondria provided that the polarity of the charge developed across the membrane was appropriate (i.e., positive outside). Conversely, Janus green B (a positively charged dye) is taken up, as might be expected, only by the mitochondria of intact cells.

A Unifying Hypothesis of Mitochondrial Energy Transduction

The experiments described in this chapter, taken in conjunction with the work of others, lead us to postulate that two modes of energy transduction exist within the mitochondria. One is essentially electrical in nature, in that fluctuating charge stored within the inner mitochondrial membrane is transduced to cause cyclical energization of the integral membrane enzymes, including the redox carriers, ATP synthase, and adenine nucleotide translocase. Voltage phase change might be achieved by looped (Mitchell, 1961, 1976) or cyclical (Mitchell et al., 1985) electron flow within the inner membrane. The effect of this energization is to adjust the activation energies of each membrane enzyme in the direction which tends to equalize the rates of forward and back reactions and the binding affinities for substrates and products. Thus, energization allows the enzymes to act not as mere catalysts but as "molecular machines" (McClare, 1971; Welch and Kell, 1986), moving the steady-state levels of reactants away from thermodynamic equilibrium and thereby providing the driving force for the multiplicity of endergonic chemical processes occurring in the cell. This driving force in electrical terms can be regarded as a voltage. It is detected as $\Delta\Psi$, the "mitochondrial membrane potential", by means of the partitioning of $TPMP^+$ and as ΔG_p, reflecting the energized steady-state value of the components of the reaction catalyzed by ATP synthase. Since a large number of enzyme-catalyzed cytoplasmic synthetic and transport processes are ATP-dependent, their mass-action ratios also would be shifted away from thermodynamic equilibrium and their Gibbs free-energy content increased, thus raising the total chemical-potential energy of the cell.

The electrical energy required to generate $\Delta\Psi$ comes from β-oxidation or pyruvate oxidation to acetyl CoA. Only a small voltage drop leads to a cessation of ATP turnover and synthetic processes, since intramembranous energy transduction acts cooperatively with the second mitochondrial energy-transducing mechanism to facilitate ATP formation. The latter mechanism is described by the now generally accepted chemiosmotic hypothesis, whereby a transmembrane electrochemical proton gradient across the inner mitochondrial membrane is generated and utilized to provide proton current for ATP synthesis and turnover. This may be measured as ΔpH. The proton current for this component of mitochondrial energy transduction is derived from the oxidations of the Krebs cycle and flows only when ATP turnover and synthesis are occurring, unless uncoupling agents are present. It is inferred that the driving force established by ΔpH in animal mitochondria is insufficient in itself to bring about ATP synthesis. The two systems, acting cooperatively, provide proton current at an adequate voltage to drive ATP-dependent synthetic processes.

On the basis of these ideas, it would appear that much of the respiration occurring in the basal state is not associated with ATP turnover but rather reflects the electrical energy flow required to maintain living systems in a steady-state far from thermodynamic equilibrium. If this conclusion proves correct, it will require a revision of presently held views concerning cellular and whole-animal efficiency. It is generally believed that differences in efficiency between individuals must relate to inherent differences in the efficiency with which chemical energy in the form of ATP is generated or utilized for metabolic syntheses. While this may be correct, it would seem equally probable that variations in efficiency may relate to intramembranous energy transduction. Acceptance of this viewpoint will obviously have considerable impact on approaches to elucidating the nature of cellular regulatory mechanisms and their disturbance in disease states.

ACKNOWLEDGEMENTS

We would like to thank Drs. G.R. Welch and M.D. Brand for valuable discussions. The skilled assistance of Mrs. S. Gay, Ms. D. Henly, Miss K. Bilyk, Miss J. Burton, Mrs. J. Fenech, Mrs. M. Grivell, Miss J. Morson, Mr. C. Strobel, Miss J. Weekley, and Mrs. E. Williams is gratefully acknowledged.

This work was supported in part by grants from the National Health and Medical Research Council of Australia.

REFERENCES

Azzi, A., 1969, Biochem. Biophys. Res. Commun., 37:254.

Beeckmans, S. and Kanarek, L., 1981, Eur. J. Biochem., 117:527.

Berry, M.N., 1974a, in: "Regulation of Hepatic Metabolism", Lundquist, F. and Tygstrup, N., eds., Munksgaard, Copenhagen.

Berry, M.N., 1974b, Methods Enzymol., 32:625.

Berry, M.N. and Friend, D.S., 1969, J. Cell Biol., 43:506.

Berry, M.N., Clark, D.G., Grivell, A.R. and Wallace, P.G., 1983a., Eur. J. Biochem., 131:205.

Berry, M.N., Gregory, R.B., Grivell, A.R., and Wallace, P.G., 1983b, Eur. J. Biochem., 131:215.

Boyer, P.D., 1975, FEBS Lett., 58:1.

Boyer, P.D., 1977, Trends Biochem. Sci., 2:38.

Brand, M.D. and Felber, S.M., 1984, Biochem. J., 217:453.

Campo, M.L., Zhang, C.-J. and Tedeschi, H., 1984, Biochem. Soc. Trans., 12:384.

Clark, A.J., Cotton, N.P.J. and Jackson, J.B., 1983, Biochim. Biophys. Acta, 723:440.

Cohen, S.M., Ogawa, S., Rottenberg, H., Glynn, P., Yamane, T., Brown, T.R., and Shulman, R.G., 1978, Nature, 273:554.

Debeer, L.J., Mannaerts, G. and DeSchepper, P., 1974, Eur. J. Biochem., 47:591.

Degenaar, C.P., 1983, Clin. Chim. Acta., 131:155.

Ferguson, S.J. and Sorgato, M.C., 1982, Ann. Rev. Biochem., 51:185.

Fillingame, R.H., 1980, Ann. Rev. Biochem., 49:1079.

Hems, R., Ross, B.D., Berry, M.N. and Krebs, H.A., 1966, Biochem. J., 101:284.

Higuti, T., 1984, Molec. Cell. Biochem., 61:37.

Hoek, J.B., Nicholls, D.G. and Williamson, J.R., 1980, J. Biol. Chem., 255:1458.

Katz, J. and Rognstad, R., 1976, Curr. Top. Cell Reg., 10:237.

Kell, D.B. and Morris, J.G., 1981, in: "Vectorial Reactions in Electron and Ion Transport in Mitochondria and Bacteria", F. Palmieri, F. Quagliariello, N. Siliprandi and E. C. Slater, eds., Elsevier/North-Holland, Amsterdam.

Kiorpes, T.C., Hoerr, D., Ho, W., Weaner, L.E., Inman, M.G. and Tutwiler, G.F., 1984, J. Biol. Chem., 259:9750.

Knox, B.E. and Tsong, T.Y., 1984, J. Biol. Chem., 259:4757.

Krebs, H.A., Dierks, C. and Gascoyne, T., 1964, Biochem. J., 93:112.

Krebs, H.A., Cornell, N.W., Lund, P. and Hems, R., 1974, in: "Regulation of Hepatic Metabolism", F. Lundquist and N. Tygstrup, eds., Munksgaard, Copenhagen.

Malpress, F.M., 1984, Biochem. Soc. Trans., 12:399.

Matlib, M.A. and O'Brien, P.J., 1975, Arch. Biochem. Biophys., 167:193.

McClare, C.W.F., 1971, J. Theor. Biol., 30:1.

Mitchell, P., 1961, Nature, 191:144.

Mitchell, P., 1976, Biochem. Soc. Trans., 4:399.

Mitchell, P., Mitchell, R., Moody, A.J., West, I.C., Baum, H. and Wrigglesworth, J.M., 1985, FEBS Lett., 188:1.

Newsholme, E.A. and Crabtree, B., 1976, Biochem. Soc. Symp., 41:61.

Nicholls, D.G., 1974, Eur. J. Biochem., 50:305.

Nordenbrand, K. and Ernster, L., 1971, Eur. J. Biochem., 18:258.

O'Shea, P.S. and Chappell, J.B., 1984, Biochem. J., 219:401.

Pietrobon, D., Azzone, G.F. and Walz, D., 1981, Eur. J. Biochem., 117:389.

Pietrobon, D., Zoratti, M. and Azzone, G.F., 1983, Biochim. Biophys. Acta, 723:317.

Plomp, P.J.A.M., van Roermund, C.W.T., Groen, A.K., Meijer, A.J., and Tager, J.M., 1985, FEBS Lett., 193:243.

Rosing, J., Kayalar, C. and Boyer, P.D., 1977, J. Biol. Chem., 252:2478.

Schlodder, E, Graber, P. and Witt, H.T., 1982, in: "Electron Transport and Photophosphorylation", J. Barber, ed., Elsevier, Amsterdam.

Scholz, R., Schwabe, U. and Soboll, S., 1984, Eur. J. Biochem., 141:223.

Seren, S., Caporin, G., Galiazzo, F., Lippe, G., Ferguson, S.J. and Sorgato, M.C., 1985, Eur. J. Biochem., 152:373.

Soboll, S. and Stucki, J., 1985, Biochim. Biophys. Acta, 807:245.

Srere, P.A., Halper, L.A. and Finkelstein, M.B., 1978, in: "Micro-environments and Metabolic Compartmentation", P.A. Srere and R. W. Estabrook, eds.,Academic Press, New York.

Sumegi, B. and Srere, P.A., 1984, J. Biol. Chem., 259:8748.

Sumegi, B., Gilbert, H.F. and Srere, P.A., 1985, J. Biol. Chem., 260:188.

Teissie, J., Knox, B.E., Tsong, T.Y. and Wehrle, J., 1981, Proc. Natl. Acad. Sci. U.S.A., 78:7473.

Welch, G.R. and Kell, D.B., 1986, in: "The Fluctuating Enzyme", G.R. Welch, ed., Wiley-Interscience, New York.

Westerhoff, H.V., Melandri, B.A., Venturoli, G., Azzone, G.F. and Kell, D.B., 1984, FEBS Lett., 165:1.

Williams, R.J.P., 1978, Biochim. Biophys. Acta, 505:1.

Williamson, J.R., Scholz, R., Browning, E.T., Thurman, R.G. and Fukami, M.H., 1969, J. Biol. Chem., 244:5044.

Witt, H., Schlodder, E. and Graber, P., 1976, FEBS Lett., 69:272.

Woods, H.F., Eggleston, L.V. and Krebs, H.A., 1970, Biochem. J., 119:501.

EXPERIMENTAL ANALYSIS OF SPATIOTEMPORAL ORGANIZATION

OF METABOLISM IN INTACT CELLS: THE ENIGMA OF "METABOLIC

CHANNELING" AND "METABOLIC COMPARTMENTATION"

E. Kohen[1], G.R. Welch[2], C. Kohen[1],
J.G. Hirschberg[3], and J. Bereiter-Hahn[4]

[1]Department of Biology, University of Miami
Coral Gables, Florida

[2]Department of Biological Sciences
University of New Orleans, New Orleans, Louisiana

[3]Department of Physics, University of Miami
Coral Gables, Florida

[4]Arbeitskreis Kinematische Zellforschung
Zoologisches Institut, J.W. Gothe Universität
Frankfurt a.M., G.F.R.

INTRODUCTION

The organization and control of multienzyme systems in the living cell raises numerous challenging questions. Based on studies of protein-protein interactions in multienzyme aggregates (Frieden and Nichol, 1981; Welch, 1977) and permeabilized cells (Shannon et al., 1977), several conclusions can be drawn concerning the intracellular behavior of complex metabolic pathways. One of the interesting concepts evolving from such studies is that of "channeling", i.e. a substrate is channeled between two or more enzymes while it is diffusionally constrained from the bulk phase. Together with hypotheses of "channeling", "metabolic compartmentation" and interactions of enzymes with cytoskeletal components, there are also differing views as to allosteric controls of multienzyme pathways. Thus, a cell viewed as a multienzyme system represents a complex structure in which the net activity of one enzyme is affected by that of all the enzymes in the system. According to Kacser and Burns (1979), detailed, or even complete, knowledge of an enzyme or all the enzymes in isolation is, therefore, insufficient to determine the role they play when they are embedded in the cell matrix and linked kinetically to each other.

Abbreviations used are as follow: FDP, fructose-1,6-diphosphate; G1P, glucose-1-phosphate; G6P, glucose-6-phosphate; 6-PG, 6-phosphogluconate; DNP, dinitrophenol, and PFK, phosphofructokinase.

This leads to the conclusion that any property of such a structure, e.g. a flux, is therefore not simply the sum of its parts. The study of enzymes or cell components in isolation neglects crucial in situ interactions which occur within the intact living cell. Therefore, no matter how accurate the information derived from isolated systems, it can be misleading with regard to intracellular fluxes and metabolite levels, which are systemic properties, and thus approach their final values by the interactions of, in principle, all the elements in the system.

The conceptual framework based on theoretical analysis presented by Kacser and Burns (1979) is combined with an experimental approach to enzyme systems in terms of modulation in vivo. The experimental analysis of the spatiotemporal organization of enzyme systems in intact living cells requires the use of optical methods. Using microspectrofluorometry for example, the result of metabolic perturbations, such as microinjections of substrates or modifiers, can be followed at different sites of the living cell by means of fluorescence changes attributable to an endogenous tracer, which is often nothing else than a coenzyme, pyridine or flavin, sensitive to redox changes. In this way a topographic mapping of dynamic processes, i.e. metabolic fluxes and redox transients, can be followed in the intact cell.

Glycolysis (and its alternate, the hexose monophosphate shunt) constitutes the central scheme for degradation of carbohydrates in living cells. This amphibolic process is vitally important in producing both energy and carbon precursors. According to the classic view, such multienzyme systems are regarded as homogeneously dissolved in the aqueous cytosol (and nucleoplasm). It is now apparent that the cytoplasm of larger eukaryotic cells is spanned not only by many membranous surfaces, but also laced with a dense array of fibrous cytoskeletal elements (e.g., the microtrabecular lattice) (Porter and Tucker, 1981). There is abundant evidence that much of intermediary metabolism takes place in association with such particulate structures (see Clegg, 1984; Wombacher, 1984; Welch, 1977, 1985; Welch and Keleti, 1981; and others cited therein), and that the enzymes of glycolysis associate with these cytological substructures (Masters, 1981; Wilson, 1980; Friedrich, 1985; Ottaway and Mowbray, 1977). This association bears many implications as regards the organization, dynamical behavior, and regulation of glycolysis (Hess, 1973).

Most of what we know about the organization of glycolytic enzymes has come from in vitro studies with isolated proteins, as well as with cellular particulate material, e.g., muscle filaments, membrane fragments, whole membranous "ghosts" (Masters, 1981). Until recently, little information was available on actual glycolytic organization within intact cells. The living cell represents in most instances a fragile entity, operating in a complex way on a time scale where enzyme kinetics and metabolic phenomena often proceed at a microsecond-millisecond pace.

Multichannel (multisite) microfluorometry, in conjunction with microinjection of metabolic intermediates and effectors, has been used for in situ topographic analysis of metabolic transients (e.g., $NAD(P)^+$-$NAD(P)H$) in correlation with structure and compartmentalization in single cells (Kohen et al., 1979, 1981a,b, 1983). Notably, these studies have revealed a mosaic-like spatial pattern for the metabolism of glycolytic phosphate esters. The form of this pattern was seen to depend on the physiological state of the cells, as well as on the nature of the glycolytic compound injected. Using Krebs cycle substrates, similar studies have been extended to a comprehensive study of metabolic responses and delays in the mitochondrial versus extramitochondrial regions. We discuss the implications of this design on current models of glycolytic spatiotemporal behavior and regulation, as well as mitochondrial-extramitochondrial compartmentation.

EXPERIMENTAL MATERIALS AND METHODS

Microspectrofluorometer

The grating microspectrofluorometer used in these studies has been described previously (Kohen et al., 1981a,b). It is designed essentially to superimpose a dynamic functional or metabolic map of the cell onto its ultrastructural map for structure and function correlation, especially when operated in one of its two modes (_viz._, the topographic or spectral options). The defined purpose is achieved by means of the following conditions:

(1) a long-working distance (10 cm), high numerical aperture phase-fluorescence system mounted on an inverted microscope, to allow free distance for cell manipulations (i.e., microelectrophoretic injections of substrates/effectors) and high efficiency in collection of fluorescence;

(2) imaging of cell organelles and sites onto detector channels of an optical multichannel analyzer (resolution element in the micrometer range); and

(3) when warranted, electron microscopy of cells premarked with a diamond objective, and thereby retrieved for ultrastructural analysis and organelle morphometry (see Fig. 1).

The apparatus is operated in two modes:

(1) topographic options: to map within the microcompartments of a single cell (or adjacent cells) the fluorescence changes associated with transient perturbations of coenzyme (pyridine, flavin) redox states triggered by microinjections of metabolites or effectors; and

(2) spectral option: to analyze, from emission spectra, the free/bound states of cellular coenzyme.

The present apparatus, based on a unilinear scan, is the forerunner of a two-dimensional system aimed at unraveling metabolic reactions throughout a single living cell--viewed as a matrix at every point of which diffusion and fluxes of metabolites occur continually.

Microelectrophoretic Injections

A dual microinjector is used for independent electrophoretic introduction of different substrates into the same cell, either sequentially or before/after injection of effector agents (Kohen et al., 1981 a,b). Cell tolerance to microcurrents in the range of 10^{-7} amp/μ has been established.

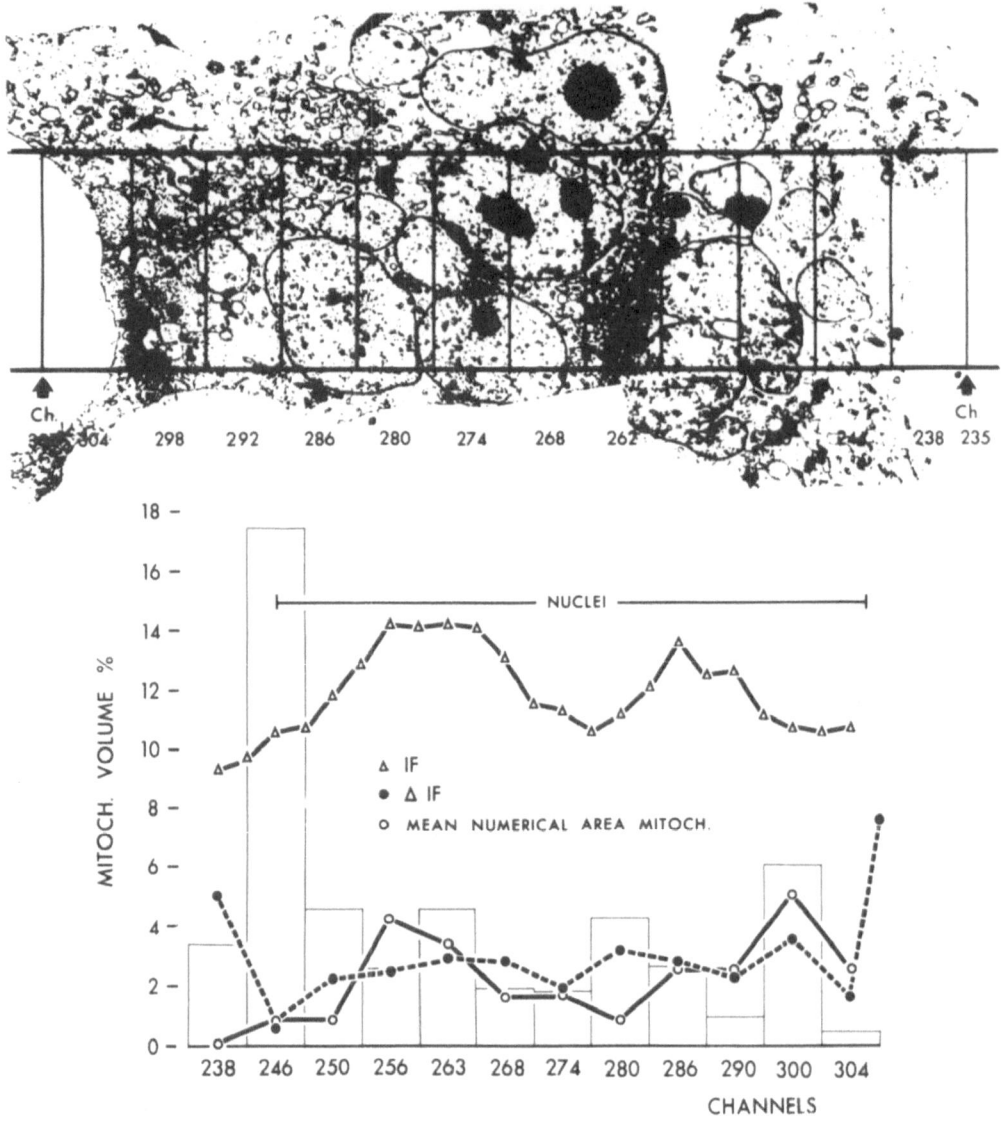

Fig. 1.   Morphometry of a cell studied by microspectrofluorometry.
A.   Electron micrograph of L cell microinjected with malate.
The detector channels on which the cell, followed by microfluorometry, was imaged, are brought to coincidence with corresponding regions of the ultrastructural image.   In this way the
metabolic map of the living cell is superimposed onto the
ultrastructural map, i.e. the metabolic response of specific
cell sites can be related to the structure, organelle density
and organelle conditions at the same sites.   B.   Comparison of
two functional (one steady state, one dynamic) parameters to
two morphometric parameters.   IF = initial fluorescence of the
cell prior to injection of malate.   ΔIF= fluorescence increase
after microinjection of malate; mean numerical area mitoch. = a
measure of mitochondrial swelling.   The columns show % mitochondrial volume over total volume for the cell region covered
by indicated channel number.   The numbers in the abscissa
correspond to numbers of detector channels viewing a given cell
region as indicated in Fig. 1A.   IF and   ΔIF are expressed in
arbitrary units (not shown in ordinate).

Upon injection of substrate/modifier, the observed transient changes in coenzyme fluorescence have two components:

(1) a change in the relative levels of oxidized/reduced coenzymes;

(2) changes in free/bound coenzyme, which can affect the fluorescence intensity and spectrum.

Studies are in progress to evaluate the contribution of this second component.

Biological Material

Well differentiated L sarcoma cells and highly malignant, poorly differentiated CCL 136 rhabdomyosarcoma cells were chosen because of their suitable morphology for the study of intracellular metabolic compartments, especially cytoplasmic/nuclear regions and subcompartments associated with glycolysis and the hexose monophosphate shunt (Kohen and Kohen, 1984). In L cells the response to substrates of glycolysis, hexose monophosphate shunt, and Krebs cycle is observed. In the case of CCL 136 cells, mitochondria [identified by use of the organelle-specific vital dye, dimethylaminostyrylmethylpyridinium iodine (DASPMI)] (Bereiter-Hahn, 1976) are evident but seem functionally deficient when challenged with injections of malate. This reinforces the fact that the cell is well suited for study of non-respiratory bioenergetic pathways.

Injection Sequences and Other Conditions

The sequential injections of two different substrates from the same, or different, multienzyme sequences were aimed at distinguishing specific localization and compartmentalization of the metabolic responses to glycolytic phosphate esters. In view of previously observed multifocal responses to fructose-1,6-diphosphate (FDP) and glucose-1-phosphate (GlP) in EL2 ascites cancer cells (Kohen et al., 1979, 1981b), these injections were repeated in CCL 136 cells. The preference given to substrates of the glycolytic and shunt pathways for CCL 136 cells is explained by the predominance in these cells of extramitochndrial pathways, compared to the activity of the respiratory pathway. However, mitochondrial activity must still be present, as the responses to glycolytic and shunt substrates were activated upon addition of oxidative-phosphorylation uncouplers (dinitrophenol [DNP], 10^{-4} M) or ATP traps (ethionine, 6×10^{-3} M), revealing an enhancement of anaerobic metabolic responses upon depletion of ATP supply (Kohen et al., 1985).

EXPERIMENTAL RESULTS

Bistable Responses

Earlier studies (unpublished) with injected FDP revealed a patchy, mosaic intracellular pattern of fluorescence. Upon injection of FDP, the NAD(P)H response tended to be initially multifocal with eventual generalization; but the observations were difficult, due to the rapid rise time (e.g., 1 second) until ultimate generalization. Also, the response to FDP may have a preliminary oxidative phase, possibly due to its action as an allosteric effector on different glycolytic enzymes or

due to partitioning between glycolysis and gluconeogenesis. Later the
substrate effect takes over, which is manifested by $NAD(P)^+$ reduction.

In L cells the sequence FDP-G6P was repeated in controls (Fig. 2A, B)
and cells maintained in a hypertonic medium, i.,e., incubation buffer
supplemented with 0.25 M mannitol (Mansell and Clegg, 1983) (Fig. 2C, D).
The purpose was to investigate the role that changes in the bulk water
level may play in the organization and activities of cell enzymes. In
the L cell maintained in isotonic medium, the response to FDP is
oxidation and that to G6P is reduction, as recognizable from the time
curve and topographic traces.

Upon incubation in the presence of 0.25 M mannitol, within minutes
the response to FDP shows a reduction following a short period of
oxidation; but this time there is only an oxidative response to G6P. The
observed patterns are consistent with the bistability of $NAD^+ \rightleftarrows NADH$
redox changes. When the response to FDP is an oxidation, a subsequent

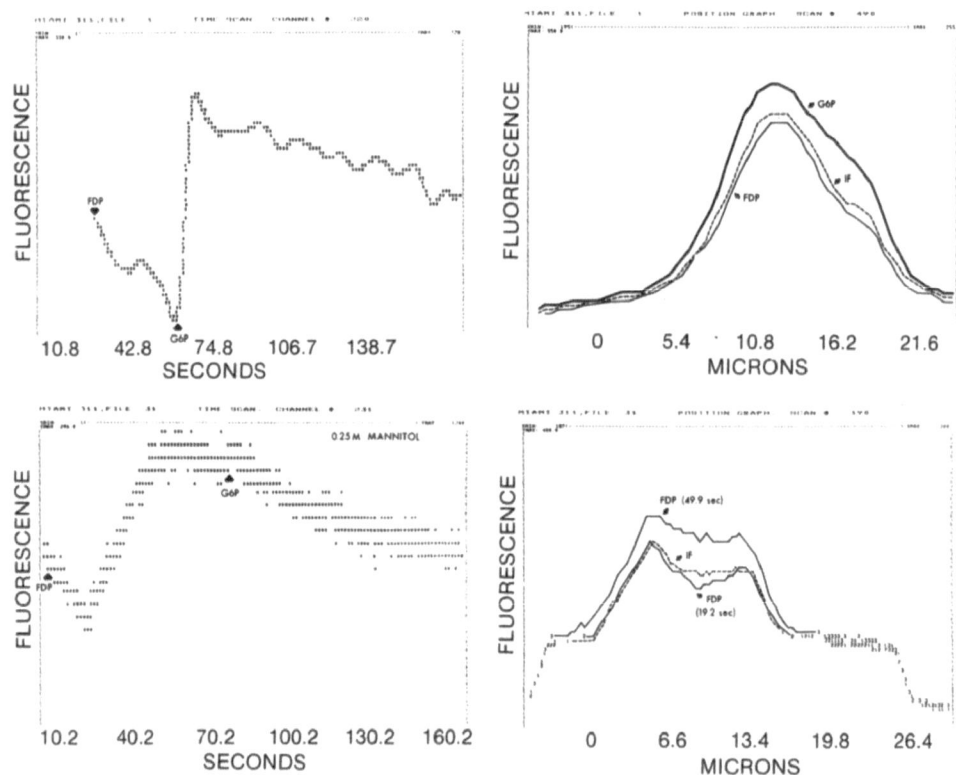

Fig. 2.   Bistability of metabolic responses. NAD(P)H transient
          fluorescence changes from sequentially injected FDP and G6P in
          L cells maintained in isotonic and hypertonic (0.25 M mannitol
          added) media. IF = fluorescence of cell prior to first
          injection. Times in parenthesis are sec after injection.
          A. Time curve (smoothed) in isotonic medium.
          B. Topographic curves of cell fluorescence along the multi-
              channel scan axis.
          C. Time curve (unsmoothed) in the presence of 0.25 M mannitol.
          D. Topographic curves under the same conditions.

injection of G6P can result in reduction; but if near maximal reduction is already observed with FDP, a swing of the pendulum will now bring forth the other bistable phase, i.e., oxidation upon subsequent injection of G6P.

Similarly with consecutive injections of G1P and G6P, a reductive response is observed only with the first intermediate, and upon injection of the second the bistable state reverses to oxidation (Fig. 3A and B).

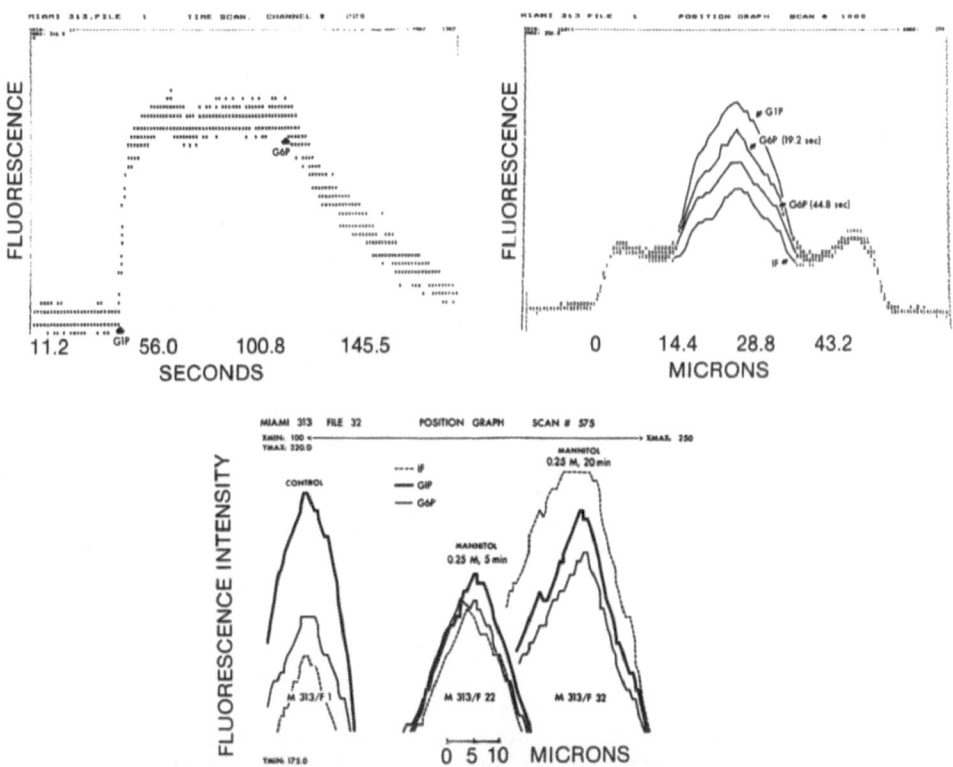

Fig. 3.   Possible evidence that two glycolytic phosphate esters (G1P and and G6P) interact with the same $NAD(P)^+$ pool. Metabolic responses (NAD(P)H transient fluorescence changes) to sequentially injected G1P and G6P. The lack of a reductive response to G6P following maximal NAD(P)H response to G1P suggests that both intermediates have access to the same NAD(P)H pool. Following near maximal NAD(P)H reduction by the first injected substrate, a subsequent injection results in an "overshoot" of the response towards oxidation (bistability) when no further reduction is possible. Thus, if channeling exists both substrates are directed to the same multienzyme channel and are in contact with the NAD(P)H pool within the same compartment. A. Time curve. B. Topographic curves. C. Topographic curves in isotonic medium, and in the presence of 0.25 M mannitol (5 and 20 min). The gradual inhibition of the reductive response to the first injected substrate is noticed as IF (prior to injection of any substrate) approximates the maximal level of reduction. The high NAD(P)H fluorescence observed 20 min after addition of mannitol could be due to activation of endogenous glycolysis under hypertonic conditions.

However, when the cell is placed in 0.25 M mannitol-possibly due to changes in the bulk water of the cell under hypertonic conditions the cells start at a high level of NAD^+ reduction (see Fig. 3C, mannitol, 20 min) and no further reduction; but only oxidation is observed upon injection of G1P and G6P.

Multifocal and Generalized Responses

In the case of G1P, the metabolic events proceed at a pace 20-30 times slower than with G6P, which circumstance facilitates the observation of unifocal, multifocal, and generalized response patterns in actual sequence. As seen in Fig. 4, small foci of increased NAD(P)H fluorescence are already observed within 0.1 second of the injection. However, the intensity of the increase is not great. Repeated observations in over 100 cells indicate that the focal change of NAD(P)H fluorescence is above the baseline level (labeled "IF" in Fig. 4A) which precedes the injection, rather than randomly above or below this line. Thus, on a statistical basis, we are observing an actual, initial multifocal response to substrate. Within another 0.1 second, the response gains prominence while remaining focal. After about 4 seconds the NAD(P)H transient covers nearly the whole cell, but a focus with no detectable activity is still noticeable. It takes more than 20 seconds to obtain the maximal and fully generalized response to G1P.

These studies were repeated with G6P, in which case the rise time of the NAD(P)H response is around 3-5 seconds. The responses to G6P were, again, initially multifocal (Fig. 5) and subsequently generalized, but on a somewhat compressed time scale compared to G1P. In the majority of cases the initial responses to G6P were coincident with the nuclear region and later spread to the cytoplasmic region. The initially nuclear localization of the NAD(P)H responses was confirmed more extensively with injections of 6-PG. The pattern of responses within the nuclear region was itself focal or multifocal (Fig. 6, Fig. 7), suggesting that different metabolic subcompartments within the nucleus join the metabolic process in a certain sequence rather than simultaneously. In other instances the responses were biphasic. After an initial increase of NAD(P)H the subsequent phase of increase was not simultaneous throughout the nucleus and cytoplasm, but again started at nuclear sites and then extended to the whole cell. There were also cells exhibiting a reversal of the response sequence, with cytoplasmic sites starting first.

Cytochalasin B (1 µg/ml) (Schliwa, 1982) was added acutely to the cell medium, to study what effect the disorganization of the intracellular microarchitecture might have on metabolic subcompartments. However, we have not determined what degree of cytoskeletal disorganization was achieved (if any) with this drug. There was little change in the multifocal/generalized response patterns. However, within an hour of the addition (Fig. 8), the tendency of the nucleus to segregate (as a prelude to its extrusion, due to disruption of the holding microfilament network) was obvious, accompanied by enhanced compartmentation of the NAD(P)H response to G6P. Within the nucleus itself, subcompartments joining the NAD(P)H response at different times were manifest.

Injection of a Krebs cycle substrate, malate, revealed time-dependent localizations, in extramitochondrial/mitochondrial regions (Fig. 9). While the site of earlier $NAD(P)^+$ reduction seems coincident with the nucleus, the response of a cytoplasmic invagination within the nucleus is strongly suspected.

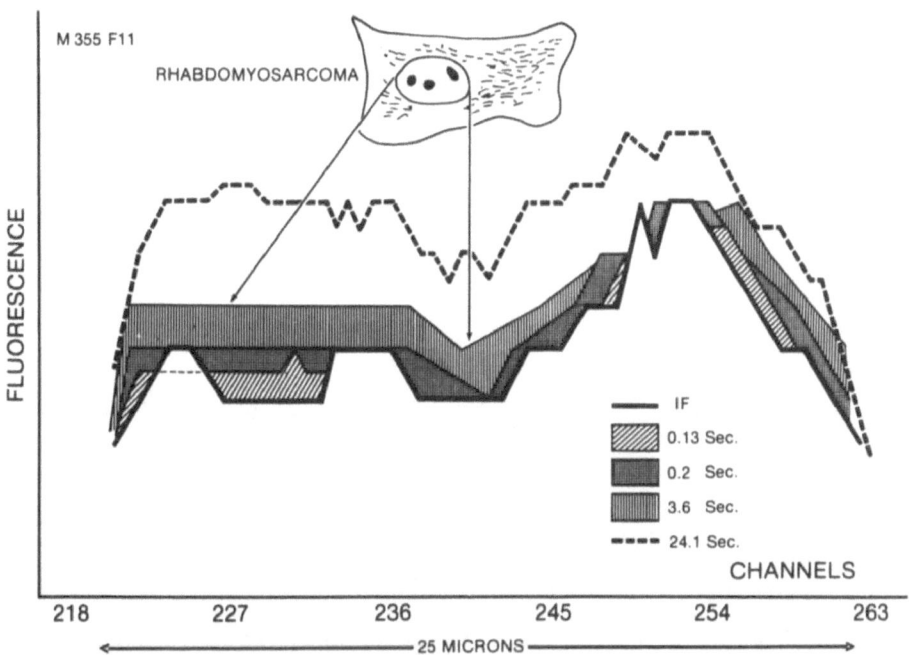

Fig. 4   Multifocal responses vs. diffusional changes. Topographic
plots of the NAD(P)H response to injection of G1P into a CCL136
cell (see drawing in middle upper part). Initial NAD(P)H level
(IF) and levels at 0.13, 0.2, 3.6 and 24.1 sec after injection
of G1P are shown. The multifocal responses at 0.13 and 0.2 sec
become generalized from 3.6 to 24.1 sec. The observed delays
are one–two orders of magnitude above substrate diffusion times
(no longer than 30-50 msec end-to-end within such mammalian
cells in culture). For position of channels in this and many
following figures, see drawing of cell with scale of image on
multichannel detector, in upper center.

Channeling(?) of Metabolic Pathways

Sequential injection of 6-PG/G1P to a cell revealed a response
pattern suggesting compartmentalization of these glycolytic phosphate
esters, i.e., the sites showing a relatively higher response to G1P
showed a relatively smaller response to 6-PG and vice-versa (Fig. 10).
Similarly, sequential injections of malate and 6PG (Fig. 11) revealed
differential behavior of malate and 6PG-induced NAD(P)H transients in
different 2-3 μm wide regions of the cell.

Localized versus Generalized Responses in the Presence of Drugs

With injections of 6-PG and Krebs cycle substrates the metabolic
transient responses observed are often the "tip of an iceberg". They show
a localized pattern (see also succinate responses in different 2-3 μm
wide regions of an L cell, Fig. 12) and relatively weak amplitude of the
NAD(P)H transients. In the case of succinate the fastest rise of NAD(P)H
fluorescence and larger amplitude of the transients are observed in
regions associated with nuclear locations of the binucleated L cells seen

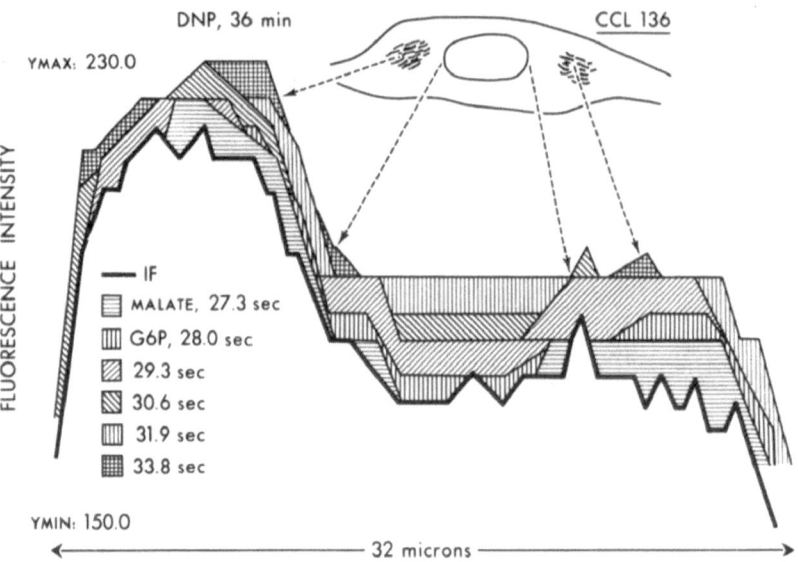

MIAMI 367, FILE 16   POSITION GRAPH

DNP, 36 min   CCL 136

YMAX: 230.0

FLUORESCENCE INTENSITY

— IF
▤ MALATE, 27.3 sec
▥ G6P, 28.0 sec
▨ 29.3 sec
▧ 30.6 sec
▥ 31.9 sec
▦ 33.8 sec

YMIN: 150.0

←——————— 32 microns ———————→

Fig. 5.   Localizations of metabolic compartments and multifocal
responses to malate and G6P. CCL136 cell maintained 36 min in
the presence of 10^{-4} M DNP prior to injection of malate.

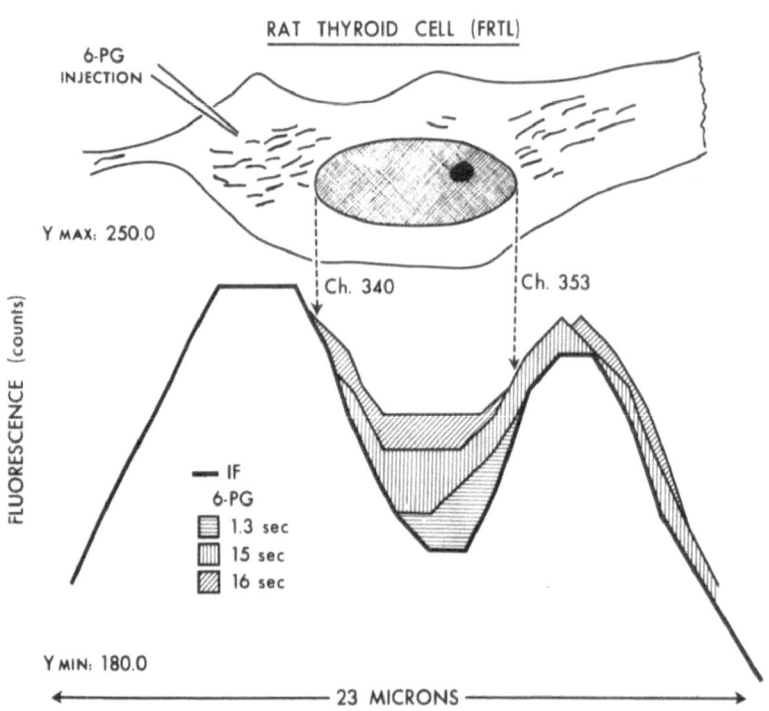

RAT THYROID CELL (FRTL)

6-PG
INJECTION

Y MAX: 250.0

Ch. 340   Ch. 353

FLUORESCENCE (counts)

— IF
6-PG
▤ 1.3 sec
▥ 15 sec
▨ 16 sec

Y MIN: 180.0

←——————— 23 MICRONS ———————→

Fig. 6.   Preferentially nuclear localization of the NADPH response to
injected 6-PG in rat thyroid cell in culture. Initial NADPH
level (IF) and topographic curves at 1.3, 15 and 16 sec after
injection of 6-PG.

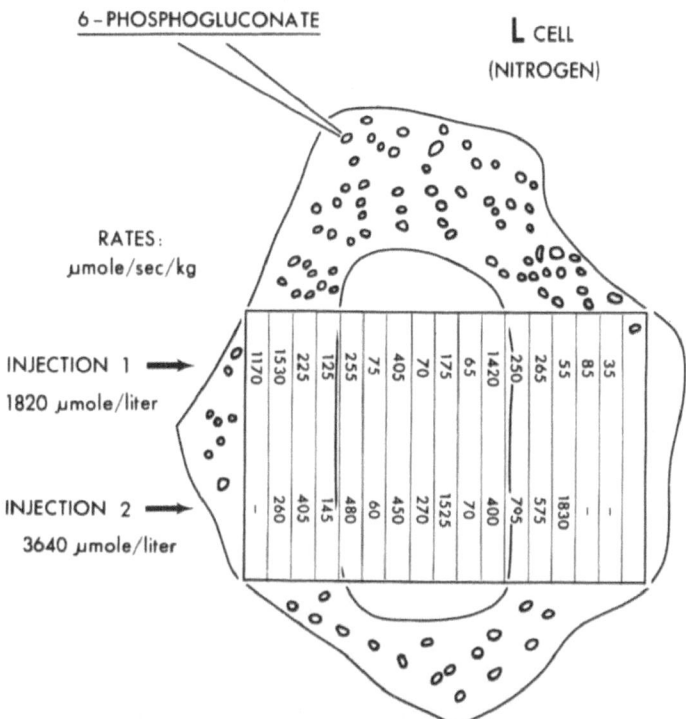

Fig. 7.   Mapping of metabolic rates (calculated from dose of injected
6-PG and NADPH transient,halftime) in an L cell maintained in a
nitrogen atmosphere (to maximize the NADPH transient
responses).   The site-by-site rates are shown for two
consecutive injections.

in Fig. 11. The weakest transient is observed in the internuclear region
(site E, Fig. 11). The association of the succinate response with the
nuclear sites does not necessarily imply an intranuclear pathway for
succinate reduction of $NAD(P)^+$. There is an energy barrier for succinate
reduction of this coenzyme, and usually flavin reduction is observed.
However, ATP-driven reverse electron transfer (Chance and Hollunger,
1961) can lead to $NAD(P)^+$ reduction by succinate. As to the nucleus
localization, it is more likely to represent the response of perinuclear
mitochondria and, possibly, also that of cytoplasmic invaginations into
the nuclear region.

    We have some evidence that localized weak responses to glycolytic,
pentose shunt, and Krebs cycle substrates are under ATP control.
Conditions leading to an ATP drop, such as addition of an ATP trap
(ethionine), an uncoupler (dinitrophenol) (Fig. 13) certain drugs
(Christopherson and Gammeltoft, 1978; Shroot et al., 1981; Kohen, E.,
unpublished results), and photosensitizers (Dougherty et al., 1978;
Santus et al., 1983; Kohen, E., unpublished results) leading to
structural damage of mitochondria, result in a "volcano-like" strong
activation of metabolic pathways. In the case of the anti-psoriatic
drug, anthralin (Christopherson and Gammeltoft, 1978; Shroot et al.,
1981; Kohen, E., unpublished results) (Fig. 14), and the photosensitizer,
hematoporphyrin (Dougherty et al., 1978; Santus et al., 1983; Kohen, E.,

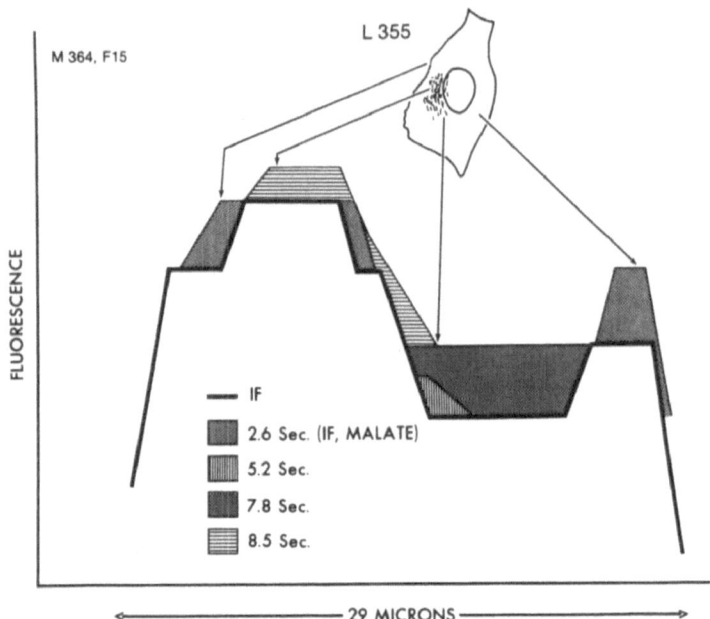

Fig. 8.   Time dependent localization of the NAD(P)H transient
           (fluorescence, see ordinate) response to injection of the Krebs
           cycle substrate malate into an L cell adapted to growth in a
           0.355 M NaCl medium (whereby there is an increased density and
           structural reorganization of mitochondria). With known
           extramitochondrial and mitochondrial malate dehydrogenases, the
           observed pattern suggests an early focal and subsequently
           generalized extramitochondrial malate dehydrogenase response,
           followed by a mitochondrial malate dehydrogenase response after
           a delay (7-8 sec) at the mitochondrial membrane. The partial
           coincidence of the extramitochondrial response with the
           nuclear region can be attributed to a cytoplasmic invagination
           at this site.

unpublished results), the structural damage in mitochondria can be
demonstrated by fluorescence micrographs using the vital probe, DASPMI
(Fig. 15) (Bereiter-Hahn, 1976). It is a striking example of the
intracellular interaction between cell organelles and metabolic pathways,
that release from mitochondrial control will lead to explosive
activations of metabolic pathways. Such interactions could be missed in
studies with isolated organelles and in vitro enzyme systems. Thus,
another justification is obtained for the pertinence of the approach
proposed by Kacser and Burns (1979), for studies in intact cell systems
rather than extrapolations from isolated enzymes and organelles.

    Another consequence of release from mitochondrial control is the
randomization throughout the cell of localized metabolic responses
observed prior to addition of agents leading to structural alterations of
mitochondria. Compare the transient responses to 6-PG in different
2-3 μm regions of a control (Fig. 16) versus an anthralin-treated
(Fig. 16B) L cell. The site-to-site delays for transient rise are
considerably decreased in the anthralin-treated cells. The near
synchronization of transient metabolic responses in the latter cells
could suggest a tuning of the different metabolic "channels" to operate

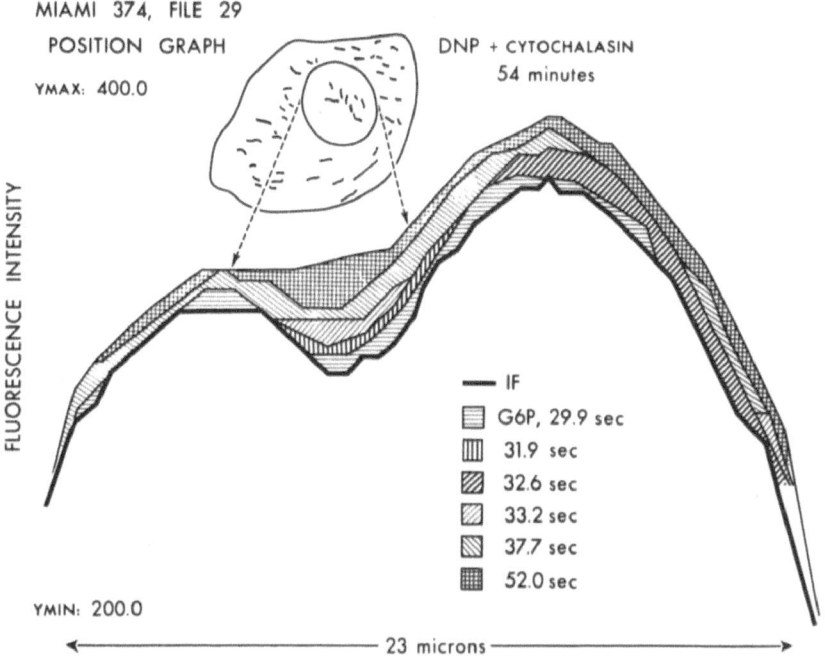

Fig. 9.   Topographic curves of the NAD(P)H response (fluorescence
          intensity) to microinjected G6P in a CCL136 cell maintained as
          indicated in presence of 10^{-4} M DNP and 1 µg/ml cytochalasin B.

on a nearly identical time clock. Assuming that the local time clocks
could be dependent on local ATP concentrations, the synchronization may
result from a global drop of the ATP in the bulk phase.

Since the cells are more sensitive to changes in ATP levels,
practically regardless of the substrate used, two possibilities are
considered:

(1) allosteric control by ATP at the level of different enzymes
    within multienzyme changes, or

(2) control by ATP of the product of bioenergetic pathways.

IMPLICATIONS OF HETEROGENEITIES IN GLYCOLYSIS AND OTHER METABOLIC
PATHWAYS

The physiological advantages of enzyme organization are numerous and
have been treated at length in the literature (for reviews see Clegg,
1984; Wombacher, 1984; Masters, 1981; Friedrich, 1985; Welch, 1977, 1985;
Welch and Keleti, 1981). Most discussions focus on such features as
"channeling" of intermediates, elimination of diffusional transit time,
and coordinate regulation of multiple related enzyme activities. Of
course, any (or all) of these features might be invoked for the case of
organization in glycolysis. There are theoretical and experimental
indications that glycolysis generates limit-cycle oscillations. Enzyme
heterogeneity can be expected to alter the character of such behavior.

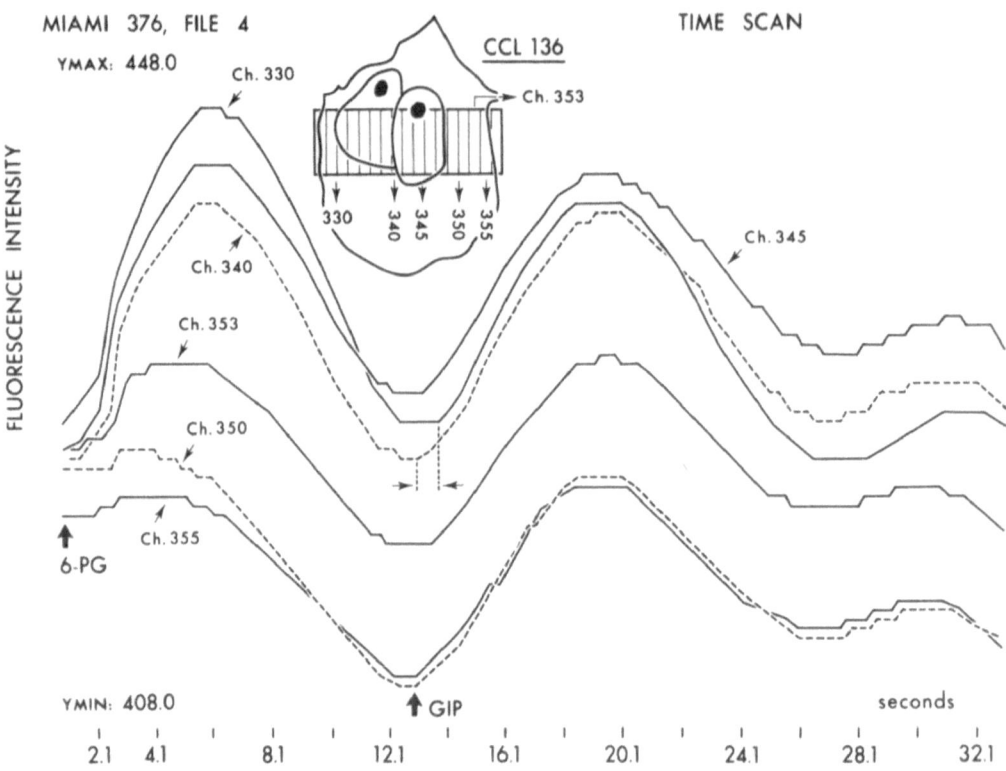

Fig. 10. Possible separate channeling and compartmentation of glycolytic
and pentose shunt pathways. Time course of the response to
consecutive injections of 6-PG and G1P, recorded as indicated
from six regions of a CCL136 cell.

The "Ambiquity" of Glycolysis In Vivo

Wilson (1980) has coined the term "ambiquity" to characterize enzymes
which reversibly associate with intracellular particulate structures, as
a function of metabolic conditions. A number of glycolytic enzymes
appear to be "ambiquitous" (Masters, 1981; Wilson, 1980). For example,
hexokinase associates with the mitochondrial membrane. The interaction
is influenced by the cellular energy status, as signaled by levels of
various metabolites and effector substances. Binding to the
mitochondrion is enhanced under conditions of glycolytic stress, as seen
with ischemia and insulin treatment. The bound form is considered the
more active species. Interestingly, this form is not subject to the
inhibition by G6P, observed with the soluble enzyme.

Also, phosphofructokinase interacts with particulate structures
(Masters, 1981). It is found associated with filamentous material in
muscle extracts. Reconstitution studies in vitro have shown that binding
involves F-actin thin filaments and is influenced by various regulatory
compounds. The interaction is enhanced by electrical stimulation of
muscle tissue. (Pyruvate kinase, another regulatory enzyme, also binds
to the filaments, but hexokinase does not). Binding of phosphofructokinase, from erythrocytes and from muscle, to erythrocyte "ghost" membranes (Karadsheh and Uyeda, 1977; Higashi et al., 1979) is enhanced by

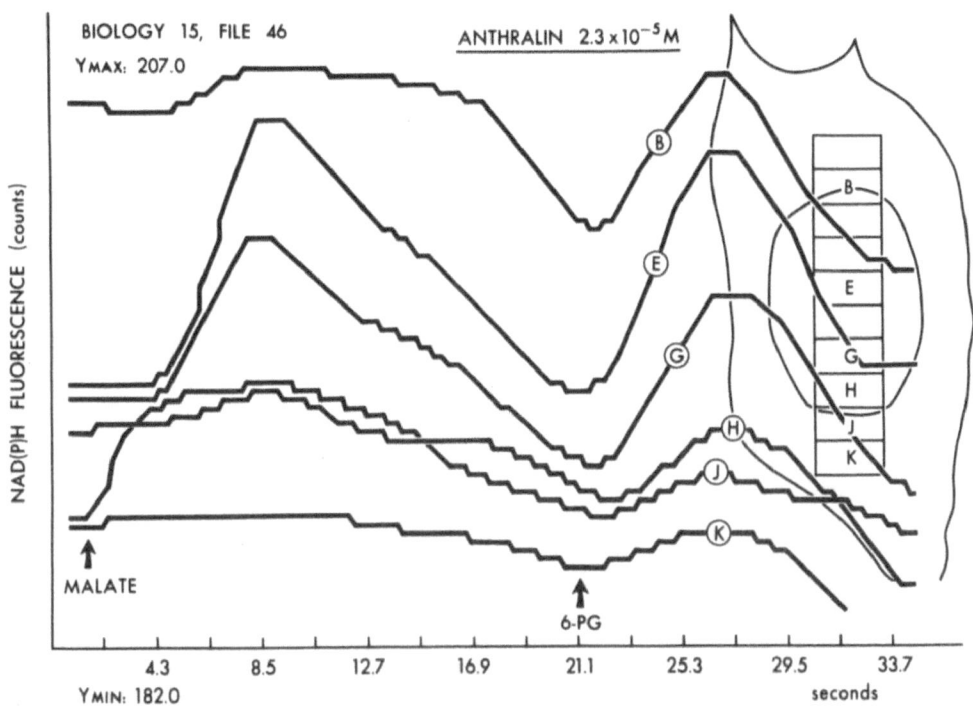

Fig. 11. As in Fig. 10 for consecutive injection of malate and 6-PG into an L cell.

effectors ATP and FDP. Importantly, the membrane-bound form of the enzyme is not inhibited by ATP; and its fructose-6-phosphate saturation curve is nonsigmoidal. This kinetic behavior is in sharp contra-distinction to the well-known properties of the "soluble" enzyme. A phosphofructokinase "activation factor" (M_r= 3000-4000) has been isolated from liver (Furuya and Uyeda, 1980). This factor, apparently removed from the enzyme during purification, prevents the inhibition of the liver kinase by ATP. (It acts synergistically with AMP in reversing the ATP inhibition). This finding correlates with the observation that purified liver phosphofructokinase is strongly inhibited by ATP, whereas the crude enzyme is only slightly inhibited. Similarly to the situation with erythrocyte-bound enzyme, the association of liver phosphofructokinase with the "activation factor" leads to nonsigmoidal kinetics.

Microfluorometric findings reported herein, as well as previous studies (Kohen et al., 1979, 1981a, b, 1983), show the patchy, mosaic nature of the fluorescence patterns obtained upon microinjection of G1P, G6P, FDP, and 6-PG. The resolution of the technique (see Figs. 15 and 16) does not allow a determination of the actual spatial extent of the localized glycolytic "generators". Studies, however, are in progress to enhance the limits of fluorescence detections in the cellular microenvi-ronment. Moreover, the temporality of the spreading fluorescence "waves" is such, that the instrumentation (with its intrinsic time delays) cannot follow the localized responses from the very moment of their inception. This mosaic picture fits the view, propounded by a number of cell biolo-gists (e.g., Siekevitz et al., 1967; Sjostrand, 1964; Tata, 1971), that specific metabolic processes in larger eukaryotic cells are topographi-

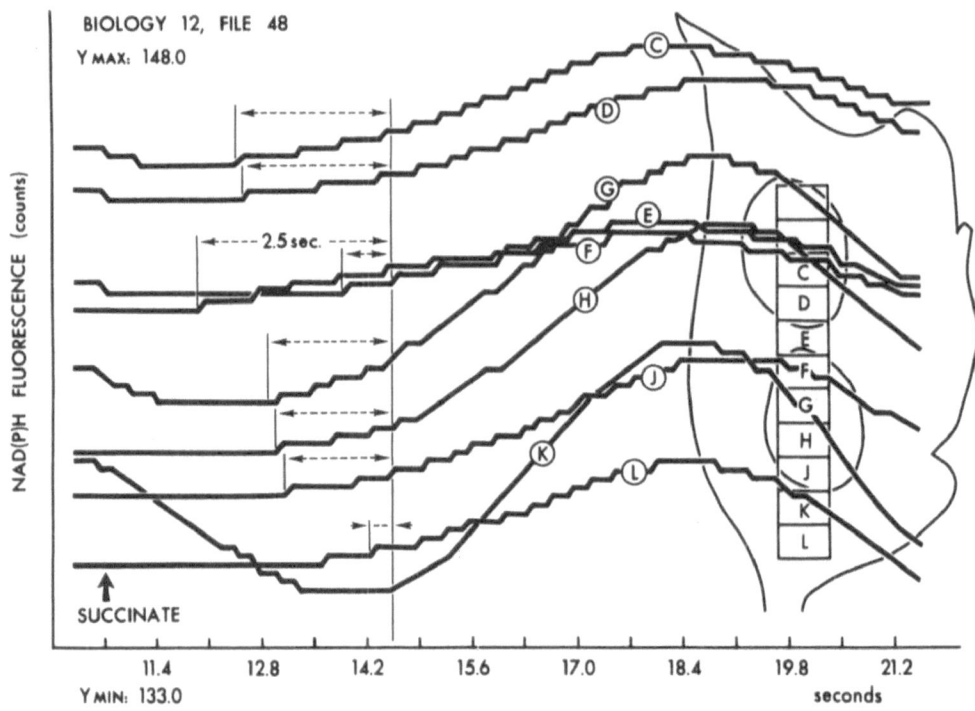

Fig. 12. Kinetics of the NAD(P)H response to succinate in different
regions of binucleated L cell. The localizations of curves C
to L corresponding to similarly identified cellular sites
viewed by the multichannel detector are shown on the right.
The delays between transient rise at different cell sites are
indicated in sec. Compare transient responses at nuclear sites
C, D, G, H, and J with response at internuclear site E. The
coincidence of the higher transient responses with nuclear
sites may be explained by cytoplasmic invaginations at such
sites or perinuclear mitochondria.

cally segregated on membranous elements such as the endoplasmic
reticulum, as well as in association with the cytomatrix.

A crucial point in the topographic evaluation of localized metabolic
responses is to determine whether they are the result of diffusional
patterns or actual segregation (compartmentation). To distinguish
between diffusion and compartmentation, the injection of "caged"
substrates (Kaplan et al., 1978) is considered. Such substrates can be
microinjected in the cell and, following intracellular diffusion, the
substrate can be released from the "caging group" by cell irradiation at
the appropriate wavelength. However, so far only caged adenine nucleotides have been synthetized, and there are currently no caged glycolytic
or pentose shunt intermediates. Nevertheless, it is apparent that the
observed site-to-site delays in the emerging fluorescence patterns upon
injection of metabolic intermediates, are one or two orders of magnitude
greater than for simple bulk diffusion (as determined from measurements
with injected dyes [Kohen et al., 1973]). This observation provides
reasonably sound evidence for compartmentation.

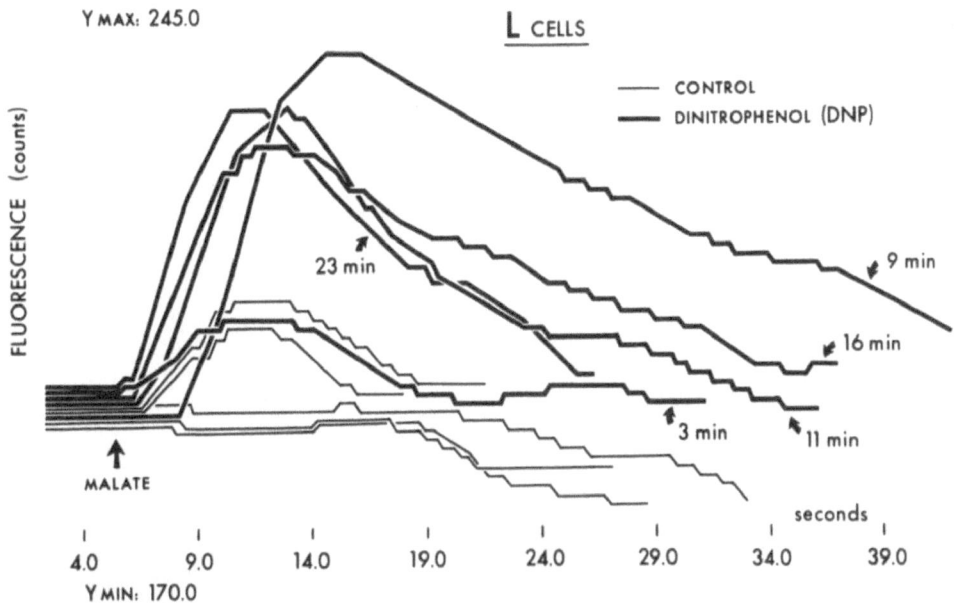

Fig. 13. Time curves of the response to malate in L cells. Enhancement of the response in the presence of DNP is shown.

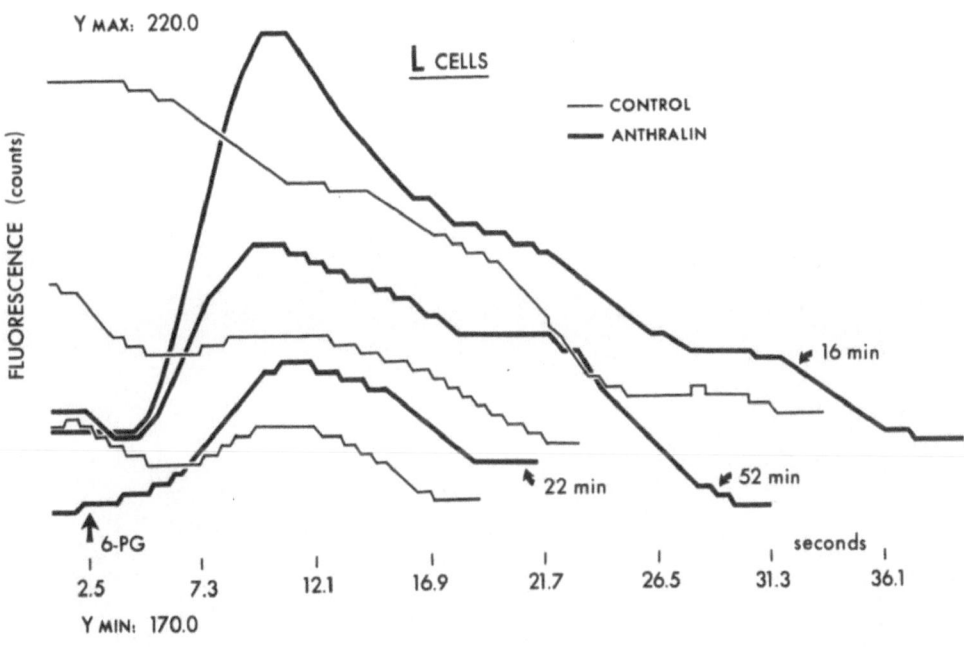

Fig. 14. Enhancement of the NAD(P)H response to 6-PG in the presence of anthralin (2.3 x 10^{-5} M).

Of course, our technique cannot discern the nature of the cytomatrix onto which the glycolytic "generators" might be organized. Although the CCL 136 cells are derived from striated muscle, they show poor differentiation in the transformed state. It is thought that actin (or

267

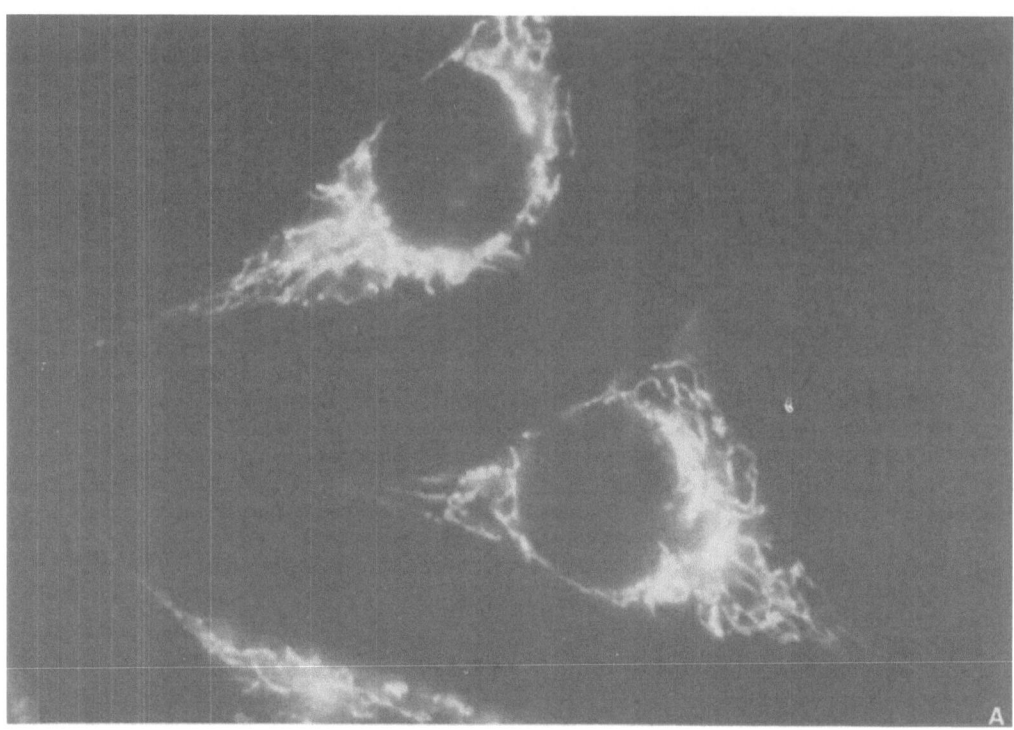

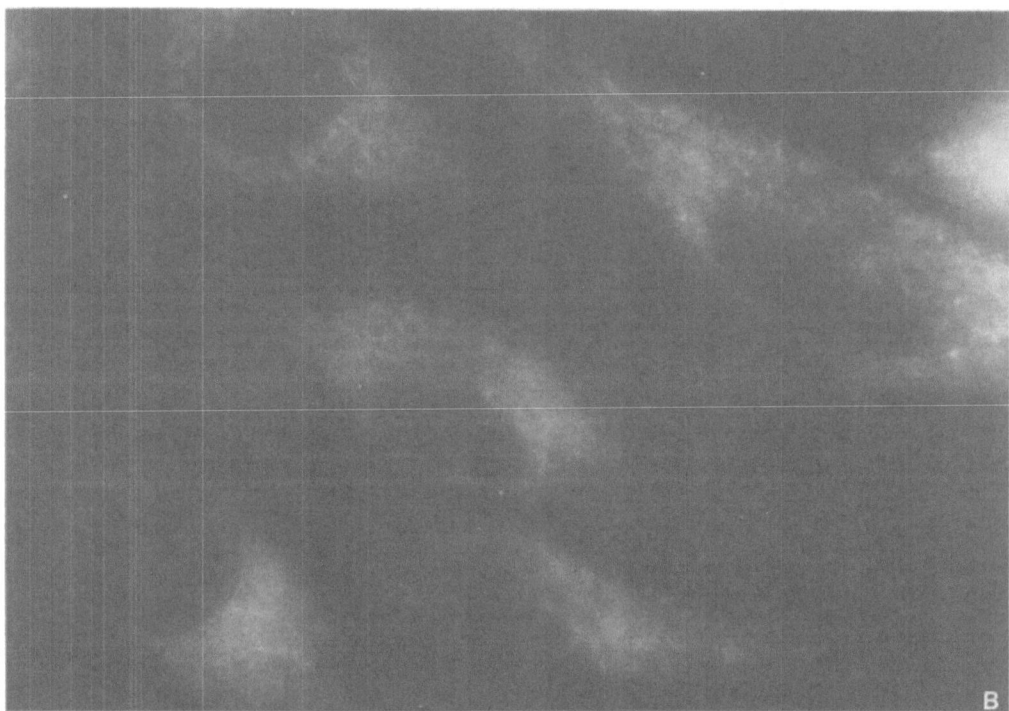

Fig. 15.   Structural alterations of mitochondria in L cells and skin
           fibroblasts incubated for 15 min in presence of hematoporphyrin
           (HP). Vital fluorescence stain with DASPMI.   A. L cell,
           control; B. L cell, 10^{-4} M HP (continued on next page).

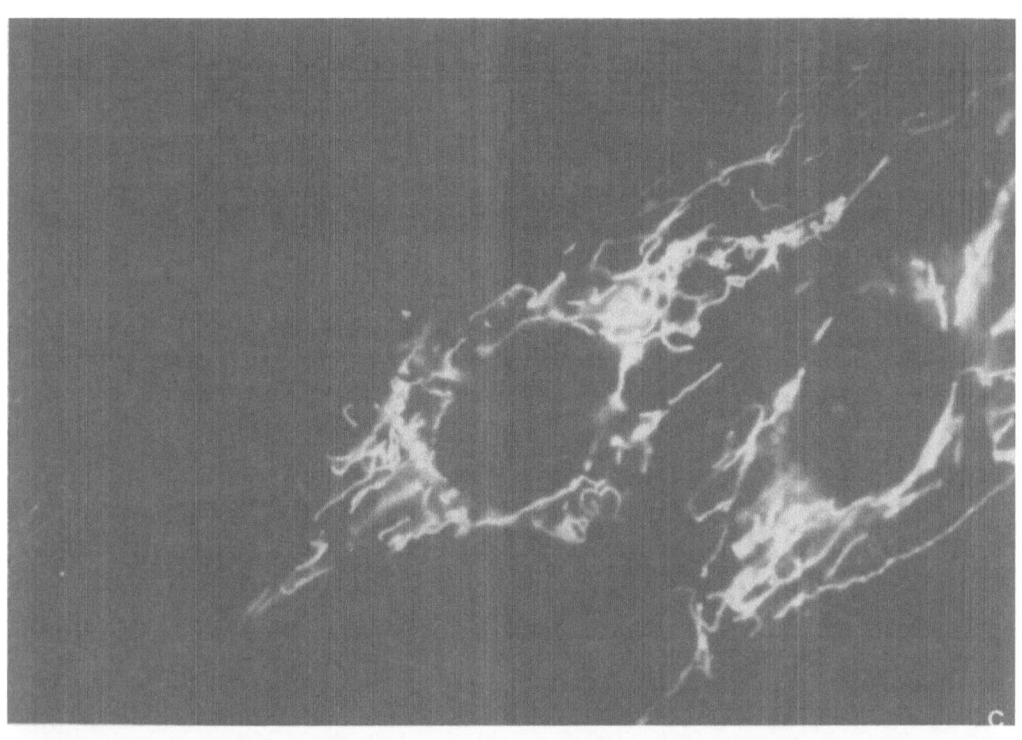

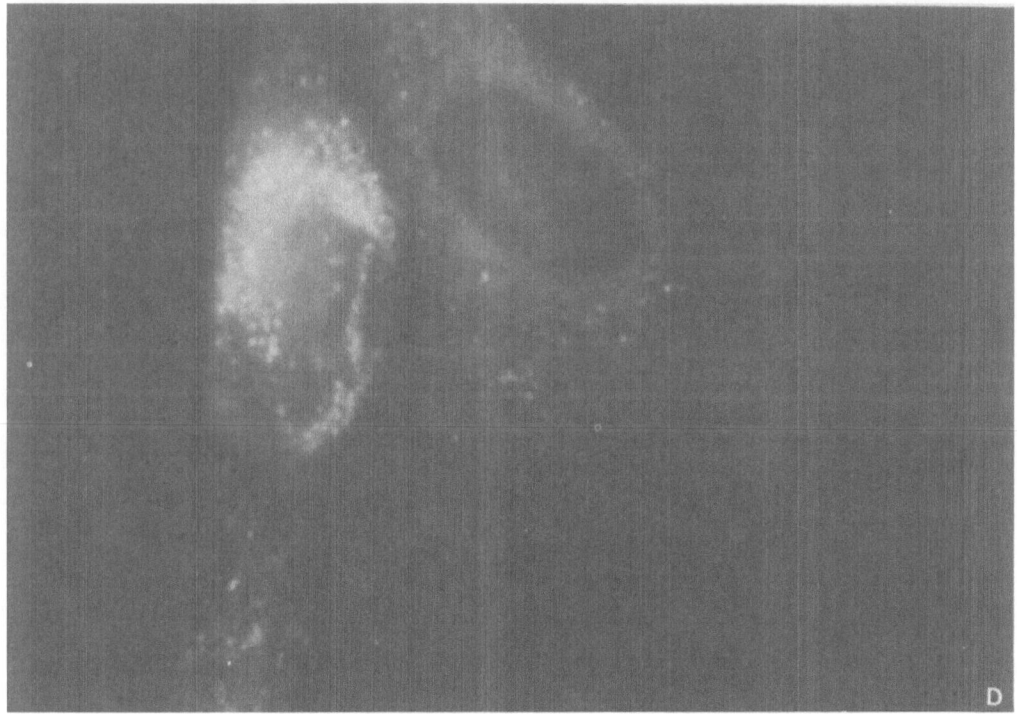

Fig. 15 (continued). C. skin fibroblast, control; D. skin fibroblast, 10^{-4} M HP.

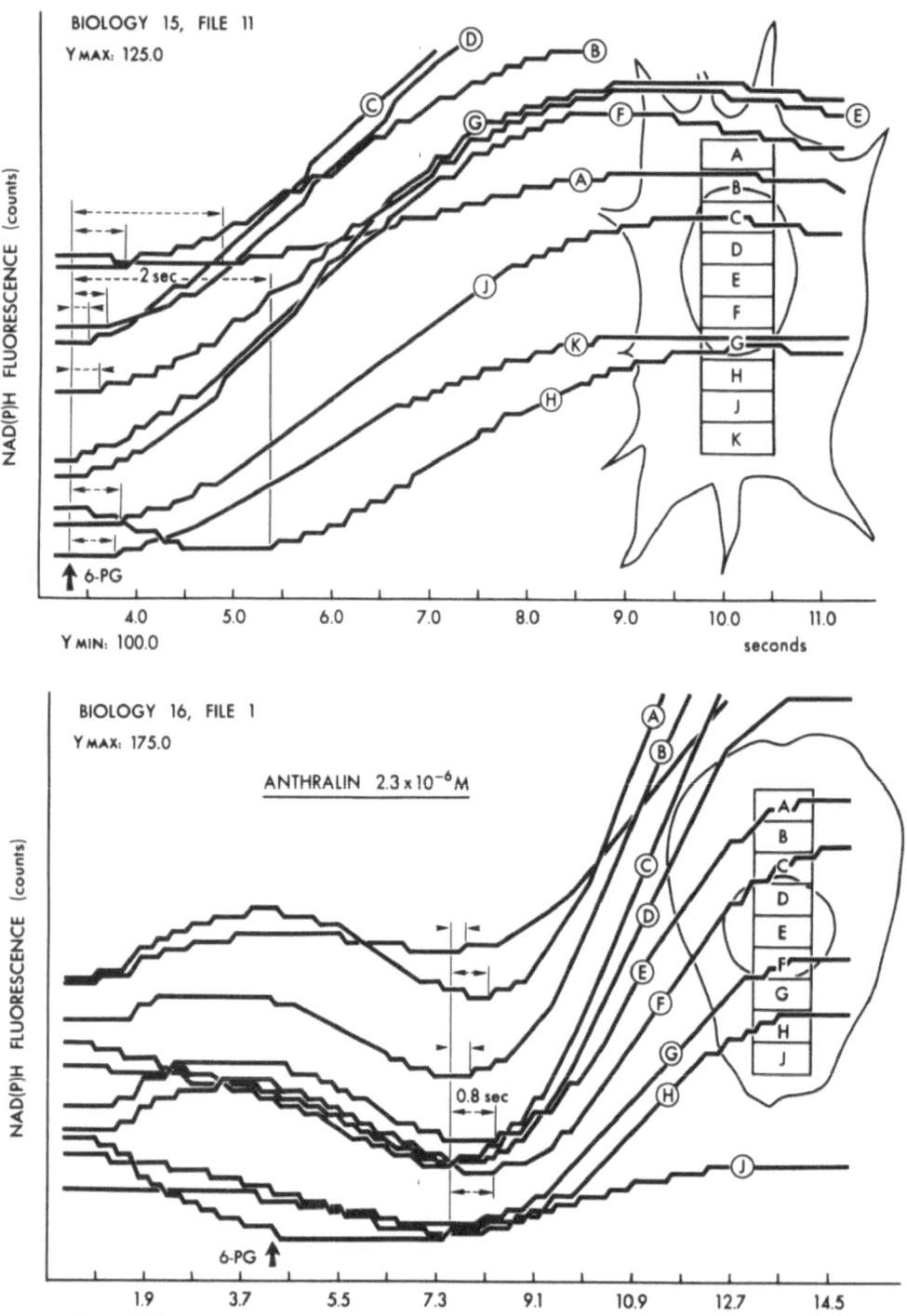

Fig. 10. Kinetics of the transient responses to 6-PG in L cells.
A. Asynchronous in the control. B. Nearly synchronous in
cells treated with anthralin (2.3 x 10⁻⁶ M).

actin-like) skeletal elements play a central role in the organization of glycolytic, and other "soluble", enzymes in a number of cell types (Clegg, 1984; Masters, 1981). It is noteworthy that cytochalasin (which disorganizes actin cytoskeletons) had little effect on the multifocal fluorescence response in the CCL 136 cells (see Fig. 9). This might suggest some other (e.g., membranous) location for the enzyme systems in these cells. Of course, the caveat must be made that we have not determined that cytochalasin actually did disrupt the cytomatrix.

Glycolytic Spatiotemporal Behavior and Regulation

Most regulatory models for glycolysis center on the allosteric properties of phosphofructokinase (PFK) (Kemp and Foe, 1983). Elaborate schemes, involving feedback and feedforward effects in combination with hexokinase and pyruvate kinase, have been constructed (reviewed in Newsholme and Start, 1973). Moreover, glycolysis is known to exhibit oscillatory behavior in a variety of systems. The best example is yeast, where oscillations have been well studied in cells and in cell extracts (Boiteux et al., 1980). The limited evidence for glycolytic oscillations in mammalian systems has been obtained from cell-free extracts of beef heart and rat muscle and from intact mouse L cells and ascites tumor cells (Lipkin et al., 1983). As discussed by Lipkin et al (1983), the lack of evidence may be due to the fact that individual cells in a population may be oscillating out of phase. Indeed, glycolysis in isolated rat fat cells was seen to oscillate, with perifusion of insulin or hydrogen peroxide serving to synchronize the oscillations. Cycles, which require a continuous regular or pulsed supply of substrate, have not been observed by microfluorometry in cells injected intermittently. However, up to 15 consecutive transients of $NAD(P)^+$ reduction/oxidation have been observed in ascites cells microinjected with G6P.

Theoretical models of glycolytic oscillations are based on the allosteric regulatory properties of PFK (Goldbeter and Caplan, 1976) and resonance coupling with pyruvate kinase (Termonia and Ross, 1981). Awareness that the enzymes of glycolysis associate with particulate structures, together with the kinds of whole-cell studies reported herein, demand that such models (hitherto constructed for a bulk-solution case) be refined to include spatial heterogeneities. The presence of heterogeneous catalytic sites can lead to oscillatory behavior under less restrictive conditions than for homogeneous cases (Ortoleva and Ross, 1975; Glass and Kauffman, 1972). Ortoleva and Ross (1975) have developed a general theoretical paradigm which seems particularly apropos. The starting point is a set of nonlinear differential equations for the time (t) evolution of chemical concentrations:

$$\frac{\partial \underline{n}}{\partial t} = \underline{D} \, \nabla^2 \underline{n} + \underline{F}[\underline{n}] + \gamma \, \underline{G}[\underline{r}, \underline{n}] \qquad (1)$$

where $\underline{n}$ is the column vector of concentrations, $\underline{D}$ the matrix of diffusion coefficients, $\underline{F}[\underline{n}]$ the chemical-reaction contribution due to the homogeneous reaction kinetics, and $\gamma \underline{G}[\underline{r}, \underline{n}]$ the reaction-rate term due to heterogenous (space-localized) effects on the kinetics (with γ a strength parameter and $\underline{r}$ the spatial coordinate). The neighboring localized sites may interact chemically via the diffusing/reacting medium between them. Membrane-localized heterogeneous catalysis (under far-from-equilibrium conditions) can generate undulatory spatial patterns in concentration, which propagate as waves along the surface due to reaction/diffusion in a boundary layer around the membrane. Our experimental results herein (e.g., Figs. 15 and 16), while not in themselves revealing glycolytic oscillation, might represent a crude indication of this wave-like phenomenon.

If the findings on the kinetics of bound vs. free PFK can be generalized, it would appear that this enzyme functions as the primary glycolytic oscillophor ideally in the solubilized state (by virtue of its allosteric, sigmoidal behavior therein). What, then, is the role of the observed population of local (kinetically nonsigmoidal) sites in this case? The heterogeneous-catalysis theory of Ortoleva and Ross (1975) provides a possible explanation. Suppose that the homogeneous bulk-phase system (defined by F[n] in Eqn. [1]) has a stable limit-cycle oscillation, determined by a soluble population of enzyme. The previous authors showed that the presence of heterogeneities (γ G[n]) leads to local renormalization (local frequency changes) and the development of phase gradients in the oscillating bulk. Defining the phase by $\phi(r,t)$, it is found that the localized sites lead to a phase-correction term, $t_1(r,t)$, such that $\phi(r,t) \sim t - \gamma t_1(r,t)$, and that $t_1(r,t)$ obeys a phase diffusion equation

$$\frac{\partial t_1}{\partial t} = D_p \nabla^2 t_1 - g(r) \tag{2}$$

where D_p is a phase diffusion coefficient (viz., a weighted average of the individual coefficients, D) and $g(r)$ is a measure of the projection of the local disturbance on the bulk kinetics. Phase waves from the local sites propagate into the medium with velocity $\sim D_p/\gamma g$. Waves from neighboring sites can interfere, with the stronger heterogeneity dominating the weaker. Thus, the localized catalytic regimes would serve as "pacemakers".

It is tempting to speculate, that the spreading fluorescence patterns observed herein relate to such "phase waves". The rate of propagation of the fluorescence fronts would suggest that the "waves" are restricted to particulate surfaces in the cell. We note that Hess (Boiteux et al., 1980) has demonstrated the existence of phase waves (of supracellular dimension) in glycolyzing yeast extracts, measured by spatiotemporal change in NAD(H) optical density. Alternatively, the dynamic fluorescence patterns in our system may represent relaxation-oscillation waves (Ortoleva and Ross, 1975). Imposition of pacemaker heterogeneities in relaxation-oscillation systems leads to the emission of trains of reaction-enhanced diffusion fronts.

Ross et al. (Termonia and Ross, 1981; Richter and Ross, 1981) have shown that the oscillatory model leads theoretically to an increase in efficiency of operation of glycolysis. Goldbeter and Caplan (1976) and Boiteux et al. (1980) have emphasized such potential features as dynamic spatial compartmentation of metabolites, excitability, and signal transmission. Excitability and speed of signal transmission may be especially important, as regards regulation of the adenylate energy charge. A qualitative model of such a regulatory design, involving heterogeneous glycolysis, was proposed previously (Welch, 1977). In this scheme it was supposed that the primary glycolytic oscillophor (viz., phosphofructokinase) is space-localized on intracellular particulate surfaces and communicates with the bulk phase by way of the adenylate kinase reaction (2ADP $\rightleftharpoons$ ATP + AMP). (In muscle, for example, it has been found that adenylate kinase does not associate with particulate material under the conditions in which glycolytic enzymes are seen to bind [Masters, 1981].) The adenylate kinase "medium" might serve to propagate regulatory signals (e.g., reaction/diffusion fronts, phase waves) among the localized sites, so as to coordinate energy metabolism as a whole by entrainment or synchronization of the local "generators". Microfluorometric studies, along lines as followed herein, have yielded results consistent with this view. It has been shown (Kohen et al., 1979), that the imposition of a general metabolic load on cells (by treatment with ethionine [an ATP trap], with an uncoupler, or with metabolizable xenobiotics) completely and rapidly randomizes the mosaic responses. In this case there is observed a uniform, blanket increase in NAD(P)H fluorescence. This observation, together with results herein, suggests that the local "generators" can communicate (and synchronize) quickly under stressful conditions.

The role of glycolytic intermediates and modifiers has also been investigated by a similar approach. When two different intermediates (e.g., G6P, F6P, FDP, ADP) are microinjected simultaneously, the NAD(P)H transients are more prominent than if a large dose of a single substrate is injected. Since this observation requires a substrate dose of each intermediate (rather than smaller doses needed for allosteric regulation), it is also conceivable that there may be compartmentalization of these intermediates in the intact living cell, with different metabolic channels. Thereby, an enhancement of the transient metabolic response may occur when all the channels are activated simultaneously (although other interpretations are equally plausible).

CONCLUDING COMMENTS

We anticipate that whole-cell techniques, such as microfluorometry, will prove very useful in the intracellular mapping of metabolic channels and compartments. The study of interactions within the metabolic continuum and dynamic microarchitectural complex, represented by the living cell, lays the basis for the establishment of an "in situ cytobiochemistry".

A number of considerations are pertinent to in vivo "cytobiochemical" studies:

1) A possibly crucial role of mitochondria is revealed in the regulation of metabolic control at extramitochondrial, including intranuclear, sites (exemplified by possible mitochondrial control of the hexose monophosphate shunt).

2) The mosaic-like spatial pattern of NAD(P)H fluorescence with sequences of local, multifocal, generalized responses, and substrate channeling, are features of cell metabolism hard to duplicate in cell-free systems.

3) Theoretical models of metabolic organization often precede actual assessment of intracellular parameters or hypothetical structures (e.g., glycolytic bodies, oscillators, channels). Close interaction between theory and experimental approaches remains essential.

4) Recourse to a growing arsenal of vital fluorescent organelle probes strengthens the approach (e.g., correlated DASPMI and coenzyme studies).

5)   Metabolic processes which occur in a dynamic cytomatrix require
     evaluation in the presence of structurally active agents which can
     alter cell microarchitecture.

In probing the intracellular microenvironment, the structural
resolution of our technique is still one or two orders of magnitude away
from hypothetical structures, such as glycolytic oscillators. Attempts
are under way to define limits attainable by fluorescence. Even at the
present resolution, topographic microfluorometry as a spatiotemporal
method provides the most befitting approach to the organization of
intracellular metabolic processes, their physiopathology, and
pharmacology.

ACKNOWLEDGEMENTS

A posthumous tribute is given to Mr. B. Rietberg who made the
drawings herein. This work was supported by grant NSF DMB 8303691 from
the U.S. National Science Foundation.

REFERENCES

Bereiter-Hahn, J., 1976, Biochim. Biophys. Acta, 423:1.
Boiteux, A., Hess, B., and Sel'kov, E.E., 1980, Curr. Top Cell. Regul.,
     17:171.
Chance, B. and Hollunger, G., 1961, J. Biol. Chem., 236:1534.
Christophersen, J. and Gammeltoft, M., 1978, Arch. Pharm. Chem., 85:566.
Clegg, J.S., 1984, Am. J. Physiol., 246:R133.
Dougherty, T.J., Kaufman, J.G., Goldfarb, A., Weishaupt, K.R., Boyle
     D.G., and Mittleman, P., 1978, Cancer Res., 38:2628.
Frieden, C. and Nichol, L., 1981, "Protein-Protein Interactions", Wiley,
     New York.
Friedrich, P., 1985, in: "Organized Multienzyme Systems: Catalytic
     Properties", G.R. Welch, ed., Academic Press, New York.
Furuya, E. and Uyeda, K., 1980, Proc. Natl. Acad. Sci. USA, 77:5861.
Glass, L. and Kauffman, S.A., 1972, J. Theor. Biol., 34:219.
Goldbeter, A. and Caplan, S.R., 1976, Annu. Rev. Biophys. Bioeng., 5:449.
Hess, B., 1973, Symp. Soc. Exp. Biol., 27:105.
Higashi, T., Richards, C.S., and Uyeda, K., 1979, J. Biol. Chem.,
     254:9542.
Kacser, H. and Burns, J.A., 1979, Biochem. Soc. Trans., 7:1149.
Kaplan, J.H., Forbush III, B., and Hoffman, J.F., 1978, Biochemistry,
     17:1929.
Karadsheh, N.S. and Uyeda, K., 1977, J. Biol. Chem., 252:7418.
Kemp, R.G. and Foe, L.G., 1983, Mol. Cell. Biochem., 57:147.
Kohen, E. and Kohen, C., 1984, Biochim. Biophys. Acta., 803:115.
Kohen, E., Michaelis, M., Kohen, C., and Thorell, B., 1973, Exp. Cell
     Res., 77:195.
Kohen, E., Kohen, C., Thorell, B., and Bartick, P., 1979, Exp. Cell Res.,
     119:23.
Kohen, E. et al., 1981a, in: "Techniques in Cellular Physiology", Part 1,
     P.F. Baker, ed., Elsevier/North Holland, New York.
Kohen, E. et al., 1981b, in: "Modern Fluorescence Spectroscopy", Vol. 3,
     E.L. Wehry, ed., Plenum, New York.
Kohen, E. et al., 1983, Cell Biochem. Function, 1:3.
Kohen, E., Hirschberg, J.G., and Rabinovitch, A., 1985, in: "Advances in
     Microscopy", R.R. Cowden, ed., Alan R. Liss, New York.

Lipkin, E.W., Teller, D.C., and de Haen, C., 1983, _Biochemistry_, 22:792.

Mansell, J.L. and Clegg, J.S., 1983, _Cryobiology_, 10:541.

Masters, C.J., 1981, _CRC Crit. Rev. Biochem._, 11:105.

Newsholme, E.A. and Start, C., 1973, "Regulation in Metabolism", Wiley, New York.

Ortoleva, P. and Ross, J., 1975, _Adv. Chem. Phys._, 29:49.

Ottaway, J.H. and Mowbray, J., 1977, _Curr. Top. Cell Regul._, 12:108.

Porter, K.R. and Tucker, J.B., 1981, _Sci. Amer._, 244:56.

Richter, P.H. and Ross, J., 1981, _Science_, 211:715.

Santus, R., Kohen, C., Kohen, E., Reyftmann, J.P., Morliere, P., Dubertret, L., and Tocci, P.M., 1983, _Photochem. Photobiol._, 38:71.

Schliwa, J., 1982, _J. Cell Biol._, 92:79.

Shannon, H.A., Matlib, M.A., Robinson, B., Hacli, M., and P. Srere, 1977, _in_: "Proc. Electron Microscopy Sol.", G.W. Bailey, ed., 5th Annual MSA Meeting, Boston, Mass.

Shroot, B., Schaeffer, H., Juhlin, L., and Greaves, M.W., 1981, _Brit. J. Dermatol._, 105, Suppl. 20:3.

Siekevitz, P., Palade, G.E., Dallner, G., Ohad, I., and Omura, T., 1967, _in_: "Organizational Biosynthesis", H.J. Vogel, J.O. Lampen, and V. Bryson, eds., Academic Press, New York.

Sjostrand, F.S., 1964, _in_: "Cytology and Cell Physiology", G.H. Bourne, ed., Academic Press, New York.

Tata, J.R., 1971, _Sub-Cell. Biochem._, 1:83.

Termonia, Y. and Ross, J., 1981, _Proc. Natl. Acad. Sci. USA_, 78:2952.

Welch, G.R., 1977, _Prog. Biophys. Mol. Biol._, 32:103.

Welch, G.R., ed., 1985, "Organized Multienzyme Systems: Catalytic Properties", Academic, New York.

Welch, G.R. and Keleti, T., 1981, _J. Theor. Biol._, 93:701.

Wilson, J.E., 1980, _Curr. Top. Cell Regul._, 16:1.

Wombacher, H., 1984, _Mol. Cell Biochem._, 56:155.

FUNCTIONAL AND STRUCTURAL HETEROGENEITY OF THE

INNER MITOCHONDRIAL MEMBRANE

Dieter Brdiczka

Faculty of Biology
University of Konstanz
D-7750, Konstanz 1, F.R.G.

and

Albrecht Reith

Laboratory for Electron Microscopy
and Morphometry
The Norwegian Radium Hospital
and Institute for Cancer Research
Montebello, 0310 Oslo 3, Norway

PERSPECTIVES AND SUMMARY

The existence in mammalian cells of two ATP-providing systems, as
well as DNA and protein synthesis, both inside and outside the
mitochondrial compartment, implies sensitive control and regulation
between the two compartments. Recent studies have revealed a complex
organization at the mitochondrial periphery responsible for this
regulation, in which the outer membrane pore, the enzymes in the
peripheral mitochondrial compartment, and the peripheral part of the
inner membrane (i.e., inner boundary membrane) are involved.

The accepted pattern of organization that characterizes mitochondria
of most known cell types is two continuous membrane systems, each
comprising a closed sac. However, it has been shown that the infoldings
of the inner membrane, termed cristae mitochondriales by Palade (1952),
are distinct in structure and function from the peripheral part.
According to Deams and Wisse (1966), the cristae can be expected to
reveal continuity with this membrane through tubes, the pediculi, which
are only a very limited part of the crista periphery. It follows that
the inner limiting membrane can be considered as a separate continuous
membrane system. In addition to structural differences, heterogeneity of
the inner membrane is exhibited in the distribution of enzymes. Negative
staining of inner membrane fragments reveals two fractions bearing
different amounts of stalked particles, which represent the ATPase
(Allmann et al., 1968; Vazquez et al., 1968). Furthermore,
glycerolphosphate dehydrogenase was assigned to the peripheral part of
the inner membrane, since in histochemical and morphometric

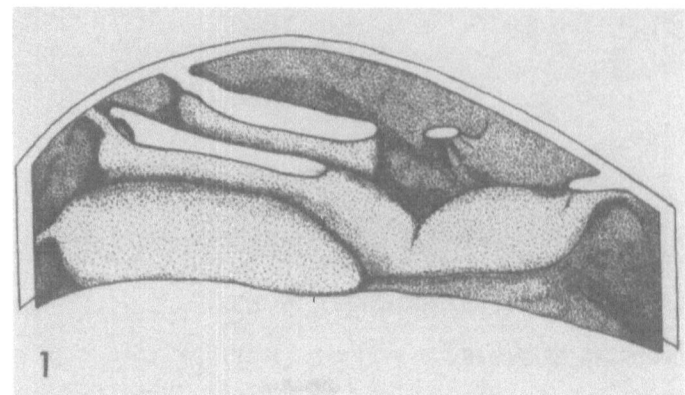

Fig. 1.   3-dimensional demonstration of a liver mitochondrion. The
           cristae are connected to the inner boundary membrane by small
           stems of tubular structure, the pediculi cristae. At the upper
           left a crista is seen which shows direct continuity with the
           inner membrane space, since the crista was sectioned on the
           level of the pediculus cristae.

Fig. 2.   Very thin section of conventionally fixed and embedded rat
           liver mitochondria. One mitochondrion contains an opaque
           crista area (asterisk) which is continuous with the inner
           membrane space through a pediculus cristae (arrow). (Inset
           higher magnification, bar = 0.2 µm). At the right a cap sec-
           tion of a mitochondrion with a cross-sectioned pediculus
           cristae (arrowhead). Further pediculi cristae (arrows). Bar =
           0.2 µm. (See next page.)

Fig. 3.   Rat heart mitochondria with fenestrated cristae (arrowheads).
           The cristae are connected by pediculi cristae (arrows) to the
           inner boundary membrane. Bar = 0.2 µm. (See next page.)

Fig. 4.   Cap sections of rat liver mitochondria under the influence of
           thyroid hormone. Several pediculi are cross sectioned
           (arrows). In (a) control, (b) treatment with thyroid hormone.
           Note the increase of pediculi cristae profiles in (b). Bar =
           0.2 µm. (See next page.)

Fig. 5.   Convex fracture face of a rat liver mitochondrion. View of the
           inner leaflet of the boundary membrane (protoplasmic face)
           covered by remnants of the inner leaflet of the outer membrane
           (exoplasmic face). Note the entrances of the pediculi cristae
           (arrowheads). Bar = 0.1 µm. (See next page.)

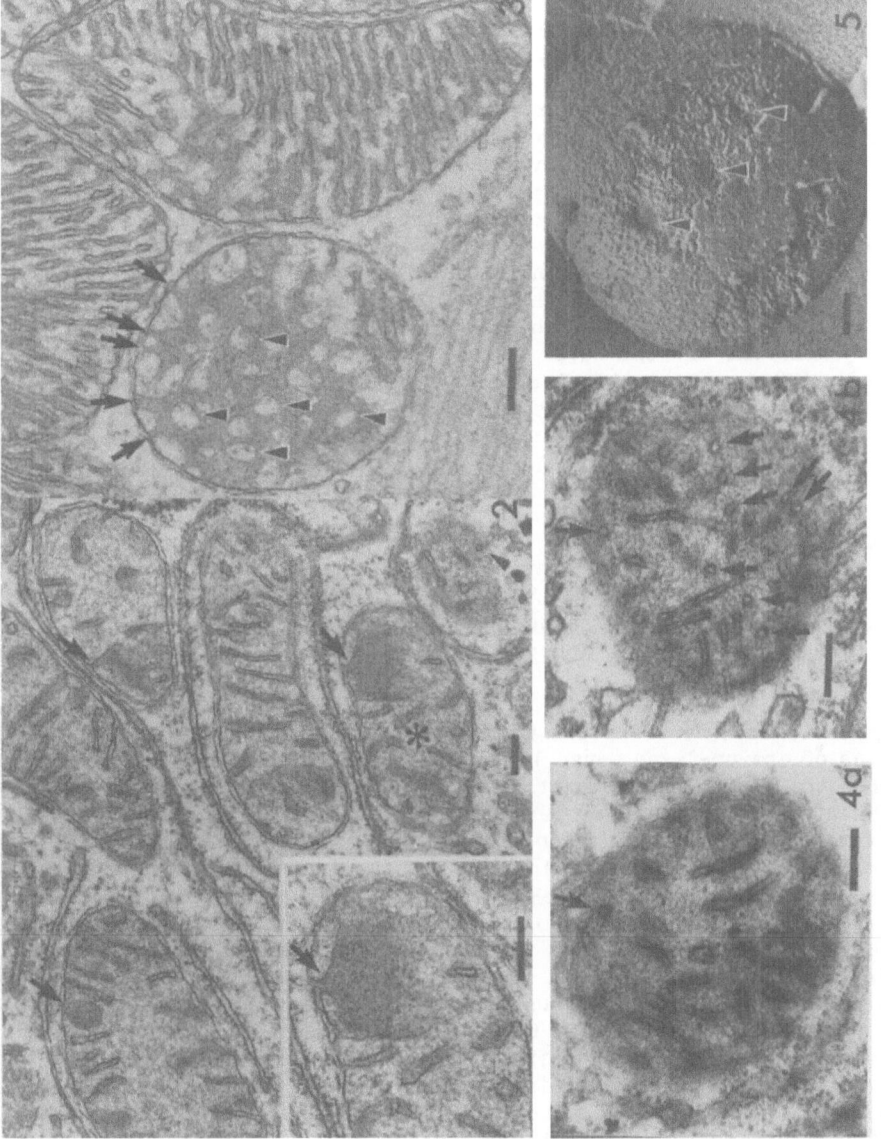

Figs. 2 – 5.

investigations of liver and muscle a positive correlation was observed
between the distribution of this enzyme and the surface density of the
inner boundary membrane of the chondriom (Reith, 1972). In view of these
findings, it seemed likely that the fraction of the inner membrane with
low activity of ATPase and high activity of glycerolphosphate dehydrogenase represents the peripheral part. Indeed, subfractionation of
the mitochondrial membranes revealed a fraction of the inner membrane
possessing these properties and a second fraction, morphologically representing the cristae membranes, which was characterized by high activity
of ATPase, succinate dehydrogenase and cytochrome oxidase (Werner and
Neupert, 1972; Brdiczka et al., 1974).

It appears that all transport and exchange processes, including
uptake of newly synthesized proteins, can be assigned to the peripheral
inner membrane, whereas oxidative phosphorylation might be restricted to
the cristae membranes. This concept of a functional heterogeneity of the
inner membrane has gained support from the postulation of mosaic
chemiosmosis, that is, individual proton spaces and proton
electrochemical potentials along the surface of the inner membrane
(Westerhoff et al., 1984), and also by the recent observation of
metabolically regulated interactions (semifusions) between the inner
boundary membrane and the outer boundary membran (vanVenetie and
Verkleij, 1982). The latter lead to a functional coupling of peripheral
kinases to the inner mitochondrial compartment and provide a one-
receptor, one-step uptake mechanism of newly synthesized proteins in the
matrix.

MORPHOLOGICAL HETEROGENEITY

The Question of Continuity between the Crista Membrane and
the Peripheral Part of the Inner Memnbrane

As early as 1966 Deams and Wisse (1966) concluded from serial
sections of osmium-fixed mouse liver mitochondria that the cristae are
attached to the peripheral part of the inner membrane through short stems
of tubular structure, called pediculi cristae, of varying length and
approximately 30 nm in diameter (Fig. 1). It follows that the cristae
can be expected to reveal continuity with the inner boundary membrane
through the pediculi (Figs. 2, 3), which are only a very small part of
the cristae. The number of pediculi which link cristae to the inner
boundary membrane can vary and is regulated by hormones. We have
developed a morphometric model for pediculi assessment (Reith, 1977).
Calculations of the pediculi surface in normal mitochondria yield 6000 to
9000 nm^2, assuming a length of 60 to 90 nm. This means that the pediculi
surface constitutes less than 1% of the cristae membrane surface. A
comparable small value is obtained by calculating the cross-section area
of pediculi, which is of the order of 700 nm^2 and represents less than 1%
of the boundary surface (assuming 1 pediculus per 10^6 nm^2 — see Table 1).
The number of pediculi relative to the cristae in normal liver
mitochondria is small, as can be seen from the tangential section of
conventionally fixed mitochondria (Fig. 4a) and a freeze fracture of the
inner boundary membrane (Fig. 5). However, under thyroid-hormone
influence a considerable increase occurs in the number of pediculi
cristae per unit inner boundary membrane (Table 1). At the same time an
increase occurs in the surface density of both inner boundary and cristae
membrane, but to a lesser extent (Reith, 1980). Therefore one may assume

Table 1. The Number of Pediculi Cristae per Inner Boundary Membrane Surface Under Thyroid Hormone[a]

| | Days of thyroid hormone treatment | | | | | |
	0	1	3	5	8	15
Number of pediculi per 1 μm^2 membrane	2	4	7	7	8	9

[a]Mild hyperthyroidism, 10 µg/100 g body weight per day, caused a doubling of the fraction of mitochondria and peroxisomes (Reith, 1980; Fringes and Reith, 1982). The same was true for total inner membranes as a consequence of an enlargement of both the cristae and the inner boundary membrane. The membrane increase was paralleled by an increase in the number of pediculi cristae.

that there is better communication between intracristae spaces and the peripheral compartment under thyroid hormone influence. If the pediculi are regulated dynamic structures and physiologically not very frequent, then we are led to the conclusion that there is either a limited proton exchange with the bulk phase of the organelle or even a possibility that some of the cristae are completely separated from the inner boundary membrane. Either way, the proton electrical potential could be restricted entirely to the surface of the cristae membrane, and each crista could be considered as a separate energy transducing unit.

Distribution of ATPase in Negatively Stained Mitochondria

The use of the negative-staining technique in the study of disrupted mitochondria led to the observation of stalked particles in association with the inner membrane (Fernandez-Moran, 1962). It has been widely assumed that the surface bearing the stalked particles is the matrix-facing surface (Muscatello and Carafoli, 1969). Close inspection of the negatively-stained broken inner membranes frequently reveals large sheets and narrow tubules. The sheets, unlike the tubules, bear relatively few stalked particles. It seems likely that the sheets represent the peripheral part of the inner membrane (Allmann et al., 1968; Vazquez et al., 1968).

QUANTITATIVE RELATION BETWEEN THE CRISTAE AND THE INNER BOUNDARY MEMBRANE AND REPRESENTATIVE MARKER ENZYMES

Crista Membrane and Succinate Dehydrogenase

The surface density of the mitochondrial inner-boundary and crista membranes has been determined by morphometric methods in a variety of tissues and species under normal and functionally and pathologically altered conditions (Reith et al., 1976).

In normal rat liver it could be shown independently by Loud (1968) and Reith (1972) that the ratio between the inner boundary-membrane surface density and crista membrane density differs significantly, depending on the location in the liver lobule. The 1.4 times higher crista density in the periportal region agrees with a 1.5 to 1.6 times higher activity of succinate dehydrogenase in the periphery that was determined by cytophotometric measurements (Nolte and Pette, 1971). This correlation between crista density and succinate dehydrogenase activity remains unchanged upon induction of the cristae by thyroid hormone in liver (Reith et al., 1973), heart (Reith et al., 1973), and muscle (Reith, 1972). Furthermore, no change of the correlation was found during development of cristae investigated in muscle during postembryonal differentiation (Vogell, 1965) and exercise (Pilstrom and Kiessling, 1972), nor under pathological conditions such as the reduction of cristae and succinate dehydrogenase in rat liver under chronic ethanol consumption (Pilstrom and Kiessling, 1972). Thus, succinate dehydrogenase appears to be a specific component of the crista membrane.

Inner Boundary Membrane and Glycerolphosphate Dehydrogenase

Comparison of the inner boundary-membrane surface density with the activity of glycerolphosphate dehydrogenase activity has led to the observation of a similar correlation between this enzyme and the inner boundary membrane under most of the conditions described above. On the basis of these data, it was postulated that glycerolphosphate dehydrogenase may be located within the peripheral part of the inner membrane (Reith, 1972). This postulate predicted that large mitochondria in heart or skeletal muscle, with many cristae, would have a lower glycerolphosphate/succinate dehydrogenase ratio than small mitochondria, such as those found in the brain. The expected relationship was indeed observed (Reith, 1972). Moreover, it could be demonstrated that the correlation between glycerolphosphate dehydrogenase and boundary surface density was maintained also under pathological conditions of ethanol consumption, in which the mitochondrial glycerolphosphate dehydrogenase is induced (Pilstrom and Kiessling, 1972).

SUBFRACTIONATION OF THE INNER MEMBRANE

Heterogeneous Protein Composition of the Subfractions

The inner membrane of osmotically shocked rat liver mitochondria is fragmented into segments which differ in enzymatic activity. The different fractions were separated on a sucrose density gradient (Werner and Neupert, 1972; Brdiczka et al., 1974). The activity profile of glycerolphosphate oxidase in such an experiment is shown in Figure 6 and reveals a fraction with a high activity of this enzyme relative to succinate dehydrogenase. ATPase and cytochrome oxidase seemed to be restricted to different gradient fractions of higher density. Although the separation of membrane enzymes was incomplete, the electron microscopic investigation showed that the fraction with high glycerolphosphate oxidase activity contained vesicles of 0.1 - 0.3 μm diameter and only a few crista-like profiles, whereas the fraction of higher density resembled peripheral and crista-forming inner membranes.

In agreement with the histochemical and structural enzyme localization in the inner membrane, these experiments suggest a heterogenic distribution of the glycerolphosphate dehydrogenase preferentially in the peripheral inner membrane and of succinate dehydrogenase, ATPase and most

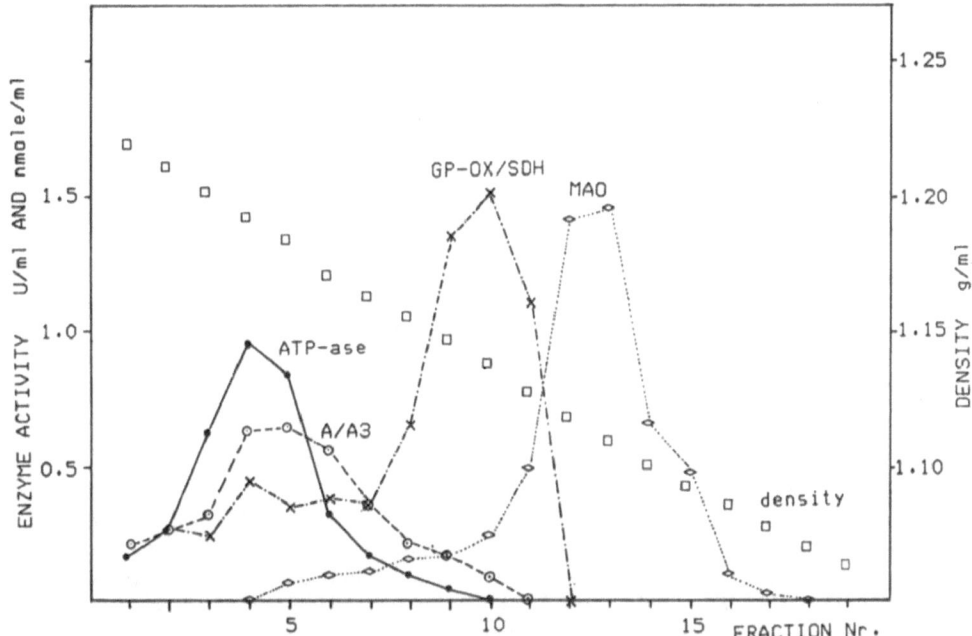

Fig. 6.   Separation of mitochondrial membranes on a density gradient
after osmotic shock. Liver mitochondria were exposed to a
swelling, shrinking and sonification procedure and were subsequently centrifuged on a linear sucrose density gradient
varying from a density of 1.22 to 1.06 g/ml. The activity of
inner and outer membrane enzymes was determined in the different fractions of the gradient: SDH = succinate dehydrogenase (EC 1.3.99.1), ATPase = oligomycin-sensitive ATP
synthetase (EC 3.6.1.3), GP-OX = mitochondrial glycerolphosphate dehydrogenase (EC 1.1.99.5), MAO = monoamine oxidase (EC
1.4.3.4), A/A3 = cytochrome oxidase (EC 1.9.3.1). Enzyme activity is given as U/ml: ATPase times 1, MAO times 10. Cytochrome a/a3 is expresed as nmol/ml. GP-OX activity is shown
relative to SDH activity in the respective fraction.

of the cytochrome oxidase in the crista membrane. The heterogeneous
distribution of flavoproteins in the inner membrane might point to the
existence of different electron transport chains linked to different
flavoenzymes. It follows that a proton electrochemical potential can
presumably be generated across the inner boundary membrane, as well as
across the crista membrane.

Heterogeneous Phospholipid Composition of the Subfractions

     The subfractions of the inner mitochondrial membrane also differ in
phospholipid composition (Brdiczka et al., 1974). A high concentration
of diphosphatidylglycerol and phosphatidyl-ethanolamine was present in
the crista membrane fraction relative to the inner boundary membrane
fraction. This non-random distribution of phospholipids agrees with the

observation that many membrane enzymes require specific phospholipids for activity, for example cytochrome oxidase, which specifically binds diphosphatidylglycerol (Racker and Kandrach, 1971).

FUNCTIONAL HETEROGENEITY

Import of Newly Synthesized Proteins

Amino acid incorporation, both in vitro and in vivo, in the presence of cycloheximide occurs predominantly into the fractions containing enriched inner boundary membrane, obtained by subfractionation as described above (Werner and Neupert, 1972). Furthermore, the inner boundary membrane became labelled first when the time-course of protein incorporation into the isolated subfractions was followed after a leucine pulse in rabbits (Brdiczka, 1974). These data suggest that extra- and intramitochondrial synthesized polypeptides are inserted at specific regions into the inner membrane, located on the inner membrane rather than on the crista membrane. The existence of "growth zones" where the enzyme complexes are assembled was postulated (Werner and Neupert, 1972). It is assumed that the incorporation of newly synthesized mitochondrial proteins is a post-translational step (Schatz, 1979; Hay et al., 1984). The precursor released in the cytosol migrates to a point on the mitochondrial surface where the two boundary membranes are in close contact. This point contains a receptor which recognizes the precursor. The interaction of the receptor with the precursor induces the uptake of the polypeptide across both membranes.

So far, specific receptors for the precursor polypeptides have been found only at the surface of the outer boundary membrane. This suggests a location of the uptake process in regions of contacts between the boundary membranes, where the protein can cross the two membranes in one step (Schatz, 1979; Hay et al., 1984).

Contacts between the Two Boundary Membranes

Pure physical fixation of mitochondria by rapid freezing and freeze fracturing exposes an irregular fracture face which is characterized by a frequent change of the fracture plane between different layers. It has become widely accepted that the fracture plane jumps between the interior of the two boundary membranes. This observation implies that the fracture plane upon deflection has to cross the outer mitochondrial compartment. The course of a fracture like this can be imagined by assuming close contacts between the two boundary membranes. Such contacts have been described by Hackenbrock (1968) in thin-sectioned mitochondria. With regard to the structural nature of the contacts, one has to consider that the fracture plane in freeze fractures follows hydrophobic regions. Therefore a semi-fusion between the two boundary membranes has been proposed in the contact zones in which non-bilayer phospholipid structures are involved (van Venetie and Verkleij, 1982). The semi-fusion model would allow dynamic changes of the contacts, which have been observed in correlation to the functional state of the mitochondria. Phosphorylating (state 3) mitochondria have a higher frequency of contacts compared to energized (state 4) mitochondria, and in uncoupled mitochondria the contacts almost disappear (Knoll and Brdiczka, 1983). The formation of the contacts changes the properties of the mitochondrial surface. For example, the binding of hexokinase to the outer membrane pore is positively correlated to the frequency of the contacts.

It has, moreover, been observed that the frequency of contacts is regulated by hormones in intact hepatocytes. Epinephrine increases, whereas glucagon decreases, the number of contacts (Riesinger and Brdiczka, unpublished).

FUNCTION OF THE BOUNDARY MEMBRANE CONTACTS IN METABOLIC REGULATION

The transport of metabolites across the outer membrane is restricted to a pore protein to which kinases (such as hexokinase) bind specifically (Fiek et al., 1982). The pores in the contact region have a higher capacity for binding hexokinase. The bound hexokinase preferentially uses internal ATP supplied by oxidative phosphorylation (Gots and Bessman, 1974; Inui and Ishibashi, 1979). In view of these findings it has been suggested that the contacts play a role in the regulation of the mitochondrial metabolism: hexokinase, when bound to the pore in the contact regions, may serve to create a microcompartment which facilitates a direct exchange of ATP and ADP between the enzyme and the compartment of oxidative phosphorylation (Fig. 7). Such a functional coupling would increase the effective concentrations of metabolites near target enzymes, and therefore maintain high activity rates. Furthermore, it may provide a mechanism for the transfer of high energy phosphate to the cytosol (Gots and Bessman, 1974; Brdiczka et al., 1985), analogous to the creatine phosphate in the creatine phosphate shuttle (Bessman and Carpenter, 1985).

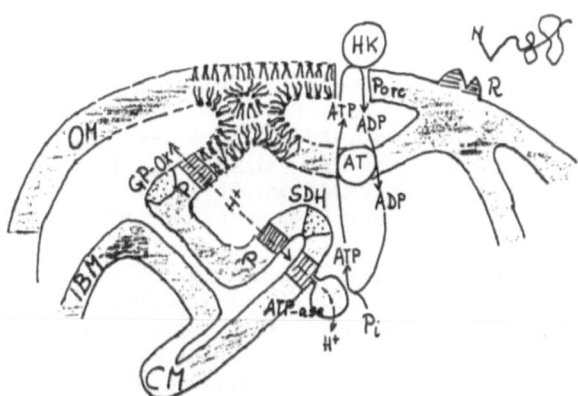

Fig. 7.   Schematic drawing of the microcompartmentation at the mitochondrial periphery. The contacts between the two boundary
membranes create a microcompartment which functionally couples
hexokinase to the inner mitochondrial compartment. The diagram
refers to the semi-fusion model in which non-bilayer phospholipid structures are involved (van Venetie and Verkleij, 1982).
Abbreviations: HK = hexokinase, SDH =succinate dehydrogenase,
GP-OX = glycerolphosphate dehydrogenase, AT = adenylate
translocator, P = proton pump representing the electron
transport chain, R = receptor for newly synthesized polypeptides, OM = outer membrane, IBM = inner boundary membrane,
CM = crista membrane.

CONCLUSIONS

The formation of reversible contacts between the two boundary membranes might play an important role in the regulation of protein uptake and metabolite exchange. It reveals a specific phospholipid and/or protein composition of the inner boundary membrane, which agrees with the assumption of a heterogenic composition of the inner membrane. According to the results described above, the inner boundary membrane is enriched in glycerolphosphate oxidase, whereas succinate dehydrogenase, ATPase and most of the cytochrome oxidase appear to be concentrated in the cristae forming part of the inner membrane. The observed difference in phospholipid composition between the two parts of the inner membrane might be a prerequisite for the proposed semi-fusion with the outer membrane in the contacts and on the other hand for a different permeability for protons in the cristae membranes, to which the ATP synthesis may be restricted. The view of the cristae as separated energy transducing units is supported by the observation that they are connected to the periphery by small pediculi. As the number of these connections is low and is subjected to regulation, the existence of a barrier to the free diffusion of protons between proton spaces of individual ATP-synthesizing systems and the bulk phase of the organelle can be assumed. This has already been postulated by the mosaic chemiosomotic hypothesis (Westerhoff et al., 1984). This idea suggests individual coupling units, in which the proton spaces of the different units can have a different proton electrical potential.

REFERENCES

Allmann, D.W., Bachmann, E., Orme-Johnson, N., Tan, W.C., and Green, D.E., 1968, Arch. Biochem. Biophys., 125:981.
Bessmann, S.P. and Carpenter, C.L., 1985, Ann. Rev. Biochem., 54:831.
Brdiczka, D., Dolken, G. Krebs, W. and Hofmann, D., 1974, Hoppe-Seylers Z. Physiol. Chem., 355:731.
Brdiczka, D., Knoll, G., Riesinger, I., Weiler, U., Klug, G., Benz, R. and Krause, J., 1985, in: "Myocardial and Skeletal Muscle Bioenergetics", N. Brautbar, ed., Plenum Press, New York.
Deams, W.T. and Wisse, E., 1966, J. Ultrastruct. Res., 16:123.
Fernandez-Moran, H., 1962, Circulation, 26:1039.
Fiek, Ch., Benz, R., Roos, N. and Brdiczka, D., 1982, Biochim. Biophys. Acta, 688:429.
Fringes, B. and Reith, A., 1982, Lab. Invest., 47:19.
Gots, R.E., and Bessman, S.P., 1974, Arch. Biochem. Biophys., 163:7.
Hackenbrock, C.R., 1968, Proc. Natl. Acad. Sci. U.S.A., 61:598.
Hay, R., Bohni, P. and Gasser, S., 1984, Biochim. Biophys. Acta, 779:65.
Inui, M. and Ishibashi, S., 1979, J. Biochem., 85:1151.
Knoll, G. and Brdiczka, D., 1983, Biochim. Biophys. Acta, 733:102.
Loud, A.V., 1968, J. Cell Biol., 37:27.
Muscatello, U. and Carafoli, E., 1969, J. Cell Biol., 40:602.
Nolte, J. and Pette, D., 1971, in: "Recent Advances in Quantitative Histo- and Cytochemistry", U.C. Durbach and U. Schmidt, eds., H. Huber, Bern.
Palade, G.E., 1952, Anat. Record, 114:427.
Pilstrom, L. and Kiessling, K.H., 1972, Histochemie, 32:329.
Racker, E. and Kandrach, A., 1971, J. Biol. Chem., 246:7069.
Reichmann, H., Hoppeler, H., Mathieu-Castello, O., Von Bergen, F. and Pette, D., 1985, Pflugers Arch., 404:1.
Reith, A., 1972, Cytobiol., 5:384.
Reith, A., 1977, Mikroskopie, 33:95.

Reith, A., 1980, *Gegenbaurs morph. Jahrb., Leipzig*, 126:327.

Reith, A., Barnard, T. and Rohr, H.P., 1976, *CRC Crit. Rev. Toxicol.*, 4:219.

Reith, A., Brdiczka, D., Nolte, J. and Staudte, H.W., 1973, *Exptl. Cell Res.*, 77:1.

Schatz, G., 1979, *FEBS Letters*, 103:201.

van Venetie, R, and Verkleij, A.F., 1982, *Biochim. Biophys. Acta*, 692:379.

Vazquez, H.J., Santiago, E., Guerra, F., and Macarulla, J.M., 1968, *Rev. Espan. Fisiol.*, 24:43.

Vogell, W., 1965, *Naturwissenschaften*, 52:405.

Werner, S. and Neupert, W., 1972, *Eur. J. Biochem.*, 25:379.

Westerhoff, H.V., Melandri, B.A., Venturoli, G., Azzone, G.F. and Kell, D.B., 1984, *FEBS Letters*, 165:1.

EXPERIMENTAL AND THEORETICAL MODELING
OF METABOLIC ORGANIZATION

COMPLEX PATTERNS OF EXCITABILITY AND

OSCILLATIONS IN A BIOCHEMICAL SYSTEM

A. Goldbeter and F. Moran*

Faculté des Sciences,
Université Libre de Bruxelles
Campus Plaine, C.P. 231
B-1050 Brussels, Belgium

INTRODUCTION

The organization of cell metabolism possesses spatial, functional, and temporal aspects; all these are closely intertwined. Spatial organization relates, for example, to the intracellular localization of biochemical processes, and to the manner in which enzymes that belong to a given pathway interact by forming multienzyme complexes. Functional organization is exemplified by the cellular response to hormonal and neural stimuli, through metabolic cascades of interconvertible enzymes controlled by intracellular messengers such as cyclic AMP. Temporal organization, on the other hand, encompasses the sequence of metabolic events along the cell cycle and their coordination according to growth and differentiation (obviously, spatial and functional organization are involved in such coordination).

More specific forms of temporal organization arise in cellular metabolism as dynamic properties associated with enzymatic and genetic regulation. Among these are autonomous oscillatory behavior (Hess and Boiteux, 1971; Goldbeter and Caplan, 1976; Berridge and Rapp, 1979), excitability, and transitions between multiple, simultaneously stable, steady states.

The product-activated enzyme reaction remains the prototype mechanism giving rise to temporal organization in the form of metabolic oscillations (Goldbeter and Caplan, 1976). Models based on this regulation have been proposed to account for glycolytic oscillations in yeast and muscle, and for cAMP oscillations in the slime mold Dictyostelium discoideum (see Goldbeter et al., 1984, for review).

Here we analyze a simple biochemical model comprising only two variables and show how more complex patterns of temporal organization may arise in regulated enzymatic systems. Among these patterns are the

---

*Present address: Department of Biochemistry, Faculty of Chemistry, Universidad Complutense, Madrid, Spain.

coexistence of two simultaneously stable periodic regimes (birhythmicity), multiple modes of oscillations for closely related conditions, excitable behavior associated with multiple thresholds, and the coexistence of up to three simultaneously stable steady states.

Beyond the dynamics of enzymatic systems, the model bears on other types of cellular organization. We show indeed how the present results throw light on the origin of multiple modes of oscillations in thalamic neurons.

THE TWO-VARIABLE MODEL: AN AUTOCATALYTIC ENZYME REACTION WITH RECYCLING OF PRODUCT INTO SUBSTRATE

The model, represented in Fig. 1, is that of an allosteric enzyme activated by its reaction product. The substrate is supplied at a constant rate, whereas the product is removed at a rate proportional to its concentration and is also recycled into the substrate. This system is governed by the following two kinetic equations for the normalized concentrations of substrate (α) and product (γ) (Moran and Goldbeter, 1984):

$$\frac{d\alpha}{dt} = v + \frac{\sigma_i \gamma^n}{K^n + \gamma^n} - \sigma_M \phi(\alpha, \gamma) \tag{1a}$$

$$\frac{d\gamma}{dt} = q\sigma_M \phi(\alpha, \gamma) - k_s \gamma - \frac{q\sigma_i \gamma^n}{K^n + \gamma^n} \tag{1b}$$

with

$$\phi(\alpha, \gamma) = \frac{\alpha(1 + \alpha)(1 + \gamma)^2}{L + (1 + \alpha)^2(1 + \gamma)^2}$$

Here, v denotes the normalized (constant) input of substrate; k_s measures the rate of product removal, e.g. by a Michaelian enzyme operating in the linear range; σ_M is the normalized maximum rate of the product-activated enzyme; σ_i denotes the normalized maximum rate of product recycling (this parameter includes the constant concentration of any cofactor needed for recycling the product into the substrate); K denotes the threshold constant for the recycling process, which is

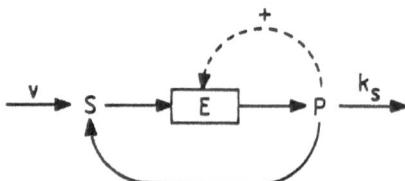

Fig. 1.   Model of an allosteric enzyme reaction with activation by the product (P) and recycling of product into substrate (S). As a function of recycling, this model exhibits simple or complex patterns of excitability and oscillations (Moran and Goldbeter, 1984, 1985).

assumed to obey a Hill equation with a Hill coefficient n. The rate
function ϕ for the product-activated enzyme obtains for a dimer obeying
the concerted allosteric model of Monod et al. (1965). Finally, q is a
constant which originates from the normalization of the substrate and
product concentrations (see Moran and Goldbeter, 1984, for further
details).

BEHAVIOR IN THE ABSENCE OF PRODUCT RECYCLING

The system governed by Eqs. (1) has been analyzed extensively in the
absence of product recycling (i.e. for $\sigma_i = 0$) as a model for glycolytic
oscillations in yeast and muscle (Goldbeter and Lefever, 1972; Boiteux et
al., 1975). These oscillations originate from the positive feedback
exerted on phosphofructokinase by a reaction product (Hess and Boiteux,
1971). Let us briefly consider the behavior of the system in such
conditions, before turning to the new behavioral modes introduced by
recycling.

Oscillatory Behavior

In the phase plane (α, γ) formed by the substrate and product
concentrations, the steady state is located at the intersection of the
two nullclines (dα/dt)=0 and (dγ/dt)=0 (see Fig. 2). The equations of
the substrate and product nullclines are obtained by setting the right-
hand side of Eqs. (1) equal to zero. Linear stability analysis of the

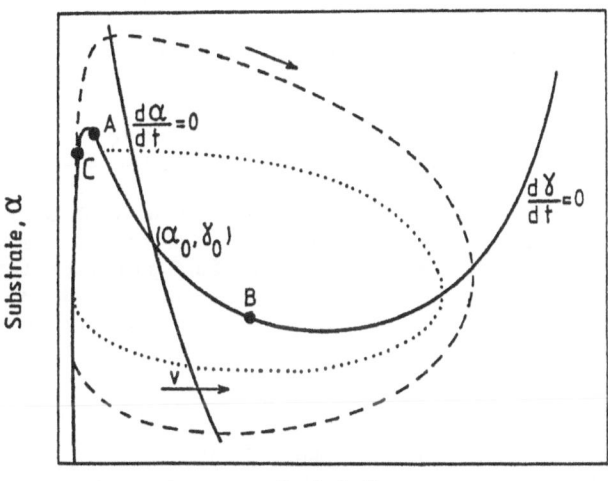

Fig. 2.   Schematic phase-plane diagram for the product-activated enzyme
reaction of Fig. 1 in the absence of product recycling. The
system is governed by Eqs. (1) with σ_i=0. As indicated by the
arrow, the substrate nullcline (dα/dt)=0 can be moved to the
right by increasing the substrate input v. The unique steady
state (α_0, γ_0) is unstable whenever it lies between points A
and B on the product nullcline; then a stable limit cycle
(dashed line) encloses the unstable steady state. In C, the
steady state is stable but excitable: a suprathreshold pulse
of product is amplified in a pulsatory manner (dotted line)
before the system returns to steady state.

steady state has shown (Goldbeter, 1980) that it is unstable whenever it lies on a region of negative slope on the product nullcline, such that

$$(d\alpha/d\gamma)_0 < -(1/q) \tag{2}$$

(the subscript zero refers to the steady state). Then the system evolves toward a stable limit cycle (Fig. 2) corresponding to sustained oscillations of the substrate and product concentrations around the unstable steady state.

In the absence of recycling, the product nullcline possesses at most one region of negative slope. As the product concentration at steady state is simply given by (qv/k_s), a continuous increase in v from a low initial value will shift the substrate nullcline from left to right. As a result, the system will move from a stable steady state into a domain of instability (located between points A and B in Fig. 2) and, finally, into a range where the system again evolves toward a stable steady state. This result accounts for the observation that glycolytic oscillations in yeast occur in a range of substrate injection rates bounded by two critical values (Hess et al., 1969).

Excitability Beyond A Single Threshold

For substrate injection rates just below those that produce oscillations, the steady state is located immediately to the left of point A on the product nullcline (e.g., in C, in the diagram of Fig. 2). The steady state is then stable, but excitable. In the phase plane, the addition of a pulse of product corresponds to a horizontal displacement to the right, away from the steady state. This perturbation is immediately damped as long as it remains below a threshold value. Above this threshold, however, the perturbation is amplified in a pulsatory manner before the system returns to the stable steady state. The dose-response curve for the magnitude of the response as a function of stimulation is a sigmoid characterized by a single, sharp threshold in which the response rises quasi-vertically from a basal to a plateau value. A similar dose-response curve has also been obtained in a model for the relay of suprathreshold cAMP pulses in the slime mold _Dictyostelium discoideum_ (Goldbeter et al., 1978). Such relay reflects the excitability of the signalling system which controls aggregation of the amoebae after starvation (Devreotes, 1982; Gerisch, 1982).

Bistability

A phase portrait such as that shown in Fig. 2 readily gives rise to bistability. Thus, when the substrate level is held constant in the range where the product nullcline possesses a region of negative slope, three steady-state values of γ are obtained. The extreme values correspond to stable steady states, whereas the third, intermediary one corresponds to an unstable state. As shown in Fig. 3, the system evolves to either one of the two stable states, depending on initial conditions.

COMPLEX PATTERNS OF TEMPORAL ORGANIZATION DUE TO PRODUCT RECYCLING

Product recycling allows for additional modes of dynamic behavior with respect to excitability and oscillations.

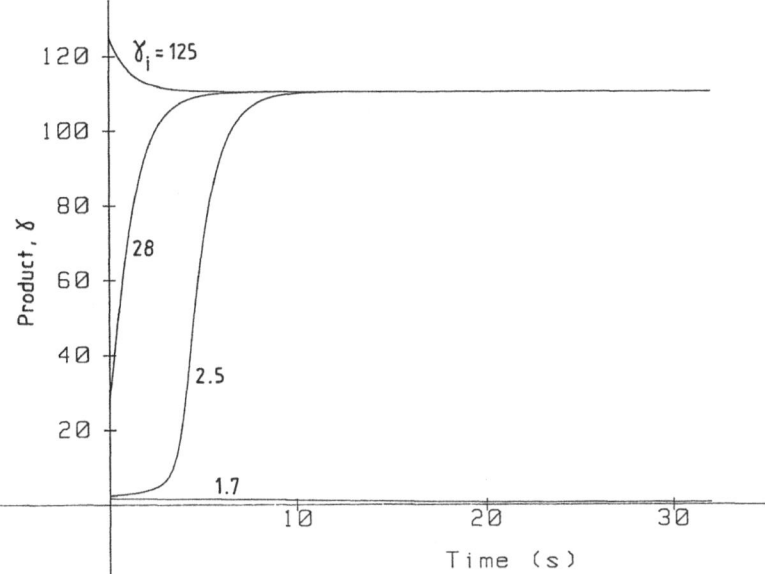

Fig. 3. Bistability. When the substrate concentration is held constant
(α =100), the system can evolve to either one of two stable
steady states, depending on initial conditions. The time
evolution is obtained by integrating numerically Eq. (1b) in
the absence of product recycling (σ_i=0) for the parameter
values: σ_M=5.8 s^{-1}, L=5x10^6, K=12, n=4, q=20, k_s=1 s^{-1}. The
initial value of γ is indicated on each curve.

Excitability With Multiple Thresholds

As in the case σ_i=0, the steady state is stable and excitable when
the value of the substrate input is taken just below that giving rise to
sustained oscillations. A small, subthreshold increase in γ is not
amplified, in contrast with a slightly larger, suprathreshold
perturbation (see the response to stimuli (1) and (2) in Fig. 4a). As
the stimulus rises, the maximum value γ_M of the response reaches a
plateau. Further increase in the perturbation fails at first to elicit
any amplification. When the perturbation exceeds a second, higher
threshold, however, the system amplifies it in a pulsatory manner before
returning to the stable steady state (see the response to stimuli (3) and
(4) in Fig. 4a).

The maximum of the peak in γ is shown in Fig. 4b as a function of
the initial displacement from steady state measured by the initial value
γ_i. Two sharp thresholds, each followed by a different plateau in the
response amplitude, can be distinguished in Fig. 4.

How does product recycling give rise to multi-threshold excitability?
This question can best be answered by resorting to phase plane analysis.
Recycling of γ into α produces a "bump" in the product nullcline around
the point of abscissa γ=K (Fig. 5). This nullcline is the locus of the
steady state as a function of parameter v. As v increases, the steady
state level of substrate, α_o, rises at first. This rise continues until
the product reaches the level (close to unity) at which autocatalysis

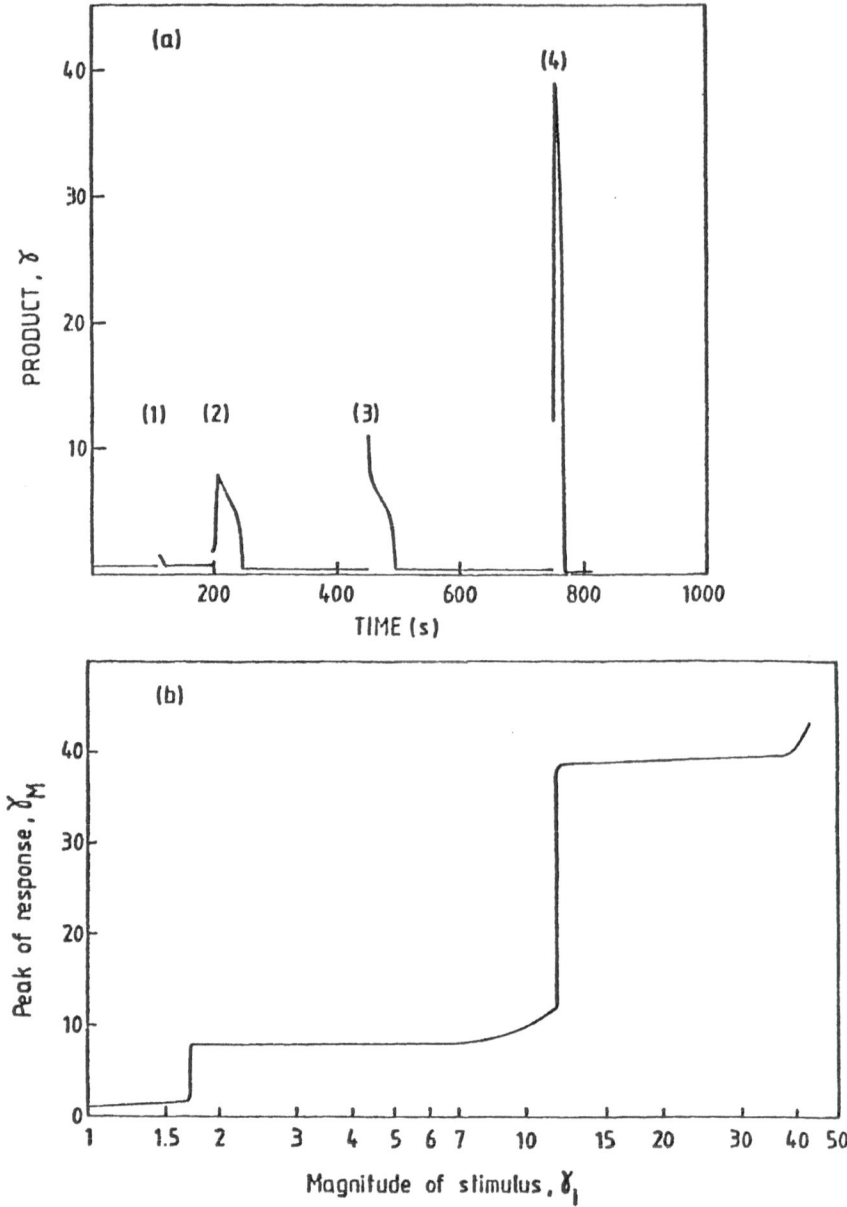

Fig. 4.   Excitability with multiple thresholds. The temporal response
to four increasing stimuli is shown in (a), whereas the full
dose-response curve is shown in (b) (redrawn from Moran and
Goldbeter, 1985). Two sharp thresholds can be distinguished,
each of which is followed by a different plateau in response
amplitude. The stable steady state is perturbed in (a) by
setting the initial value γ_i equal to 1.6, 1.8, 11, and 12,
successively. Parameter values are v=0.04 s^{-1}, σ_M=4 s^{-1},
L=5x10^6, K=8.7, n=4, q=50, k$_s$=3 s^{-1}, σ_i=1 s^{-1}.

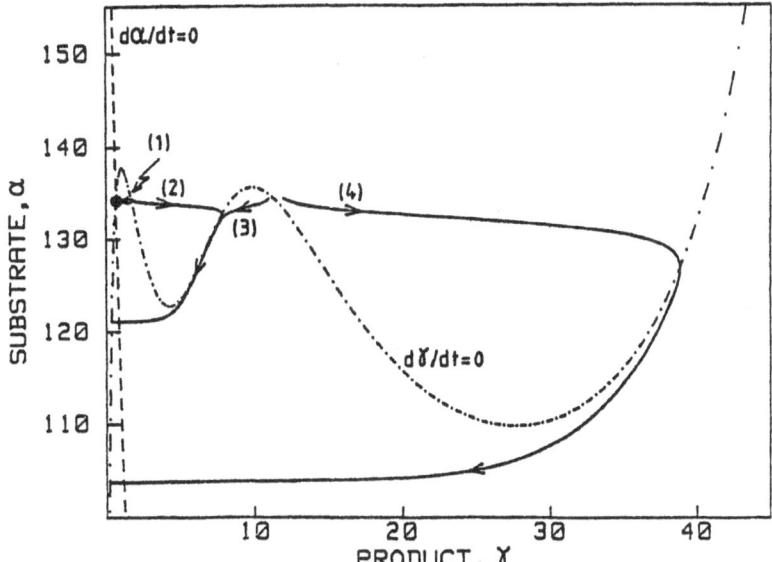

Fig. 5.   Explanation of multithreshold excitability by phase plane
analysis. The trajectories followed in response to the four
stimuli of Fig. 4a are shown, together with the substrate and
product nullclines (redrawn from Moran and Goldbeter, 1985).

becomes effective. Then, any increase in v will lead to an increase in γ
such that the rate of the product-activated reaction will be enhanced
owing to autocatalysis; the substrate level at steady state will decrease
as a result of this enhanced consumption, and the slope $d\alpha/d\gamma$ on the
product nullcline becomes negative.

Around $\gamma = K$, however, product recycling becomes effective and
counterbalances the decrease in α_o due to substrate consumption; hence
the local maximum in the nullcline. At higher rates v, autocatalysis
once again overrides recycling until the enzyme is fully activated by the
product at sufficiently large values of γ; then, further increse in v
again leads to an increase in the substrate level at steady state.

Such phase portrait in which the product nullcline possesses two
regions of negative slope readily accounts for multi-threshold
excitability. In Fig. 5, the trajectories following the four increasing
stimuli of Fig. 4a are indicated. When the steady state is located on
the left limb of the sigmoid nullcline, the dynamics of the system is
dictated by the position of the initial value of γ relative to the two
portions of the curve where $d\alpha/d\gamma$ is negative. These portions of the
nullcline indeed determine two thresholds, each of which separates a
region of amplification of the perturbation ($d\gamma/dt>0$) from a region of
immediate decay ($d\gamma/dt<0$) since the sign of the derivative of γ changes
upon crossing the nullcline $d\gamma/dt=0$.

Birhythmicity: Coexistence of Two Stable Periodic Regimes

The stability analysis in the presence of product recycling shows
that the instability condition is still given by Eq. (2), as in the case

σ_i=0. Oscillatory behavior is, however, significantly affected by the recycling process, as shown by the bifurcation diagrams of Fig. 6 where the maximum amplitude α_M of the substrate in the course of oscillations, as well as the steady state level α_0, are plotted as a function of the substrate input v for different values of the maximum recycling rate σ_i (Moran and Goldbeter, 1984).

For σ_i=0, a single domain of oscillations exists in which the amplitude passes through a maximum as v increases (this behavior is accounted for by the phase portrait of Fig. 2 in which the steady state is unstable between A and B). At large values of σ_i (Fig. 6h), two domains of oscillations are separated by a domain in which the steady state is stable; this is due to the stabilizing shoulder induced in the product nullcline around γ=K by recycling. For intermediate values of the recycling rate, the bifurcation diagrams indicate the occurrence of hard excitation, i.e. the coexistence between a stable steady state and a stable periodic regime separated by an unstable limit cycle.

In a narrow range of v values, the coexistence between two stable periodic regimes can also be observed (Fig. 6d-f). Such phenomenon of birhythmicity (Decroly and Goldbeter, 1982) is represented in the phase plane in Fig. 7, where arrows indicate how the system can switch from one periodic regime to the other upon chemical perturbation. The actual switch from the low-amplitude to the large-amplitude limit cycle is shown to occur in response to a substrate pulse in Fig. 8a, whereas the reverse transition is shown in Fig. 8b.

As observed in a recent study of birhythmicity in a model for the cAMP signalling system of D. discoideum (Goldbeter and Martiel, 1985), the transition from the large to the small limit cycle requires much finer tuning of the perturbation with respect to both the magnitude and the phase at which it is applied. The reason for such asymmetry in the sensitivity of the two cycles with respect to perturbations becomes clear when considering Fig. 7. Switching from the small to the large limit cycle occurs whenever the perturbation brings the system across the boundary set by the unstable cycle. The basin of attraction of the larger limit cycle is much more extended than that of the small-amplitude oscillatory regime. Hence the increased difficulty of reaching the latter regime by chemical perturbation.

Tristability

The existence of two regions of negative slope on the product nullcline gives rise to the possibility of a coexistence between three stable steady states. This occurs when the substrate level is held constant in the range in which the nullcline possesses the two regions of negative slope. For the parameter values which produced bistability in the absence of product recycling in Fig. 3, an additional stable steady state appears when recycling takes place. Thus, for σ_i=2.2 s^{-1}, the system can evolve toward either one of the three simultaneously stable steady states γ_0=55.8, 8.9 or 0.59, depending on initial conditions (Fig. 9).

Here, two sharp thresholds, located near γ_i=25 and γ_i=2, separate the three basins of attraction; these thresholds correspond to the two unstable steady states which separate the three stable ones. A horizontal line corresponding to a constant level of substrate indeed

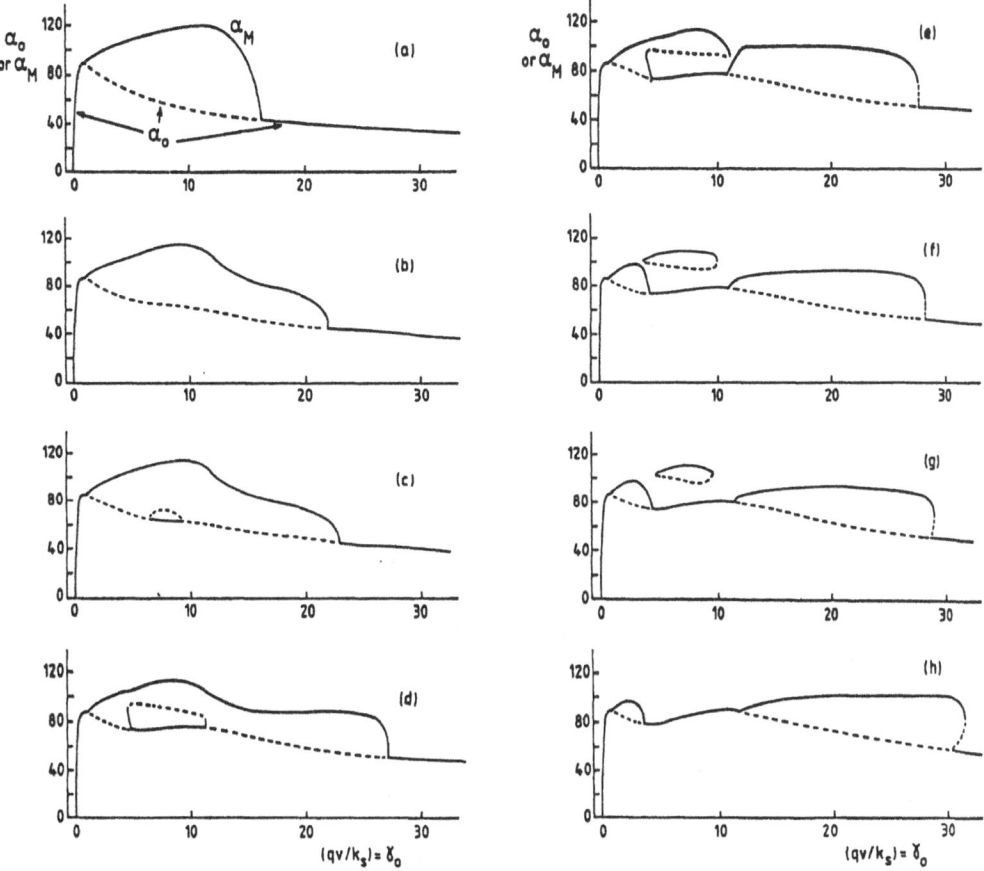

Fig. 6. Effect of product recycling on the appearance of birhythmicity
and of two domains of oscillations. Shown are a series of
bifurcation diagrams as a function of v for different values of
the maximum recycling rate σ_1 (in s^{-1}); (a) 0, (b) 0.5, (c)
0.6, (d) 1.2, (e) 1.3, (f) 1.4, (g) 1.5, (h) 2.0. As indicated
in panel (a), each diagram shows the steady-state level of
substrate (α_0) (lower curve), as well as the maximum amplitude
α_M of substrate oscillations (upper curves). Stable and
unstable regimes are represented by solid and dashed lines,
respectively. Parameter values are $\sigma_M=10$ s^{-1}, $L=5 \times 10^6$, $K=10$,
$n=4$, $q=1$, $k_s=0.06$ s^{-1} (redrawn from Moran and Goldbeter, 1984).

intersects the product nullcline in up to five points corresponding to as
many steady states.

Multiple Oscillatory Domains

For appropriate values of the recycling rate, two domains of
oscillations are encountered upon variation of the control parameter v
(see Figs. 6f-h). This property differs from birhythmicity which refers
to the coexistence of two distinct, simultaneously stable oscillatory
regimes. Here, two distinct periodic regimes are observed for closely
related, but different, parameter values.

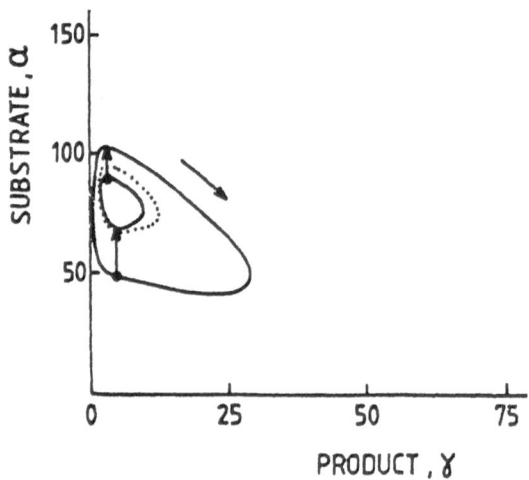

Fig. 7.   Birhythmicity. The behavior of the model is represented in the
phase plane (α, γ) for the situation of Fig. 6e, with v=0.255
s^{-1}, i.e. (qv/k_s)=4.25. The two stable limit cycles (solid
lines) are separated by an unstable cycle (dashed line). The
vertical arrows indicate how the system may switch from one
stable cycle to the other upon addition of substrate (Moran and
Goldbeter, 1984).

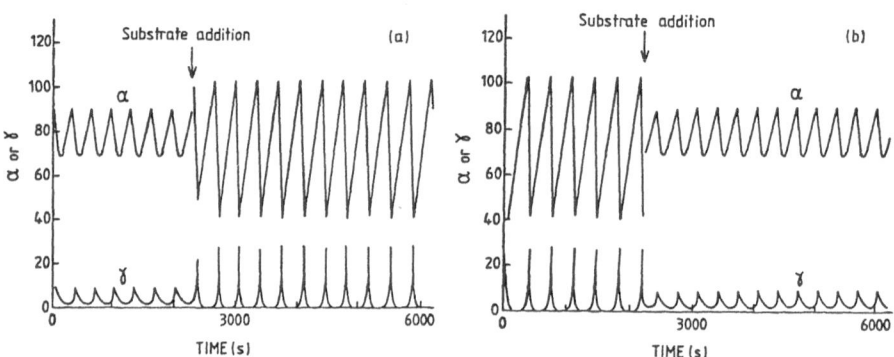

Fig. 8.   Switching between two simultaneously stable periodic regimes.
The passage from the small-amplitude to the large-amplitude
oscillations of Fig. 7 is shown in (a), whereas the reverse
transition is achieved in (b) (Moran and Goldbeter, 1984).

An interesting property associated with such behavior is illustrated
in Fig. 10. Starting at a value of v=0.05 s^{-1} corresponding to a stable
state, we apply square-pulse perturbations bringing the substrate input
to the values 0.15, 0.6 and 1.5 s^{-1}, successively. The first increase
produces small-amplitude oscillations; the second brings the system into
a stable steady state; whereas the last gives rise to large-amplitude,
high-frequency oscillations.

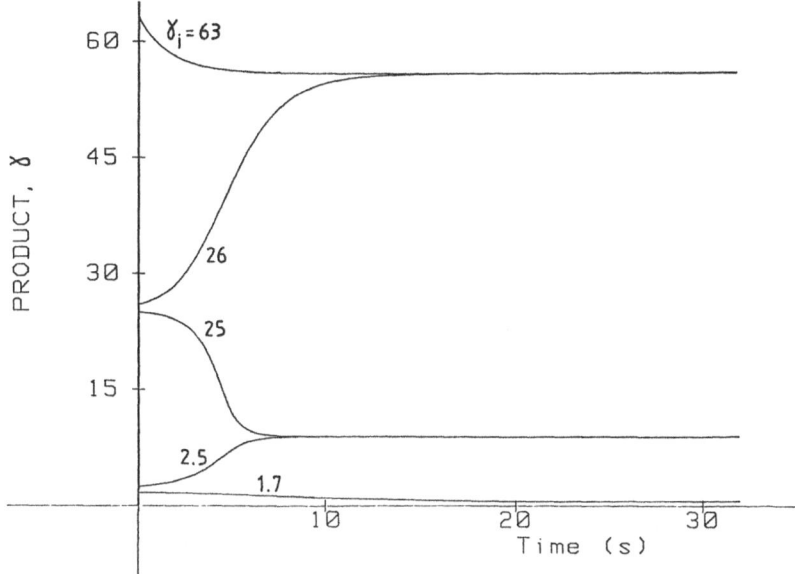

Fig. 9.   Tristability. For the parameter values of Fig. 3, at the same
fixed concentration of substrate, the system evolves to either
one of three stable steady states in the presence of product
recycling (σ_i=2.2 s^{-1}). The initial product concentration is
indicated on each curve.

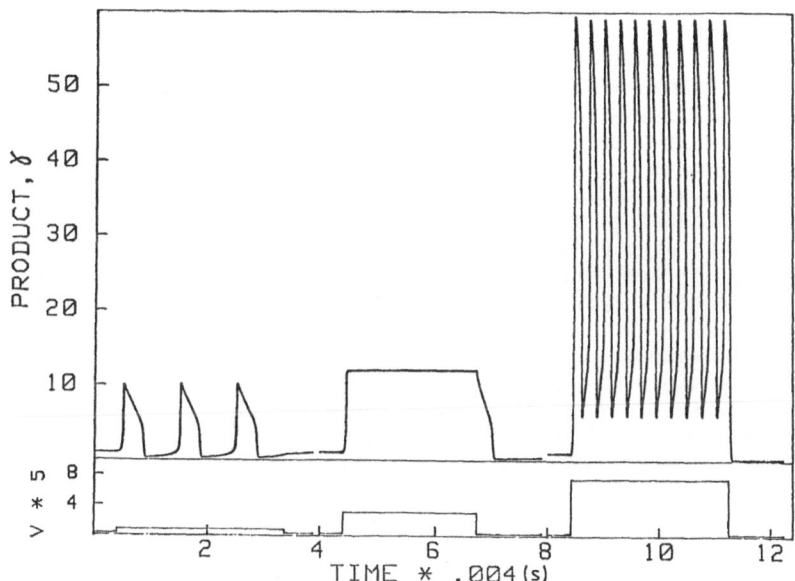

Fig. 10. Multiple modes of oscillations. Shown is the time evolution of
the product concentration following successive square-pulse
increases in the substrate input v from the initial value 0.05
s^{-1}. A stable steady state is established at the intermediate
value v=0.6 s^{-1}, whereas the upper and lower values v=1.5 s^{-1}
and 0.15 s^{-1} produce two different modes of rhythmic behavior
(parameter values are as in Fig. 9). This phenomenon is
similar to that observed by Jahnsen and Llinas (1984) in
thalamic neurons, where a hyperpolarizing and a depolarizing
current pulse produce distinct modes of rhythmic behavior when
cells are initially in a nonoscillatory state (see Discussion).

301

DISCUSSION

We have explored the patterns of temporal organization in a biochemical system with multiple domains of instability. Given that the model considered comprises only two variables, we used phase plane analysis to explain the origin of a variety of complex dynamic phenomena. The peculiarity of the phase portrait of the model is that one of the nullclines possesses two regions of negative slope, or, in other words, admits up to five intersections with a straight line.

In the absence of recycling, the model of Fig. 1 is characterized by a phase portrait in which the product nullcline (i.e., the nullcline of the variable that exerts autocatalysis) possesses but a single region of negative slope. The system is then characterized by a single domain of limit-cycle oscillations, and by excitability associated with a single threshold. Such analysis has previously been proposed to account for glycolytic oscillations in yeast and muscle, and for the oscillations and relay of suprathreshold cAMP pulses in _Dictyostelium_ amoebae (see Goldbeter, 1980, and Goldbeter et al., 1984, for review). A similar phase plane analysis was first used by Fitzhugh (1961) to account for the excitable and oscillatory properties of the nerve membrane. Finally, we showed that in these conditions, the system is capable of bistability when the substrate concentration is "frozen" so that the dynamics of the model is governed by a single variable.

Recycling of product into substrate modifies the dynamic properties of the model of Fig. 1, by introducing a second region of negative slope on the product nullcline. The consequences of this change in phase portrait are at least fourfold. First, excitability can now be characterized by two distinct thresholds and two different plateaus, so that the dose-response curve takes a staircase appearance (Fig. 4). Second, the system becomes capable of birhythmicity: two simultaneously stable periodic regimes with different amplitudes and frequencies may coexist _in the same conditions_. The evolution toward either one of these regimes depends on initial conditions, i.e. on the history of the system. The reversible switching from one mode of oscillation to the other can be induced by chemical perturbations applied at the appropriate phase, with the appropriate magnitude (Fig. 8); alternatively, an increase followed by a decrease in the control parameter (here, the substrate input v) can be used to create a hysteresis loop between the two periodic regimes.

The third consequence is that the system becomes capable of exhibiting two markedly different modes of oscillation in slightly different conditions (Fig. 10). This property, which differs from birhythmicity, may, however, be of greater physiological significance as it occurs in a much larger domain in parameter space.

Finally, when the substrate level is maintained constant, a third stable steady state is added to the two already present in the absence of recycling. Here also (Fig. 9), the choice of the asymptotic state is governed by the initial conditions since each of the stable steady states possesses its basin of attraction.

Phase plane analysis readily accounts for all the above modes of dynamic behavior, and helps to predict the conditions in which the various patterns of temporal organization will occur. It also suggests that further complexity would arise from the presence of additional regions of negative slope on the product nullcline.

Previous analyses of biochemical models comprising three or more variables showed that birhythmicity occurs there in the neighborhood of complex periodic (bursting) or aperiodic (i.e., chaotic) oscillations (Decroly and Goldbeter, 1982; Goldbeter et al., 1984; Martiel and Goldbeter, 1985). Here, bursting or chaos may not occur as these phenomena require at least three interacting variables. The conditions for the occurrence of birhythmicity are therefore less stringent.

The road to birhythmicity illustrated in Fig. 6 shows that the phenomenon accompanies the splitting of a single domain of instability of the steady state into two such domains. It is at the edges of the intermediate zone of stability that birhythmicity occurs in the present model, as was the case in a three-variable system of coupled autocatalytic enzyme reactions (Decroly and Goldbeter, 1982) and in a model for the cAMP signalling system of D. discoideum (Goldbeter and Martiel, 1985). This road to birhythmicity may therefore be generic.

Birhythmicity has not yet been observed in biological systems, but has recently been demonstrated in a number of chemical oscillatory reactions (Alamgir and Epstein, 1983; Lamba and Hudson, 1985; Roux, 1983). Whereas the phenomenon analyzed here is autonomous, multiple periodic regimes have also been obtained in a model for glycolytic oscillations subjected to a periodic input of substrate (Markus and Hess, 1984). Chaotic behavior has been demonstrated experimentally in the latter system (Markus et al., 1984), whereas autonomous chaos has been reported for the peroxidase reaction (Olsen and Degn, 1977), and may well underlie the aperiodic aggregation of the D. discoideum mutant Fr17 (Durston, 1974), as suggested by a recent theoretical study (Martiel and Goldbeter, 1985).

The phenomenon of bistability has been observed in a variety of models and demonstrated experimentally in chemical as well as biochemical systems. To the latter belong the peroxidase reaction (Degn, 1968), membrane-bound enzymes under pH control (Naparstek et al., 1974), as well as a reconstituted, partial glycolytic system containing phosphofructokinase and pyruvate kinase (Eschrich et al., 1980). The phenomenon of tristability is much less common and has been reported for chemical systems only (Epstein, 1984). All-or-none transitions between multiple steady states provide switch mechanisms for the control of metabolic processes (e.g., Boiteux et al., 1980). On the other hand, steep transitions may arise in metabolic pathways in the absence of multiple steady states, either as a result of allosteric regulation or of zero-order ultrasensitivity in covalent modification (Goldbeter and Koshland, 1982).

With regard to temporal organization, the present analysis of a relatively simple biochemical system shows the richness of behavioral modes, even in the absence of multiple feedback processes. Some of these modes have been observed in cellular metabolism, e.g., oscillations, excitability with a single threshold, and bistability. Some other modes have not yet been observed, e.g., multi-threshold excitability, birhythmicity, and the coexistence between three stable steady states. The prediction of their occurrence, corroborated by observations in chemical systems, may prompt their investigation at the cellular level.

Beyond their bearing on the patterns of temporal organization in enzymatic systems, the present results throw light on the dynamics of some neurophysiological processes. In addition to excitability, rhythmic activity is one of the most conspicuous properties of neurons (Fessard,

1936). Jahnsen and Llinas (1984) have recently shown that neurons of the
thalamus are capable of two distinct modes of rhythmic behavior which
occur at different levels of the membrane potential, with a frequency
close to 6 Hz and 10 Hz, respectively. At an intermediate level of the
potential corresponding to the absence of rhythmic behavior, a
hyperpolarizing current pulse gives rise to one mode of oscillation,
whereas a depolarizing current pulse produces the second kind of
periodicity.

Such results may be interpreted in the light of the present analysis
(see Fig. 10, and Moran and Goldbeter, 1985; Goldbeter and Moran,
manuscript in preparation). The occurrence of multiple modes of thalamic
oscillations can indeed be accounted by an underlying phase portrait
similar to that of the biochemical model of Fig. 1: the multiplicity of
oscillatory modes reflects the existence of two instability domains; in
the phase plane, the latter property should be linked to the existence of
a nullcline with two regions of negative slope. The fact that the
variables here are biochemical instead of current and voltage in the
neuronal system should not prevent relating the above results to thalamic
behavior. From a dynamic point of view, the key fact is that similar
dynamic phenomena are associated with similar phase portraits.

The above view is corroborated by a recent study of Rose and
Hindmarsh (1985) who proposed a model for thalamic cells, on the basis of
a nullcline structure similar to that of the present model. At the core
of their two-variable model, Rose and Hindmarsh indeed consider that the
nullcline for the membrane potential possesses two regions of negative
slope; to this end, the equation for this nullcline is taken as a combination of two cubic polynomials. This nullcline shape gives rise to two
oscillatory domains which are then related to the multiple modes of
oscillations observed in thalamic neurons.

In contrast to this approach, the present model provides a molecular
mechanism that gives rise to a nullcline with two regions of instability.
The continuous deformation of the nullcline due to product recycling
leads to patterns of dynamic behavior which range from a unique periodic
regime to birhythmicity and multiple modes of oscillations. In neurons,
similar nullcline deformations due to a variety of ionic currents
produce, in the same manner, increasingly complex patterns of rhythmic
activity.

ACKNOWLEDGMENTS

Fruitful discussions with Dr. R. Llinas are gratefully acknowledged.
One of the authors (F.M.) was supported by a short-term EMBO fellowship
and by a fellowship from Fundacion Juan March during completion of this
work.

REFERENCES

Alamgir, M. and Epstein, I.R., 1983, J. Am. Chem. Soc., 105:2500.
Berridge, M.J. and Rapp, P.E., 1979, J. Exp. Biol., 81:217.
Boiteux, A., Goldbeter, A., and Hess, B., 1975, Proc. Nat. Acad. Sci.
    USA, 72:3829.
Boiteux, A., Hess, B., and Sel'kov, E.E., 1980, Curr. Top. Cell. Regul.,
    17:171.
Decroly, O. and Goldbeter, A., 1982, Proc. Nat. Acad. Sci. USA, 79:6917.
Degn, H., 1968, Nature, 217:1047.

Devreotes, P.N., 1982, in: "The Development of Dictyostelium discoideum",
    W.F. Loomis, ed., Academic Press, New York.
Durston, A.J., 1974, Devel. Biol., 38:308.
Epstein, I.R., 1984, in: "Chemical Instabilities", G. Nicolis and F.
    Baras, eds., Reidel, Dordrecht.
Eschrich, K., Schellenberger, W., and Hofmann, E., 1980, Arch. Biochem.
    Biophys., 205:114.
Fessard, A., 1936, "Proprietes Rythmiques de la Matiere Vivante,"
    Hermann, Paris.
Fitzhugh, R., 1961, Biophys. J., 1:445.
Gerisch, G., 1982, Ann. Rev. Physiol., 44:535.
Goldbeter, A., 1980, in: "Mathematical Models in Molecular and Cellular
    Biology", L.A. Segel, ed., Cambridge Univ. Press, Cambridge.
Goldbeter, A. and Caplan, S.R., 1976, Ann. Rev. Biophys. Bioeng., 5:449.
Goldbeter, A., Erneux, T., and Segel, A., 1978, FEBS Lett., 89:237.
Goldbeter, A. and Koshland, D.E. Jr., 1982, Quart. Rev. Biophy., 15:555.
Goldbeter, A. and Lefever, R., 1972, Biophys. J., 12:1302.
Goldbeter, A. and Martiel, J.L., 1985, FEBS Lett., in press.
Goldbeter, A., Martiel, J.L., and Decroly, O., 1984, in: "Dynamics of
    Biochemical Systems", J. Ricard and A. Cornish-Bowden, eds., Plenum,
    New York.
Hess, B. and Boiteux, A., 1971, Ann. Rev. Biochem, 40:237.
Hess, B., Boiteux, A., and Kruger, J., 1969, Adv. Enzyme Regul., 7:149.
Jahnsen, H. and Llinas, R., 1984, J. Physiol., 349:227.
Lamba, P. and Hudson, J.L., 1985, Chem. Eng. Commun., 32:369.
Markus, M. and Hess, B., 1984, Proc. Nat. Acad. Sci. USA, 81:4394.
Markus, M., Kuschmitz, D., and Hess, B., 1984, FEBS Lett., 172:235.
Martiel, J.L. and Goldbeter, A., 1985, Nature, 313:590.
Monod, J., Wyman, J., and Changeux, J.P., 1965, J. Mol. Biol., 12:88.
Moran, F. and Goldbeter, A., 1984, Biophys. Chem., 20:149.
Moran, F. and Goldbeter, A., 1985, Biophys. Chem., in press.
Naparstek, A., Romette, J.L., Kernevez, J.P., and Thomas, D., 1974,
    Nature, 249:490.
Olsen, L.F. and Degn, H., 1977, Nature, 267:177.
Rose, R.M. and Hindmarsh, J.L., 1985, Proc. R. Soc. Lond., B 225:161.
Roux, J.C., 1983, Physica, 7D:57.

REACTION-DIFFUSION COUPLING IN IMMOBILIZED ENZYME

SYSTEMS: IONIC INTERACTIONS[*]

Alain Friboulet, Gilles Cauet,
Emmanuelle Mege and Daniel Thomas

Laboratoire de Technologie Enzymatique
UA 523 du CNRS
Université de Technologie de Compiègne
BP 233
60206 Compiègne, France

INTRODUCTION

Little attention has been paid to the influence of diffusion phenomena on nonlinear enzyme kinetics, although there is strong experimental evidence that living cells are highly compartmentalized and structured. Within a cell, the majority of enzymes are fixed reversibly or irreversibly onto membrane structures. Therefore, it is highly likely that diffusional limitations, mass transfer, and microenvironmental effects play an important role in the global behavior of cellular enzyme systems.

From an experimental point of view, the in situ accessibility of parameters governing the behavior of these microenvironments has proved to be limited. In order to overcome partly these difficulties, enzymologists have attempted to reproduce artificial cellular conditions in such a way that diffusional constraints and local concentrations can be accurately taken into account. At this level, the so-called "heterogeneous enzymology" is certainly an important step toward a better understanding of underlying laws governing metabolic pathways. In particular, the feasibility of manufacturing membranous supports with well-defined geometrical and physicochemical properties facilitates mathematical modeling, taking into account the coupling between metabolite diffusion (and/or compartmentation) and enzymatic reactions. The method developed in our laboratory (Broun et al., 1973) consists of a co-crosslinking between an inert protein (e.g., albumin, gelatin, hemoglobin) and the enzyme protein, with the help of a bifunctional agent (viz., glutaradelhyde). This method allows an homogeneous distribution of active sites throughout the thickness of the membrane. By the mere fact of their membranous shape, our supports can be used for kinetic studies with either symmetrical or asymmetrical boundary conditions. Moreover, compartmentalized experimental schemes can be designed.

---

[*]This paper is dedicated to the memory of our friend and colleague Yves Malpièce.

With this experimental system, we have developed both theoretical and experimental investigations along three directions:

1) Multienzyme systems which can be designed to obtain information on the efficiency of crosslinked regulation, channeling effects in linear or branched sequences, etc.

2) The effects of the physicochemical properties of the support, and more precisely the influence of fixed and mobile charges, on kinetic modulation.

3) The sophisticated regulatory properties exhibited by allosteric and substrate-inhibited enzymes, which give rise to complex behavior such as multistability or spatiotemporal structures.

The first point was reviewed recently by Hervagault and Thomas (1985). We will focus mainly on the latter two points with simple monoenzymatic immobilized systems.

GENERAL THEORETICAL TREATMENT OF DIFFUSION-REACTION

Kinetic formulations derived from studies in homogeneous and isotropic media are no longer valid when describing overall kinetics of a membrane system, in which substrate and product concentrations (as well as effector concentrations) may vary at each point of the thickness of the membrane. However, the kinetic formulation remains valid for any volume element small enough to be considered homogeneous with regard to the substrate concentration. The rate of change of substrate concentration with time in such a local volume element, $\delta S/\delta t$, depends on two simultaneous phenomena: metabolite diffusion in the membrane matrix and enzymatic activity, as follows:

$$\delta S/\delta t = (\delta S/\delta t)_{diffusion} + (\delta S/\delta t)_{reaction} \tag{1}$$

Considering variations in concentrations as taking place only in the x direction, perpendicular to the plane of the membrane, and introducing Fick's second law for the diffusion process, Eq. (1) becomes:

$$\delta S/\delta t = D_S (\delta^2 S/\delta x^2) - f(S) \tag{2}$$

where the local diffusion coefficient of the substrate, D_S, is assumed to be independent of concentration and space, and $f(S)$ is the enzymatic kinetic relation. Similarly we can write for the product:

$$\delta P/\delta t = D_p(\delta^2 P/\delta x^2) + f(S) \tag{3}$$

Thus the system is ruled by nonlinear partial differential equations of second-order. When the enzyme reaction obeys the Michaelis-Menten relation, Eq. (2) can be written in dimensionless form:

$$\delta S/\delta t = (\delta^2 S/\delta x^2) - \sigma[S/(1+S)] \tag{4}$$

with e (thickness) as space unit, e^2/D_S as time unit and K_m as concentration unit, with:

$$\sigma = (e^2 \cdot V_m)/(D_S \cdot K_m) \tag{5}$$

Thus the system is ruled by the dimensionless parameter σ, similar to the square of the Thiele modulus. The σ parameter as defined by Eq. (5) contains the three factors that determine the substrate concentration profile within an enzyme membrane: the membrane thickness, the facility of the substrate to diffuse through the support, and the intrinsic activity of the catalyst. The overall diffusion limitations in the membrane increase with the σ value (Engasser and Horvath, 1976).

For steady-state conditions $(\delta S/\delta t) = 0$, analytical solutions of Eq. (4) were obtained for zero-order (Selegny et al., 1968) or pseudo first-order kinetics (Goldman et al., 1968; DeSimone and Caplan, 1973). The solutions are obtained as expressions for the fluxes and concentration profiles of substrate and product within the membrane.

For non-steady-state conditions, the evolution of concentration profiles as a function of time can be obtained by a finite-difference implicit method (Kernevez and Thomas, 1975).

CYTOCHEMICAL DEMONSTRATION OF LOCAL CONCENTRATION IN A MONOENZYMATIC

IMMOBILIZED SYSTEM

It is possible, in very simple systems, to study the influence of structure and of diffusional constraints on the overall enzyme membrane properties. There is experimental evidence for the existence of local concentrations inside artificial enzyme films (Barbotin and Thomas, 1974; Malpièce et al., 1980). Cytochemical methods were used to visualize the microenvironmental concentrations. No similarities were found between the heterogeneous distribution of the electron-dense reaction product and the homogeneous distribution of the enzyme.

In view of the physiological importance of acetylcholinesterase, it was of interest to evaluate theoretically and experimentally the behavior of acetylcholinesterase with a model system (Malpièce et al., 1981). The critical point in the histochemistry of cholinesterases is the introduction of an appropriate capturing agent, which reacts with the primary reaction product to form an insoluble compound. The specific and widely used procedure of Koelle and Friedenwald (1949) is based on acetylthiocholine as substrate and Cu^{2+} as capturing agent. Copper-thiocholine solubility under basic conditions and enzyme inhibition by cupric ions impose severe limiting constraints on the incubation conditions. Experiments have therefore usually been carried out at pH 5-6, well away from the optimum pH of the enzyme, known to be close to 8.5. Succinate, acetate, or maleate buffers were used with glycine to form a chelate with Cu^{2+}. The working conditions of the model were similar to those usually used for the histochemical demonstration of cholinesterases.

Homogeneous active films were studied numerically by computer recording, with the scheme:

Acetylthiocholine $\text{-------}>$ thiocholine + CH_3-COOH

2 Thiocholine + Cu^{2+} $\text{---}^k\text{---}>$ precipitate

309

with S_1, S_2, S_3, S_4 as concentrations of acetylthiocholine, thiocholine, Cu^{2+}, and precipitate, respectively. Defining D_{Si} as the diffusion coefficient of substance i in the film, the governing balance equations are:

$$\frac{\delta S_1}{\delta t} - D_{S1} \frac{\delta^2 S_1}{\delta x^2} + V_1 = 0$$

$$\frac{\delta S_2}{\delta t} - D_{S2} \frac{\delta^2 S_2}{\delta x^2} - V_1 + 2kS_2^2 S_3 = 0$$

$$\frac{\delta S_3}{\delta t} - D_{S3} \frac{\delta^2 S_3}{\delta x^2} + kS_2^2 S_3 = 0 \tag{6}$$

$$\frac{\delta S_4}{\delta t} - kS_2^2 S_3 = 0$$

$$V_1 = V_M \frac{S_1}{S_1 + (S_1^2/K_{ss}) + K_M(1 + I/K_i)}$$

Complete kinetic studies of the system are required to generate a computer simulation of the system. These studies deal with enzyme kinetic parameters (Michaelis constant, maximum rate, and rate equation), diffusion coefficients, inhibitory properties of copper(II), and complexation of copper(II), which we asume to take place with glycine and with Tris-maleate buffer alone. The resulting experimental parameters are as follow:

$K_m = 0.13$ mM, $K_{ss} = 50$ mM, $K_i = 1.2 \times 10^{-5}$ M

$k = 10^{15}$ mol^{-2}h^{-1}; $D_{s1} = 2 \times 10^{-3}$ cm^{2}h^{-1}

$D_{s2} = 2.5 \times 10^{-3}$ cm^{2}h^{-1}; $D_{s3} = 1.5 \times 10^{-5}$ cm^{2}h^{-1}

The mean effective diffusion coefficient of Cu^{2+}, that is to say the penetration, is very low. This is the result of electrostatic interactions between the net negative charge of the membrane and the positive charge of cupric ions. The set of equations was solved by a numerical method (Kernevez and Thomas, 1975).

A membrane containing enzyme was incubated for 2 hrs in 1 mM acetylthiocholine and 2 mM copper sulphate. In the first case, the membrane contained 5 IU (5 µg, $V_m = 0.7$ mM/min) of acetylcholinesterase. The precipitate repartition is a direct visualization of the thiocholine distribution in the insoluble phase at the stationary state (Fig. 1a). The precipitate profile calculated by computer (Fig. 1b) is in reasonable agreement with the electron micrograph.

In the second case, with a greater activity (25 IU, $V_m = 3.5$ mM/min) in the membrane, the local concentration of free cupric ions becomes the limiting parameter for the visualization. The precipitate profile is no longer convex and presents an intermediate maximum (Fig. 2a). The theoretical profile is presented in Fig. 2b.

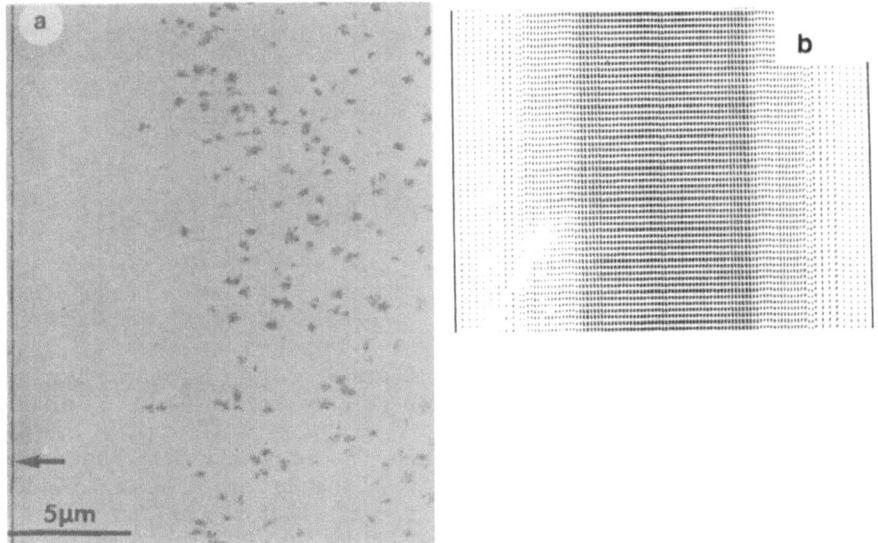

Fig. 1.   Convex repartition of the crystalline precipitate of copper
sulfide.   The membrane contained 5 IU of enzyme.   (a) One-half
a membrane is shown, the arrow indicates the edge.   (b)
Simulation of the precipitate repartition in the membrane.

Fig. 2.   Intermediate repartition of the crystalline precipitate in a
membrane containing 25 IU of enzyme.   (a) Electron micrograph
of one-half of the membrane.   (b) Calculated precipitate
repartition.

In the third case, the demonstration of diffusional constraints is
obvious.   The membrane contained 50 IU (V_m=7 mM/min) of enzyme.   The
local free Cu^{2+} concentration is a strongly limiting parameter of the
system.   The precipitate is therefore only visible near the edges of the
membrane (Fig. 3a), as predicted by computer simulation (Fig. 3b).

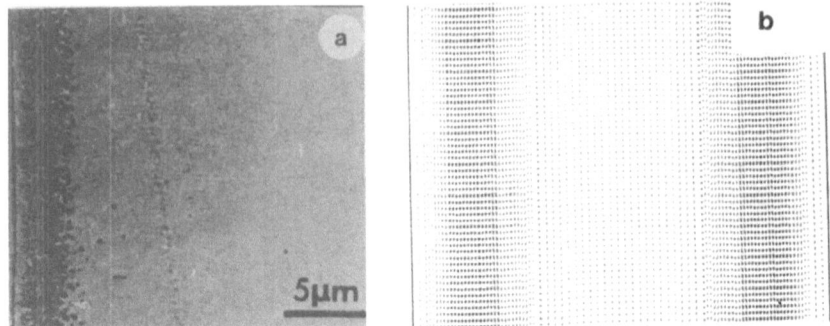

Fig. 3.   (a and b) Crystalline precipitate is only visible on the edges
of the membrane. The proteic film contained 50 IU of enzyme.

The study of this model system leads to the conclusions that, first,
there is no geometrical similarity between the active-site distribution
and the precipitate repartition, and, second, that the diffusion reaction
creates a focalized precipitation line (Figs. 1a, 2a, 3a). The phenomena
are explained by the existence of a concentration peak, resulting from
the spatiotemporal behavior of the system.

MODIFICATION OF THE KINETIC PROPERTIES INDUCED BY AN IONIC

MICROENVIRONMENT

Shifts in pH-Activity Profiles

The earliest observations of a displacement in the pH-activity
profile after immobilization of an enzyme within a membrane were
described by Goldman et al. (1965, 1968). Using a series of synthetic
substrates, the profiles for a papain-membrane were found to differ
significantly from the usual bell-shaped curves obtained in solution.
These differences can be accounted for by the assumption that the
immobilized enzyme acts in a microenvironment different from that
prevailing in the external solution. This situation is somewhat
different to that in which enzymes are bound to charged carriers, such as
the attachment of trypsin to a polyanionic carrier (Goldstein et al,
1964), or enzymes bound to biomembranes (Silman and Karlin, 1967; Maurel
et al., 1978). Goldstein et al. (1964) proposed the "polyelectrolytic
theory" to explain these deviations, by the unequal distribution of
hydrogen and hydroxyl ions and of charged substrates between the
polyelectrolyte phase within which the enzyme is embedded and the outer
solution. However, the treatment developed by Katchalski and co-workers
was based on restrictions, e.g., that the electrostatic potential
generated within the membrane is assumed to be homogenous and pH-
independent, and that the contribution of the total ionic strength
brought about by substrates and products (when these are charged
molecules) is assumed to be negligible. Whenever acid or base is
liberated during an enzymatic reaction, one may expect the occurrence of
local pH values which differ from those in the outer medium. The extent
of deviation of the pH-activity profiles is obviously dependent on the
rate of hydrolysis of the substrate in the membrane. Typical profiles
exhibiting these shifts are shown in Fig. 4.

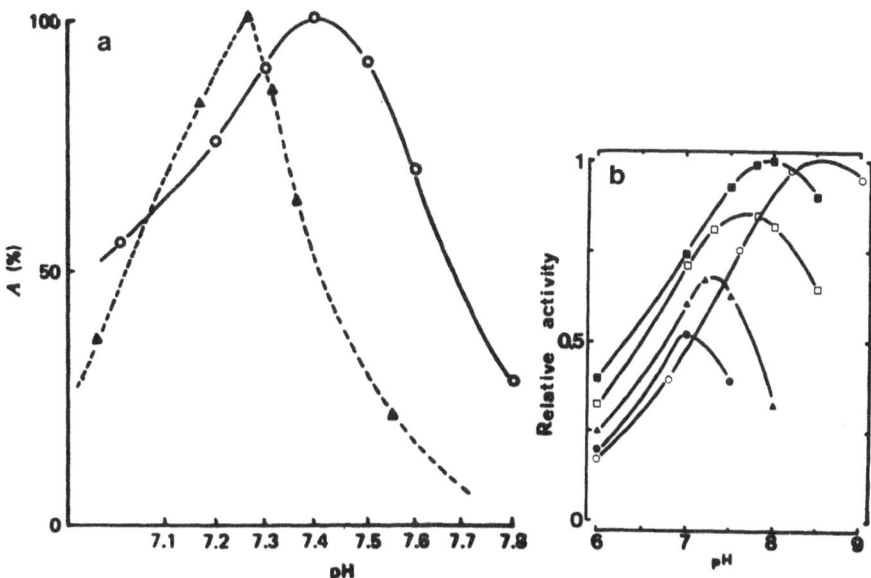

Fig. 4. pH-Activity profiles. (a) Urease-albumin membrane acting on
urea (▲). The profile of native ureseae is given for
comparison (o). (b) Acetylcholinesterase-albumin membrane in
the presence of 5 (●), 10 (▲), 20 (□) and 40 mM (■)
acetylcholine. The pH dependence of soluble acetylcholinesterase (o) is not influenced by the substrate concentration.

In the case of the urease membrane studied by Thomas et al. (1972), a
base is liberated and the apparent pH optimum is shifted to lower pH
values. The steady-state local pH is more alkaline than that of the
external solution. The acetylcholinesterase reaction produces protons
and, due to the amphoteric properties of the proteinaceous matrix, the
enzymatic reaction modifies the ion mobilities. So, the kinetic behavior
of the acetylcholinesterase system does not obey the polyelectrolytic
theory; the pH optimum values are displaced towards acidic values and
depend on the external substrate concentration. These results cannot be
explained without taking into account the mutual involvement of the
diffusion of charged molecules, the fixed charge density, and the
enzymatic reaction.

Modulation of the Kinetic Parameters

The effect of diffusion and/or ionic microenvironment on an
immobilized enzyme system can be established using both enzyme grafted on
soluble proteic polymers and then gelified to form membranes by
ultrafiltration (Wang et al., 1976). This method not only allows the
study of the effect of a chemical modification of the enzyme during the
insolubilization procedure, but also the insolubilization effects on the
kinetic parameters. The experimental demonstration of the influence of
diffusion and ionic microenvironment were obtained with three enzymatic
systems: urate oxidase, hexokinase, and cholinesterases (Remy et al.,
1978).

The urate oxidase system. Urate oxidase (EC 1.7.3.3) catalyzes the

313

oxidation of an uncharged substrate, uric acid, to give an uncharged product:

$$S \longrightarrow P$$

The results obtained with the enzyme in solution, immobilized on soluble polymers and gelified in a proteic film, are given on Fig. 5.

Activity as a function of substrate concentration (Fig. 5a), as well as linear transformation (Fig. 5b), display a two-fold K_m increase after grafting on soluble polymers and a five-fold increase after gelification. The insolubilization effect can be explained by diffusion limitation. So, the kinetic modifications can be fitted just by taking into account diffusion and reaction parameters in computer simulations, without any charge effect.

The hexokinase system. Hexokinase (EC 2.7.1.1), with a neutral substrate (glucose), a negatively-charged co-substrate (ATP), and a positively charged activator (Mg^{2+}), allows study of the influence of fixed charges on kinetic behavior:

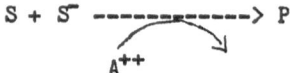

The polymers are produced by polymerization of albumin molecules. At pH 8 the polymer is a negatively-charged polyelectrolyte. The activities of the free hexokinase, bound hexokinase on a soluble polymer, and gelified in a membrane were studied as a function of glucose, ATP, and Mg^{2+} concentrations. In each case the concentrations of the two other molecules were chosen in order to give zero-order kinetics with respect to their concentrations. Results obtained with glucose are given in Fig. 6. An increase of 1.3-times the K_m value was observed after binding on

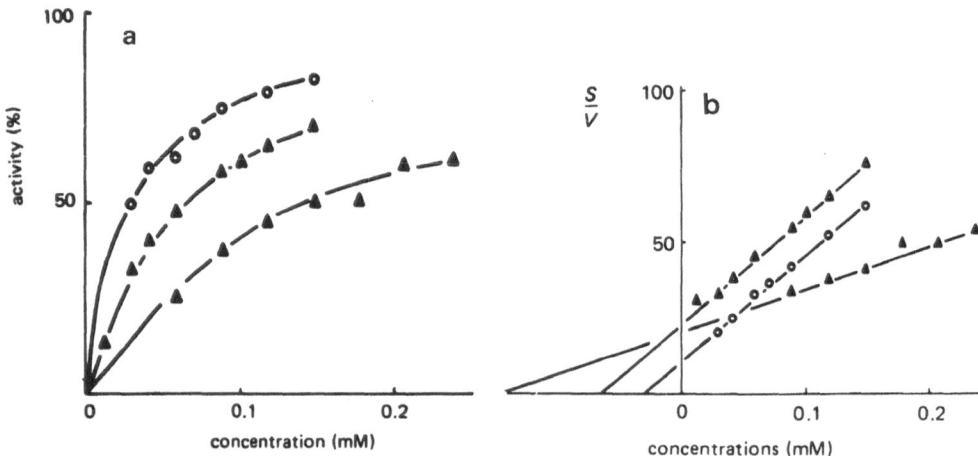

Fig. 5. (a) Urate-oxidase activity free in solution (o), bound to soluble proteic polymers ($\triangle$) and to the same polymers gelified as a membrane by ultrafiltration ($\blacktriangle$) as a function of uric acid concentration. (b) Linear transformations for free enzyme (0), soluble polymer ($\triangle$), membrane ($\blacktriangle$).

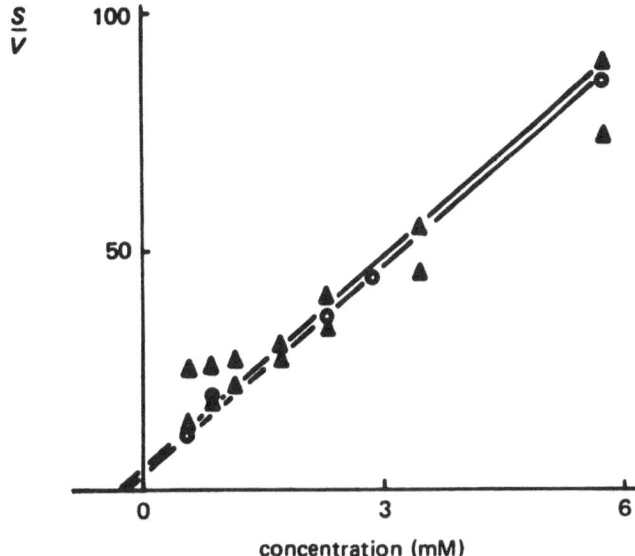

Fig. 6. Ratio of glucose concentration to velocity as a function of
glucose concentration for the free enzyme (o), bound to soluble
polymers ($\triangle$) and to the gelified membrane ($\blacktriangle$). The system is
studied under zero-order kinetic conditions for ATP and Mg^{2+}.

the soluble polymers, and 3.5-times on the gelified membrane. Results
obtained with ATP are given in Fig. 7.

The K_m value was not modified after the binding of the enzyme in the
soluble polymer, but an increase of 6.2-times was observed with the
gelified membrane. It is possible to explain the results by a charge
effect. ATP is strongly negatively charged, and an exclusion effect is
introduced. Inside the negatively charged membrane, due to the Donnan
effect, the concentration of anions would be much smaller than outside.
The concentration of ATP was measured outside, and the apparent affinity
between the enzyme and ATP was found smaller than in the solution.

The results dealing with Mg^{2+} (Fig. 8) show the opposite effect. Due
to the ion-exchange properties of the membrane, the local concentration
of Mg^{2+} is higher inside the membrane than outside in the bulk solution.
The value of the K_a was decreased by a factor 0.38 when the enzyme was
insolubilized.

The cholinesterase system. Acetylcholinesterase (EC 3.1.1.7) and
butyrylcholinesterase (EC 3.1.1.8) catalyze the hydrolysis of choline
esters. The main kinetic difference between these two enzymes is that
acetylcholinesterase activity is inhibited, while butyrylcholinesterase
is activated, by excess substrate. In both cases the activity as a
function of pH exhibits a bell-shaped curve. The reaction can be
written:

$$S^+ \longrightarrow P^+ + P^- + H^+$$

The behavior of the two cholinesterase systems is given in Fig. 9.
The native enzyme and that grafted on soluble polymers behave similarly

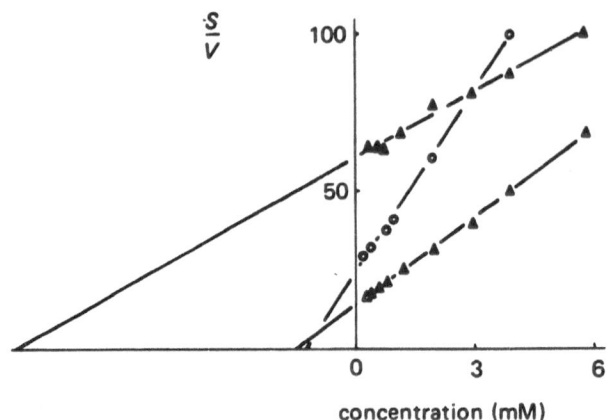

concentration (mM)

Fig. 7.   The ratio of ATP concentration to hexokinase reaction velocity
as a function of ATP concentration for the enzyme free in
solution (0), bound to soluble proteic polymers (△) and to the
polymers gelified as a membrane (▲).   Zero-order kinetic
conditions for glucose and Mg^{2+}.

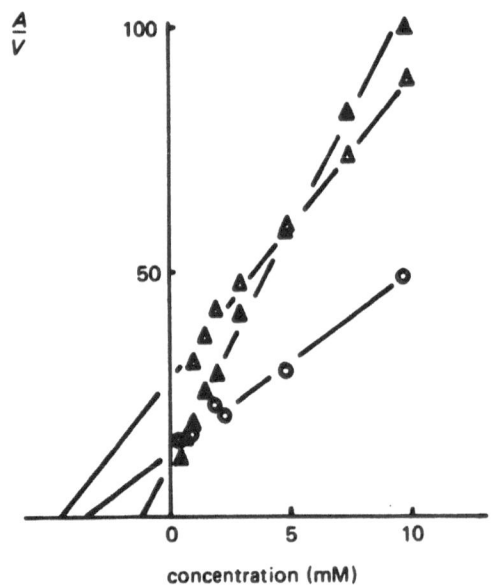

concentration (mM)

Fig. 8.   The ratio of Mg^{2+} concentration to hexokinase reaction velocity
as a function of Mg^{2+} concentration for the enzyme free in
solution (0), bound to soluble polymers (△) and to the same
polymers gelified as a membrane by ultrafiltration (▲).   Zero-
order kinetic conditions for glucose and ATP.

in both cases. After insolubilization in a membrane, the apparent enzyme
affinities are shifted. These results cannot be explained without taking
into account both the production of protons inside the membrane and the
modification of the fixed and mobile charge densities in the microenvironment of the enzyme.

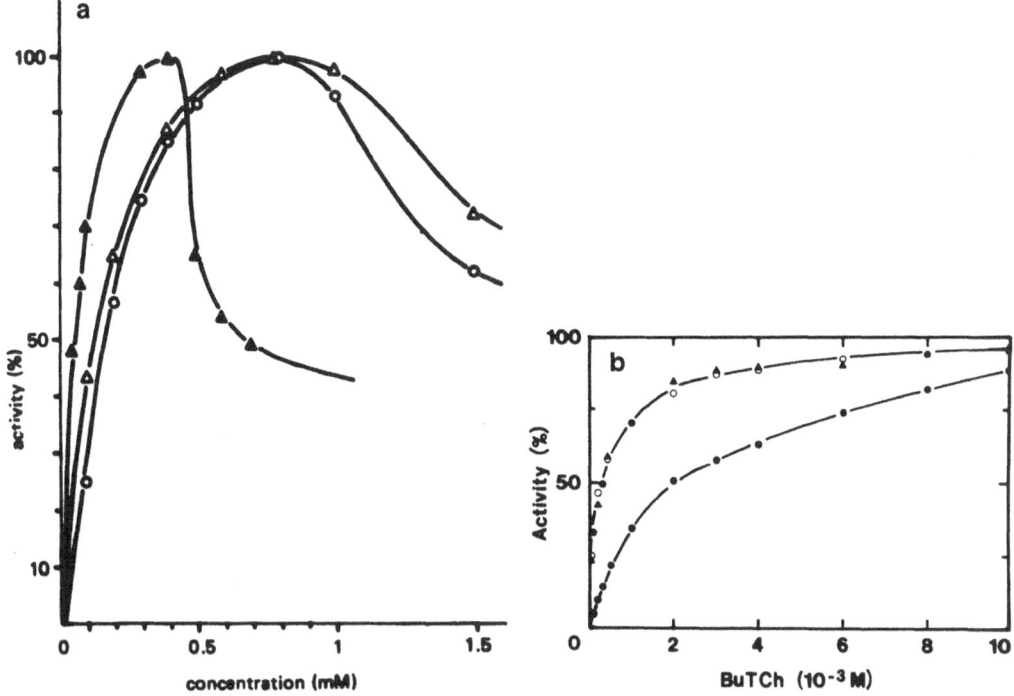

Fig. 9.   (a) Acetylcholinesterase activity as a function of
           acetylcholine concentration for the enzyme free in solution
           (0), bound to soluble polymers (△) and insolubilized in a
           membrane (▲).   (b) Butyrylcholinesterase activity as a
           function of butyrylthiocholine concentration: (0) free enzyme,
           (▲) soluble polymers and (●) membrane.

     Moreover, the apparent affinity of the enzyme after insolubilization
is greatly influenced by the ionic strength.   This is clearly
demonstrated by coating a pH glass electrode with a butyrylcholinesterase
membrane and by measuring the internal pH as a function of the external
substrate concentration (Fig. 10). For increasing external ionic
strength at a constant buffer concentration, the apparent affinity of the
insolubilized enzyme is increased.

     All these results show the role of local concentrations of
metabolites in the microenvironment of an enzyme. Local concentrations
can be modulated, not only by the diffusion limitations but also by the
fixed charges in the insoluble phase.

INTERACTIONS BETWEEN ENZYME ACTIVITY AND THE POTENTIAL DIFFERENCE AT THE

BOUNDARIES OF THE MEMBRANE

     The use of artificial enzyme membranes allows study of the
interaction between enzyme activity and membrane potential in a well
defined context.   Before the recent progress in manufacturing artificial
membranes bearing immobilized enzymes, Blumenthal et al. (1967) described
a system in which a papain solution was sandwiched between two cation and
anion exchange membranes.   Under short-circuit conditions the system was

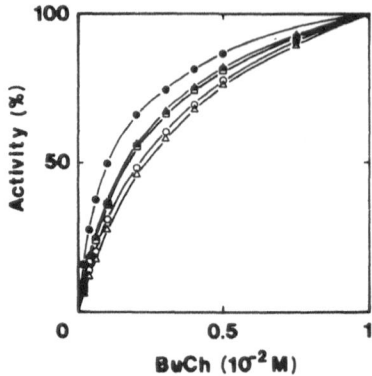

Fig. 10. Butyrylcholinesterase activity as a function of butyrylcholine
concentration for different NaCl concentrations in the bulk
solution. NaCl 0 (●), 5×10^{-3} (▲), 10^{-2} (□), 5×10^{-2} (○) and
10^{-1} M (△).

able to generate a current.

When a proteic membrane with fixed charges separates two compartments
containing electrolyte solutions of monovalent ions (I^+, I^-), these ions
diffuse through the membrane, creating an electric field within it (Fig.
11).

The determination of the electrical potential, concentrations, and
flux of electrolyte liquids can be formulated, so as to require the
solutions of fluid dynamical equations for the liquid convection velocity
$\underline{v}$ simultaneously with the classical Nernst-Planck flux equations and the
flux-continuity equation. With the assumption that the liquids flowing
through the membrane are convectionless ($v=0$), the above system would be
defined mathematically by the following equations:

- the flux:

$$J_{I^+} = -D_{I^+} \left(\frac{\delta[I^+]}{\delta x} + \lambda[I^+]\frac{\delta\Psi}{\delta x} \right)$$

$$J_{I^-} = -D_{I^-} \left(\frac{\delta[I^-]}{\delta x} - \lambda[I^-]\frac{\delta\Psi}{\delta x} \right)$$

(7)

- the mass-balance equation:

$$\frac{\delta J_{I^+}}{\delta t} + \frac{\delta J_{I^+}}{\delta x} = 0$$

$$\frac{\delta J_{I^-}}{\delta t} + \frac{\delta J_{I^-}}{\delta x} = 0; \quad 0 < x < L; \ t > 0$$

(8)

- the Poisson equation:

$$-[\varepsilon]\frac{\delta^2\Psi}{\delta x^2} = [I^+] - [I^-] \pm [S^{\pm}]$$

(9)

where D_{I^+} and D_{I^-} are the diffusion coefficients of the positive and negative ions, respectively; Ψ is the potential of the electric field; $\lambda = F/RT$ (F: Faraday's constant; R: gas constant; T: absolute temperature); $[\varepsilon] = D \cdot \varepsilon_0 / F$ (D: dielectric constant of the electrolyte liquid; ε_0: permittivity of a vacuum); t: time; L: thickness of the one dimensional membrane; and $[S^{\pm}]$: concentration of the fixed univalent charges in the membrane.

The system can be solved by a numerical method using correctors (Trubuil et al., 1985). Typical results are shown in Fig. 12.

When the system is studied under symmetrical conditions for I^+, I^-, and when a charged substrate is injected on one side of the membrane, the theoretical potential difference can be calculated (Table 1). The results obtained are in good agreement with the experimental data.

Moreover, the corrector method allows the simulation of the system when an enzyme reaction (e.g., acetylcholinesterase) modifies locally the fixed charge density in the membrane (Friboulet, Trubuil, Thomas, and Kernevez, in preparation).

The immobilized urease system can be used to study these inter-actions. The artificial membrane was put between two compartments containing, respectively, 1 and 10 mM sodium phosphate at pH 5.1 (David et al., 1974). The potential is recorded at the boundaries of the membrane. After introducing the substrate in both compartments, the reaction induces both an increase of the membrane mobile-ion concentration and an increase of the local pH, modulating the enzyme reaction and the fixed charge density. So, a jump of the potential difference is recorded at the boundaries of the membrane (Fig. 13a). After introduction of an enzyme inhibitor, the potential difference returns to the initial value. The phenomenon of potential modification is clearly linked to the enzyme reaction.

A similar behavior is obtained when a membrane containing acetyl-cholinesterase is put between two compartments containing both 1 mM sodium phosphate, pH 7.5, and when the substrate is added only on one side of the membrane (Fig. 13b). In both cases the steady-state membrane potential studied as a function of substrate concentration exhibits a

Table 1. Experimental and Theoretical Potential Differences Obtained when Acetylcholine is Injected in One Compartment.

Acetylcholine mM	Experimental ΔV mV $\pm$ S.E.M.	Calculated ΔV mV
2.5	-15.75 ± 0.44 (4)	-16.01
5	-21.75 ± 0.59 (6)	-21.97
10	-26.20 ± 0.58 (6)	-26.49
20	-29.00 ± 0.92 (5)	-28.88
30	-29.90 ± 1.07 (6)	-29.25
40	-29.40 ± 1.23 (6)	-29.20
60	-28.15 ± 1.06 (5)	-28.70

S.E.M.: standard error of the mean. The number of independent experiments is indicated in brackets.

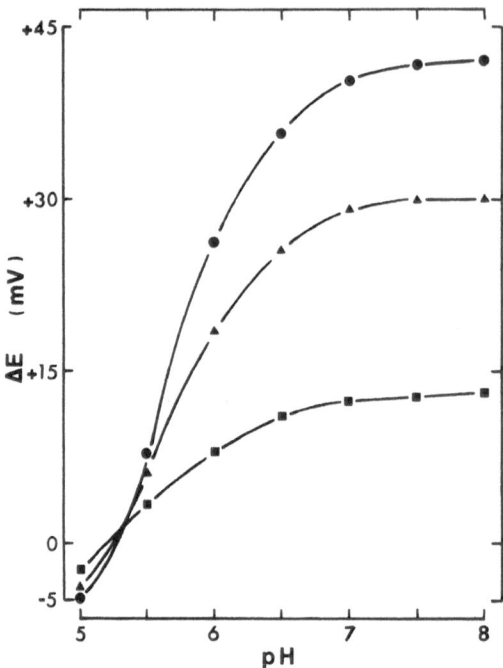

Fig. 11. Steady potential difference _vs_ buffer solution pH. Sodium
phosphate concentration: compartment 1, constant, 1 mM;
compartment 2, (■) 2 mM, (▲) 5 mM, (●) 10 mM. Membranes are
generated from serum albumin.

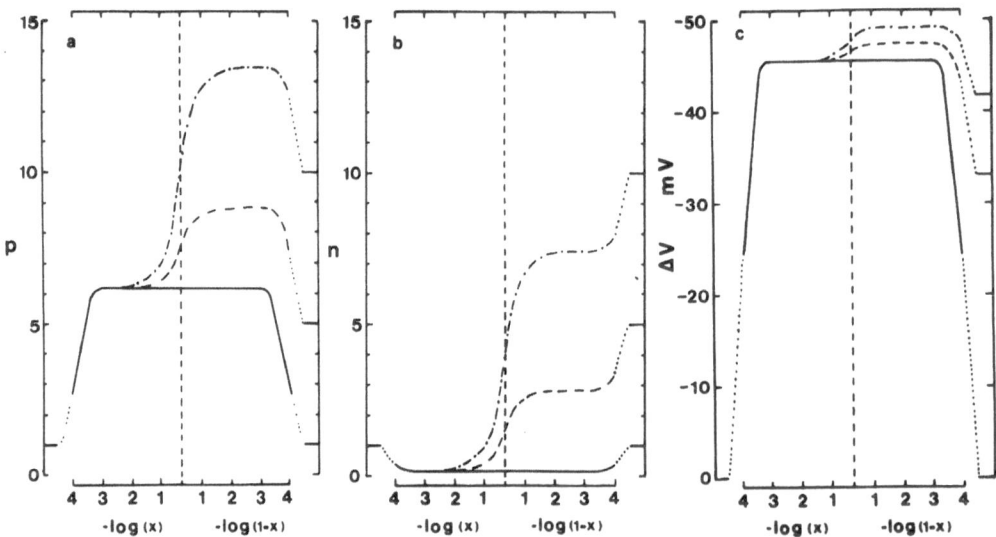

Fig. 12. Calculated intramembranous concentrations of I^+ (a), I^- (b) and
electrical potential (c) as a function of $-\log(x)$ for $0 < x \leq$
0.5 and $-\log(1-x)$ for $0.5 \leq x < 1$ where $x = X/L$. Compartment 1:
$[I^+] = [I^-] = 1$ mM; compartment 2: $[I^+] = [I^-] = 1$ mM (———), 5 mM
(– – –) and 10 mM (–·–·–). Concentration of negative fixed
sites: 6 mM; $D_{I^-}/D_{I^+} = 0.77$.

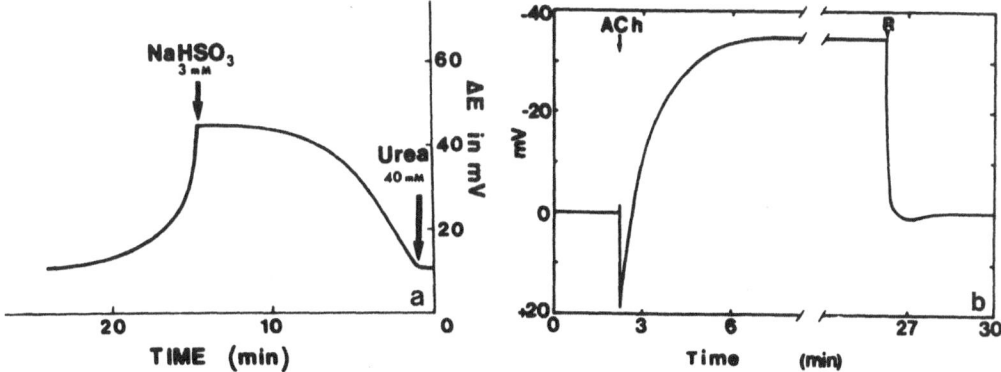

Fig. 13. Membrane potential difference recorded as a function of time.
(a) Artificial urease membrane when adding substrate (urea) and
inhibitor (NaHSO$_3$) into both compartments. (b) Acetylcholinesterase membrane when adding acetylcholine (arrow) and
rinsing into one compartment.

sigmoid shape. However, the steady-state potential difference strongly
depends on the physicochemical properties of the protein molecules used
for the matrix (Fig. 14). Three different proteins were used: serum
albumin and ossein gelatin (which have cation exchange properties for pH
> 5 but differ in the charge density), and pig skin gelatin (which has
anion-exchange properties for pH < 7). In equilibrium with an external
solution, the distribution of the solute between the membrane and the
external solution is governed by a Donnan equilibrium operating at the
interface.

MEMORY AND OSCILLATORY BEHAVIOR IN ARTIFICIAL MEMBRANES

The first immobilized enzyme system used to study hysteresis
phenomena and spontaneous oscillations was papain (Naparstek et al.,
1973, 1974). The reaction studied is the hydrolysis of the substrate
benzoyl-L arginine ethyl ester (BAEE):

$$S \longrightarrow P^- + H^+$$

The enzyme exhibits maximal activity at neutral pH. By coating papain on
the surface of a glass pH electrode, it is possible to follow the reaction inside the membrane. In a weakly buffered medium and for pH values
greater than the optimum, the ester hydrolysis decreases the pH. The pH
inside the membrane was studied under zero-order conditions for the
substrate, as a function of the pH in the bulk solution in which the
electrode was immersed (Fig. 15).

An hysteresis effect is observed, such that the enzyme reaction rate
depends not only on the metabolite concentration but also on the history
of the system. The decrease of the local concentration of substrate is
compensated by the production of hydroxyl ions, which diffuse rapidly in
the outer medium, thus creating a feedback effect on the enzyme activity.
The association between autocatalysis and feedback may result not only in
memory phenomena, but also in oscillatory behavior. Oscillatory
experiments using the papain electrode and the substrate, BAEE, were
carried out at room temperature. On introduction of the substrate into

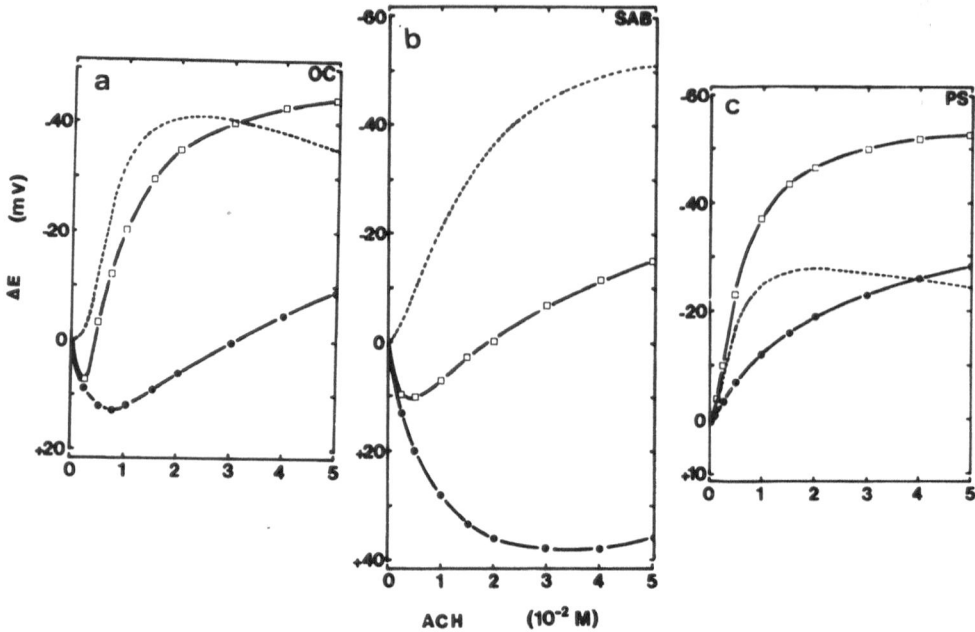

Fig. 14. Membrane potentials under steady-state conditions vs
acetylcholine concentration for active (□) and inactive (●)
membranes. (---) Curves due to acetylcholinesterase activity
obtained by difference. (a) ossein gelatin, (b) serum albumin
and (c) pig skin gelatin membranes.

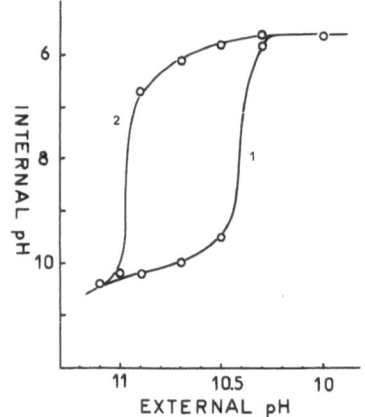

Fig. 15. Stationary-state description of internal pH as a function of
increasing (1) and decreasing (2) the external pH values. The
experiment started with decreasing the pH and then increasing.

the solution, the pH within the membrane decreases immediately, while the
external pH is kept constant at a predetermined value by addition of
NaOH. For a well defined range of pH and substrate concentration values
(pH 9.3 and 4.5 mM BAEE) in the bulk solution, spontaneous time
oscillations of the intramembranous pH occur, as shown in Fig. 16.

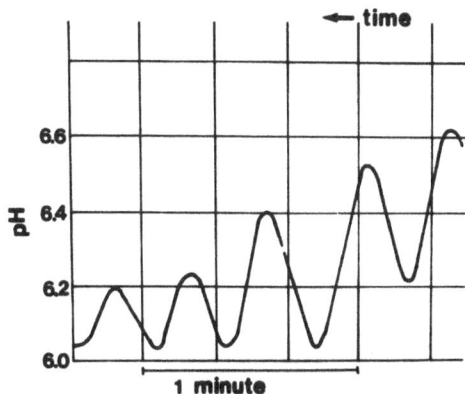

Fig. 16. Time-dependence of pH at the membrane-glass electrode interface. Nominal membrane thickness 10 μm, gentle stirring.

By addition of 0.1 mM $HgCl_2$, a papain inhibitor, in the bulk
solution, both the oscillations and steady pH differences vanish. The
overall system has been simulated, taking into account both the
autocatalytic properties of the enzyme reaction and the diffusion of the
different species in the membrane (Caplan et al., 1973; Chay, 1979).

Similar behavior was obtained with the acetylcholinesterase system.
Steady potential difference as a function of substrate concentration
exhibits a hysteresis loop when the membrane bears acetylcholinesterase
(Fig. 17).

Moreover, when the artificial membranes are produced with serum
albumin, and for a narrow range of enzyme concentration insolubilized in
the membrane, instabilities and oscillations occurred. The oscillations
obtained are given in Fig. 18. These oscillations were sustained for at
least 3 hrs. Their amplitude remained constant (3-5 mV), and the period
depends on substrate concentration. For $1x10^{-2}$ and $2x10^{-2}$ M, the
period was 13 and 4 min, respectively. Upon addition of an enzyme
inhibitor, oscillations vanished.

It is noteworthy that both papain and acetylcholinesterase, which
have been studied exhaustively, have never shown metastable states nor
periodic behavior in homogenous solution.

The behavior of the acetylcholinesterase system cannot be explained
simply by the autocatalysis due to the local decrease of pH, since the
external pH (7.5) is lower than the optimum pH of the enzymatic reaction.
Theoretical simulations of this behavior, taking into account all the
ionic interactions due to the charged sites in the membrane (as described
above) are now in progress.

CONCLUSIONS

Our purpose is not to mimic biological reality, but simply to
demonstrate that the coupling between enzymatic reactions and mass
transfer may lead to the occurrence of specific kinds of behavior. We
can manufacture synthetic enzyme membranes, constructed in such a way

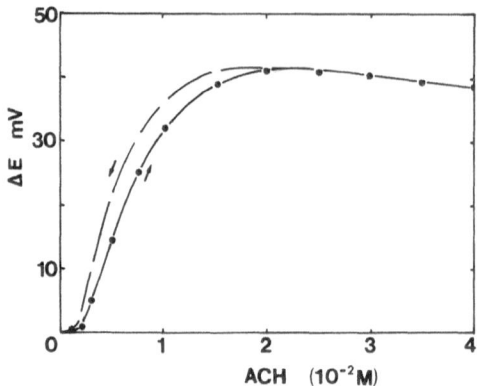

Fig. 17. Membrane potential due to acetylcholinesterase activity in ossein gelatin membranes as a function of increasing (↗) and decreasing (↙) acetylcholine concentrations.

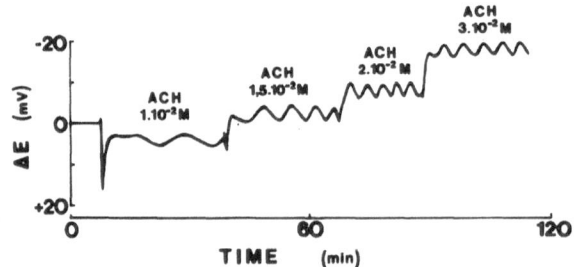

Fig. 18. Membrane potential recorded as a function of time when oscillations occurred with an acetylcholinesterase/albumin membrane.

that the distribution of the active sites is homogenous throughout the entire membrane structure. Application of such membrane-bound enzymes allows us to demonstrate that the internal microenvironment, as determined by the local concentrations of reactants and products, exerts a profound effect on the mode of action of enzymes. Due to the amphoteric properties of our proteic matrices, the reciprocal interactions between diffusion of ionic species and enzymatic reaction can be studied. The local concentrations of substrates, products, or effectors not only modulate enzyme kinetic properties (K_m values, feedback effects, pH-activity curves, etc.), but also introduce totally new phenomena, such as asymmetrical behavior, memory, and oscillations. Moreover, it is only with an enzyme membrane that it becomes possible to study the influence of an electrical potential difference on enzyme kinetics, which may well be of crucial importance in many biological systems.

REFERENCES

Barbotin, J.N. and Thomas, D., 1974, J. Histochem. Cytochem., 22:1048.
Blumenthal, R., Caplan, S.R., and Kedem, O., 1967, Biophys. J., 7:737.

Broun, G., Thomas, D., Gellf, G., Berjonneau, A.M. and Guillon, C., 1973, Biotechnol. Bioeng., 15:359.

Caplan, S.R., Naparstek, A., and Zabuski, N., 1973, Nature, 245:364.

Chay, T.R., 1979, J. Theor. Biol., 80:83.

David, A., Metayer, M., and Thomas, D., 1974, J. Membrane Biol., 18:113.

DeSimone, Y., and Caplan, S.R., 1973, J. Theor. Biol., 39:523.

Engasser, J.M., and Horvath, C., 1974, Appl. Biochem. Bioeng., 1:127.

Goldman, R., Silman, I., Caplan, S.R., Kedem, O. and Katchalski, E., 1965, Science, 150:758.

Goldman, R., Kedem, O., Silman, I., Caplan, S.R., and Katchalski, E., 1968, Biochemistry, 7:486.

Goldstein, L., Levin, Y. and Katchalski, E., 1964, Biochemistry, 3:1913.

Hervagault, J.F. and Thomas, D., 1985, in: "Organized Multienzyme Systems", G.R. Welch, ed., Academic Press, New York.

Kernevez, J.P. and Thomas, D., 1975, J. Appl. Math. Optim., 1:222.

Koelle, G.B. and Friedenwald, G., 1949, Proc. Soc. Exp. Biol. Med., 70:617.

Malpièce, Y., Sharan, M., Barbotin, J.N., Personne, P., and Thomas, D., 1980, J. Biol. Chem., 255:6883.

Malpièce, Y., Sharan, M., Barbotin, J.N., Personne, P., and Thomas, D., 1981, J. Histochem. Cytochem., 29:633.

Maurel, P., Douzou, P., Waldmann, J., and Takashi, Y., 1978, Biochim. Biophys. Acta, 525:314.

Naparstek, A., Thomas, D., and Caplan, S.R., 1973, Biochim. Biophys. Acta, 323:643.

Naparstek, A., Romette, J.L., Kernevez, J.P., and Thomas, D., 1974, Nature, 249:490.

Remy, M.H., David, A., and Thomas, D., 1978, FEBS Letters, 88:332.

Sélégny, E., Avrameas, S., Broun, G., and Thomas, D., 1968, C. R. Acad. Sci., Paris, 266:1431.

Silman, I. and Karlin, A., 1967, Proc. Natl. Acad. Sci. USA, 58:1664.

Thomas, D., Broun, G. and Sélégny, E., 1972, Biochimie, 54:229.

Wang, D., Wilson, G., and Moore, S., 1976, Biochemistry, 15:660.

ON PARTS AND WHOLES IN METABOLISM

Henrik Kacser

Department of Genetics
University of Edinburgh
West Mains Road
Edinburgh, Scotland

> *Trace Science then, with modesty thy guide;*
> *First strip off all her equipage of pride;*
> *Deduct what is but vanity or dress,*
> *Or learning's luxury, or idleness,*
> *Or tricks to show the stretch of human brain,*
> *Mere curious pleasure, or ingenious pain;*

Alexander Pope, An Essay on Man

INTRODUCTION

Historically, the progress of science has been driven by its analytical approach. Although the biological sciences have been lagging somewhat behind the physical sciences, the last fifty years or so have seen an increasing avalanche in the identification and description of the components of living organisms. Refined instrumentation and ingenious techniques have gone hand in hand with sophisticated mathematical models of molecular properties. The method of choice has been to cut up the organism into ever smaller pieces and look at the pieces in ever greater detail. We can now detect a single specific nucleotide among millions and measure the presence of a few molecules in a whole cell. We can describe the dynamics of internal molecular motions and calculate the thermodynamics of their transitions.

Side by side with these pursuits there have been studies of the organization of cells, organs and tissues. We know of membranes, organelles, compartments and molecular aggregates. Superimposed or underlying these spatial organizations is one which is functional in nature. This is symbolized by the metabolic map which reflects the kinetic organization of the cell. Its units are the individual molecular transformations, overwhelmingly catalyzed by enzymes (or other quasi-catalytic entities), which are linked to each other by sharing the metabolites and other effectors. The familiar wall charts are just superficial sketches, and the complete map of all interactions (even with the inadequate knowledge of today) would soon run out of colors, space and dimensions. What we would have in such a map, however, is a bare skeleton. It tells us which parts are connected, but not what the quantitative outcome of the interactions are. To put the muscles on this

327

skeleton we must turn to the kinetic and thermodynamic equations which describe the individual parts. In particular, enzymological studies have provided us with a mass of information and, although by no means complete, appears to give a basis for synthesizing the whole. In attempting to do this we face an immediate dilemma. The equations describing each rate expression (even if we believe the constants, determined _in vitro_, to be correct for the _in vivo_ conditions) do contain the concentrations of the molecules which make up the reaction scheme of each step in the system. Now, the product of one reaction is the substrate for the next, and the rate expression for this next reaction, therefore, enters implicitly the first expression. As we proceed along a pathway we will involve further reactions, and so on right through the system. We shall end up with many thousands of simultaneous nonlinear equations which are algebraically insoluble. The concentrations of the intermediate pools and the fluxes through any part of the system are variables which, say at the steady state, have settled to their particular values by the joint effects of all the reactions. To solve one equation, we have to solve all.

The problem is not a new one, nor is it peculiar to the study of metabolism. Take a single enzyme from the cell and try to "synthesize" its catalytic activity from the properties of its parts, the amino acids. We believe the activity to be fully determined by the amino acid sequence and hence by the DNA sequence. Even if we solve the "folding pathway" into the globular form, the specificity and catalytic activity is a joint property of the van der Waals, polar and electrical forces of all the amino acids, and particularly those lying somewhere in that soggy marsh known as the "active site". No useful calculation could predict the "emergent" properties. Instead, we look at the whole and probe the three dimensional structure in order to identify "major groups" and hope that by ignoring the rest we do not seriously distort the result. We work at the "higher level".

Going one step back, we can inquire whether the properties of the individual amino acids can emerge from those of their atoms and their arrangements. Again, in principle we can, but in practice we do not attempt to, solve the Schrödinger equations.

In spite of this realization, that for systems of reasonable complexity there is no way to understand or predict their behavior in terms of their parts, there is a persistent tendency to pursue the analytical method as if the problem did not exist. There are, of course, distinguished studies which deal with interaction problems and have yielded fascinating insights into some aspects of molecular behavior. The areas studied are, however, restricted to specific problems that "can be handled", i.e., a small number of components. The success of such studies has, in some sense, seduced us into believing that this is the way forward. "If we only worked hard enough, and if we were ingenious enough, and if there were enough of us — then the sum of all these endeavors would, somehow, enable us to generate the behavior of the whole". I believe this view to be fallacious, for reasons sketched out above. Let me hasten to add that I am not suggesting that such studies are trivial, nor that they could not have important practical consequences, nor indeed that they should not be pursued for their own sake. What I am saying is that they will not yield the answers to questions which they purport to ask. If the questions are to address the problems of control and regulation of metabolism, then a radically different approach is required. The gas laws are understood because we can handle the kinetics of millions of identical molecules, and the same goes for the treatment of individual chemical reactions. In metabolism we have the presence of thousands of reactions, each uniquely different from one another. Such a system is more than the sum of its parts. It will have emergent properties not possessed by its components. We therefore must develop methods - of thought and experiment - which reveal the action of each part within the whole. In what follows I will sketch out how this can be approached. I shall not go into detailed algebraic treatment and the proofs of various theorems, which can be obtained in a number of publications (Kacser, 1983; Kacser and Burns, 1973, 1979, 1981; Heinrich and Rapoport, 1974, 1975; Westerhoff and Chen, 1984; Westerhoff et al., 1984; Fell and Sauro, 1985; Hofmeyer et al., 1986). I shall be mainly concerned with the conceptual framework which allows us to view the problems of control in intact biological systems in a new way. Such a treatment will also have experimental consequences, insofar as different types of experiments have to be done to obtain quantitative insight.

THE FUNCTIONAL RESPONSES OF THE PARTS

Since we cannot solve the large set of simultaneous equations, how then does the cell "solve" them? We can think of the process as not very different from a computer simulation where, having specified the "structure" by specifying the shared variables and having set the values of the enzyme parameters, we ask for the integration routine to achieve the steady state. Starting from an initial condition of the variables, i.e., the metabolite pools and fluxes, these will begin to move because the rates across each step generally start being unequal. These unbalanced rates will produce increases and decreases in substrates, products and other effectors. Such changes will, in turn, change the rates and so on, until, for most systems, the successive rates through any particular section of the system become equal to one another and the pools settle to time-invariant values. The steady state has then been achieved.

It is this response of the individual reaction rate which is the relevant function for every transformation step in the system. Although we shall describe, for convenience, what follows in terms of free enzymes in solution, the treatment is also applicable to membrane bound permeases and translocators, to quasi-catalytic entities such as receptors, RNA's, or anything else which, in a concentration dependent manner, influences the rate of a process.

Consider then the rate of one of these elementary processes, catalyzed by an enzyme, in a system at steady state (s.s.). Any perturbation, transient or permanent, somewhere in the system will reverberate through the system by changing metabolite concentrations and hence fluxes, until it has either returned to the steady state (if the perturbation was transient), or will have settled to a new neighboring steady state (if the change was permanent). The particular rate on which we are focussing our attention will have been determined by the steady state concentrations of all the participating molecules — substrates, products, cofactors, effectors, and the enzyme itself. The kinetic constants of the enzyme will equally enter the determination of the rate:

$$(v_j)_{s.s.} = f(E_j, S_i, S_j \ldots I, A, \ldots k, K)$$

This may be a very complex function and, in general, will be

different for each step. We need not concern ourselves with such
details. When we consider the effect of the perturbation, this can only
change the value of the rate by changing one or several of the
concentrations of the participating molecules. Such a change, say δS_i,
leads to δv, or using normalized changes to eliminate units of
measurement,

$$\frac{\delta S_i}{S_i} \quad \text{leads to} \quad \frac{\delta v_j}{v_j}$$

The ratio

$$\frac{\delta v_j}{v_j} \Big/ \frac{\delta S_i}{S_i}$$

i.e., the "effect"/"cause", is a measure of the response of the rate due
to the change in <u>one</u> of the factors entering the rate expression. Since
this relationship is, in general, nonlinear, we must consider the limit

$$\delta S_i \longrightarrow 0$$

when we obtain

$$\left(\frac{\partial v_j}{v_j} \Big/ \frac{\partial S_i}{S_i} \right) S_j, S_k \ldots \quad = \quad \varepsilon\, ^{v_j}_{S_i} \qquad (1)$$

or $\quad \left(\frac{\partial \ln v_j}{\partial \ln S_i} \right) S_j, S_k \ldots \quad = \quad \varepsilon\, ^{j}_{i} \qquad (1a)$

i.e., the partial derivative of the rate with respect to S_i, while all the
other variables are held constant at their steady-state values. This is
designated the <u>elasticity coefficient</u>, ε. There are, of course, similar
elasticity coefficients of the same rate with respect to each of the
participating molecules.

$$\left(\frac{\partial \ln v_j}{\partial \ln S_j} \right) S_i, S_k \ldots \quad = \quad \varepsilon\, ^{j}_{j}$$

$$\left(\frac{\partial \ln v_j}{\partial \ln S_k} \right) S_i, S_l \ldots \quad = \quad \varepsilon\, ^{j}_{k}$$

$$\left(\frac{\partial \ln v_j}{\partial \ln E_j} \right) S_i, S_j \ldots \quad = \quad \varepsilon\, ^{j}_{E_j}$$

including the elasticity coefficient with respect to its "own" enzyme
concentration.

Two important points should be noted. The algebraic procedure
described above has "isolated" the step in question, has "frozen" it in
the steady state, so that we are investigating, by infinitesimal
modulations, the potential response of the step to each of its

participants. It is the equivalent of a real isolation or extraction of the enzyme and a consequent reconstruction of its molecular milieu. The method of determining the elasticities would then follow the familiar enzymological procedures, since we can rewrite, for example (1) as:

$$\varepsilon_i^j = \frac{\partial v_j}{\partial S_i} \times \frac{v_j}{S_i} \qquad (2)$$

i.e., the slope of v_j versus S_i relationship multiplied by the scaling or normalizing factor S_i/v_j. The particular value so obtained will depend on the absolute value of S_i, as well as on all the others. It reflects the kinetic behavior of the isolated rate in one particular milieu. The elasticity coefficients are "local" properties, albeit that their values are also dependent on the steady-state concentrations of the "globally" determined metabolites. These values may be +ve (e.g., for substrates) or -ve (e.g., for products), and may go from zero (e.g., saturation) to infinity (e.g., near-equilibrium). The elasticity coefficient of the rate with respect to the enzyme, $\varepsilon_{E_j}^j$, is usually unity, since, for most enzymes, the rate is proportional to enzyme concentration. This would not apply to, e.g., monomer-oligomer systems if the activities/monomer differed in the two states and if the equilibrium changed when changes in the total concentration took place. Similarly, if there are associations between different enzymes which affect the rates (e.g. channelling), the change in the concentration of one enzyme could affect the equilibrium of the complex and hence affect the rate in a nonproportional manner.

In practical terms extraction/reconstruction experiments are difficult and full of question-begging assumptions. There are methods of obtaining the values from in vivo experiments (Kacser and Burns, 1979). For the moment, however, we shall not be concerned with these problems.

We thus have each rate defined by m elasticity coefficients, where m is equal to the number of molecular species interacting directly in the rate equation. The actual change in rate, then, is the sum of all the individual effects:

$$\frac{\delta v_j}{v_j} = \varepsilon_i^j \frac{\delta S_i}{S_i} + \varepsilon_j^j \frac{\delta S_j}{S_j} + \ldots + \varepsilon_{E_j}^j \frac{\delta E_j}{E_j} \qquad (3)$$

provided the δS values are small.

Unlike classical enzymology, which attempts to describe a particular rate in terms of all the fundamental kinetic constants, its mechanism, the thermodynamics of the reaction process, and so on, the establishment of elasticity coefficients selects only those properties of the catalyzed rate which are relevant to its control behavior. While each rate represents the elementary functional part of the kinetic system, the array of elasticity coefficients, the "elasticity profile", represents the response characteristics of the individual part. Every rate has its own local profile, and the matrix of all elasticity coefficients in the system will describe the potential response of the whole (Kacser and Burns, 1979; Fell and Sauro, 1985; Hofmeyer et al., 1986). This matrix will be determined by the structure, i.e., which rates are coupled to each other by sharing metabolites. By "unfreezing" the system and therefore allowing every metabolite and flux to "float", changes around

each part will be transmitted to the adjacent ones. The quantitative
results of such transmission will be determined by the quantitative
values of the elasticity coefficients.

THE RESPONSES IN THE WHOLE

In a complete system, we can now describe the changes in the
individual rates by a set of equations of type (3) which, it will be
noted, are linear equations and, for certain conditions, are soluble
(Kacser, 1983; Kacser and Burns, 1973; Heinrich and Rapoport, 1975; Fell
and Sauro, 1985). At steady state a number of such rates are equal to
one another (in a simple pathway), or are fixed by stoichiometric
relations, or are constrained by flux conservation relations (divided
pathways), or other constraints (moiety conserved cycles [Hofmeyer et
al., 1986]). We shall describe the rates through each portion of the
system (and hence through many enzymes) as system "fluxes", to distinguish them from the "rates" of the isolated reactions. At steady state
they are, of course, equal.

We began by using the device of considering an unspecified perturbation. When asking questions about control we must now specify such
perturbation. Indeed, we must consider what we mean by "control". We
often find discussions on "which is the rate limiting enzyme in a
pathway?", or "which is the control site?". On consideration, we may
find such questions rather metaphorical than rigorous. They do not arise
from an analysis of system kinetics but appear to imply preconceptions.

Take the first of these questions. We have argued that the flux
through one enzyme in a pathway must be determined not only by its
kinetic parameters but equally by its neighboring enzymes, insofar as
they "supply" the substrate(s) and "remove" the product(s). In a simple
pathway all the individual rates must be _equal_ to one another and equal
to the pathway flux when the system is at steady state. What meaning (if
any) can we then give to "rate limiting"? In such a pathway, embedded in
the whole system, _each_ of the enzymes can be made to limit the flux by
reducing their activity to zero in turn (metabolic blocks). The flux
will then be zero. While all the enzymes in a pathway are therefore in
some sense equal, they will, in general, have different kinetic
characteristics. In the past attempts have been made to use certain of
these as criteria for "control". Such attempts are bound to fail,
because they do not take into account the effects of neighboring enzymes
(and their neighbors, etc.) on the rate sustained through one of them.
The flux through a pathway is a systemic property, and only a definition
of control which takes account of this can be meaningful. Such a
definition will now be given.

Although the flux through each step is the same for all the steps at
steady state, we can inquire whether _changes_ in the activity of each step
will have the same effect on the flux. Clearly we must compare equal
fractional changes:

$$\frac{\delta e_1}{e_1} = \frac{\delta e_2}{e_2} = \ldots\ldots\ldots = \frac{\delta e_n}{e_n}$$

where e_i = "activity" of the enzyme E_i and is of the form

$$e_i = [E_i] k_{cat} \, K_{eq} / K_m$$

for a simple Michaelis-type enzyme, or more complex expressions for others. The changes which we consider can therefore be in any or all the kinetic parameters of the enzyme. The effect on the flux is similarly expressed as $\delta J/J$. If we now made equal fractional changes (say 1%) to each of the enzymes in turn, are the fractional changes in the flux the same?

$$\frac{\delta J}{J} \bigg/ \frac{\delta e_i}{e_i} = ? \tag{4}$$

This will obviously not be so. Take, for example, two adjacent enzymes in a pathway, one highly saturated by the steady-state concentrations of its substrate, preceded by one with very little saturation. A 1% change in the activity of the preceding enzyme, whatever the changes in its product (the substrate to the next) is brought about by the change, will have little effect on the flux. Changing the saturated enzyme, on the other hand, may give an almost proportional response. It should not be taken that the above arguments are rigorous, since many more considerations apply and hand-waving is not the best approach. Nevertheless, it will be agreed that our expectations are, in general, for unequal consequences. It will be noted that such differences are intimately connected with the transmission of metabolite movements, which were discussed in the section on elasticity coefficients. Since the ratio in Eqn. (4) is nonlinear in e_i, we consider $\delta e_i \longrightarrow 0$ and obtain a new definition:

$$\left(\frac{\delta J}{J} \bigg/ \frac{\delta e_i}{e_i} \right)_{e_j, e_k \cdots} = C^J_{E_i} \tag{5}$$

This is designated the control coefficient of flux J with respect to enzyme E_i (Kacser and Burns, 1973; Burns et al., 1985). It is a "global" coefficient, defining how the systemic flux is influenced by the activity of one particular enzyme. It takes account of all the movements of the variables, which will settle to new values in the neighboring steady state. There are as many flux-control coefficients as there are enzymes in the whole system, since every enzyme, even if it is not "in the pathway", will have some effect. The different values of the control coefficients will give a measure of the "importance" of each enzyme. The relative values will indicate how control is distributed. In practice, it is found, that if determinations are carried out (Kacser and Burns, 1981; Groen et al., 1982; Flint et al., 1981; Salter et al., 1986; Middleton and Kacser, 1983), the major contribution to the control is vested in a number of enzymes and never in a single one. The idea of a "rate-limiting enzyme" should therefore be abandoned and replaced by the assignation of the magnitude of the control coefficients.

The distribution of control is constrained by the flux summation theorem (Kacser and Burns, 1973; Heinrich and Rapoport, 1974; Westerhoff et al., 1984):

$$\sum_{i=1}^{n} C^J_{E_i} = 1 \tag{6}$$

For any one flux in the system, the sum of all the flux control coefficients is unity. The majority of coefficients are positive, but there are some negative ones (e.g., divided pathways) (Kacser, 1983). Because the number of enzymes, n, is large, the magnitude of most coefficients will be very small; and this is borne out experimentally (Kacser and Burns, 1981; Groen et al., 1982; Salter et al., 1986).

The other class of variables, the metabolites, also has control coefficients

$$\left(\frac{\partial S_j}{S_j} \Big/ \frac{\partial e_i}{e_i}\right)_{e_j, e_k} = C_{E_i}^{S_j} \tag{7}$$

designated the "concentration control coefficient". They also obey a concentration summation theorem

$$\sum_{i=1}^{n} C_{E_i}^{S_j} = 0 \tag{8}$$

One can determine the control coefficients experimentally _in vivo_ (Kacser and Burns, 1973, 1979, 1981; Flint et al., 1981; Groen et al., 1982; Salter et al., 1986; Middleton and Kacser, 1983; Dean et al., 1986). Some of the methods involve direct manipulation of the activity by the use of mutants, heterokaryons, or induction/repression. Other methods use externally applied inhibitors, yet others involve the measurement of certain elasticity coefficients.

The relationship between the "global" control coefficients and the "local" elasticity coefficients can be made explicit by means of a simple example. In a system consisting of only three enzymes and two external (constant) source and sink metabolites, the three flux-control coefficients turn out to be

$$C_{E_1}^{J} = \frac{\varepsilon_2^3 \varepsilon_1^2}{\varepsilon_2^3 \varepsilon_1^2 - \varepsilon_2^3 \varepsilon_1^1 + \varepsilon_2^2 \varepsilon_1^1} \tag{9}$$

$$C_{E_2}^{J} = \frac{-\varepsilon_2^3 \varepsilon_1^1}{\varepsilon_2^3 \varepsilon_1^2 - \varepsilon_2^3 \varepsilon_1^1 + \varepsilon_2^2 \varepsilon_1^1} \tag{10}$$

$$C_{E_3}^{J} = \frac{\varepsilon_2^2 \varepsilon_1^1}{\varepsilon_2^3 \varepsilon_1^2 - \varepsilon_2^3 \varepsilon_1^1 + \varepsilon_2^2 \varepsilon_1^1} \tag{11}$$

This shows that all the elasticity coefficients in the system jointly determine the flux control coefficients. It follows that not only are the fluxes and metabolites system variables, but their coefficients themselves are systemic in nature. Any change, and particularly any large change, in the system will affect the values of all coefficients. What may be an "important" enzyme in one condition or species may not be so in another. The distribution will change but will always be constrained by relation (6) or (8).

Finally, enzymes interacting with a common substrate, S, will be related by the connectivity theorems (Kacser and Burns, 1973; Westerhoff and Chen, 1984):

$$\sum_{i=1}^{n} C_{E_i}^{J} \cdot \varepsilon_{S}^{i} = 0 \tag{12}$$

and

$$\sum_{i=1}^{n} C_{E_i}^{S_j} \cdot \varepsilon_{S_k}^{i} = -\delta_{jk} \tag{13}$$

This has important applications in the design of experiments to determine the control coefficients. One consequence of relationship (12) is that, in a pathway, the ratio of the control coefficients of two adjacent enzymes is simply the ratio of the elasticity coefficients with respect to the shared metabolite. This is also evident if we take, e.g., the ratio of (9)/(10)

$$C_{E_1}^{J} / C_{E_2}^{J} = -\varepsilon_1^2 / \varepsilon_1^1 \tag{14}$$

The _relative_ values of the kinetic responses of the two enzymes determine their _relative_ effect in the whole system. The absolute ε values are uninformative about their importance in metabolism. A good deal of confusion in the literature is caused by attempts to conclude something about control by detailed investigations of single enzymes. Inspection of equations (9)-(11) re-enforces this point. Each control coefficient contains not only the elasticity coefficients of its "own" enzyme but those of all the others. The control which a particular enzyme exercises can therefore be changed by changes in any of the other enzymes.

Great emphasis is laid in many studies on "regulatory" enzymes, i.e., those whose activity depends on some metabolites (effectors) other than substrates and products. It will be realized that this means that additional elasticity terms will appear in the equations of type (3) and will therefore occur in some combination in expressions of the type (9)-(11). Bigger and more complex systems than the example chosen may result in huge and impenetrable expressions which are experimentally not useful. Their merit principally lies in exposing the interactive nature of our system.

Negative feedback loops are well known examples of effector interactions. In the simple system of three enzymes discussed previously, a feedback loop from the third metabolite on the first enzyme would add an additional term to the denominator of (9)-(11). It is evident that the net effect of the feedback will depend on the magnitude of that term relative to the sum of the others. Identification of such a "control site" only informs us of the possible importance of the loop. Quantitative assessment of other factors is required to make assertions about its metabolic significance. Enzymology would give a detailed description of the mechanism of inhibition, while control analysis attempts to quantify the effect of its operation.

THE MESSAGE

Control analysis, as surveyed here, attempts to answer the question: How does the organism work? This is what the early physiologists asked before the divisions into Biochemistry, Genetics, and Molecular Biology (and all their own subdivisions) developed. This specialization resulted in enormous progress in solving the many particular problems which such studies uncovered. It was, of course, realized that each discipline would have to become more quantitative in its description. In keeping his nose to the grindstone of his own problem, each worker is dimly aware of the hum of all the other stones grinding away. In spite of inumerable conferences and workshops, it rarely happens that two or more diverse specialists can put their findings together. This is no reflection on their abilities or will to succeed. It is simply in the nature of such analysis that synthesis, in practice, is not possible. The best we can, and have, achieved is a qualitative understanding. Quantitative insight is, however, still the aim. As we have seen, there is an approach which promises to give the answers. It is essentially a method which probes the intact organism by modulating each part. In so doing we must be prepared to ignore much of the sophisticated and intimate knowledge which will fill the pages of our journals and concentrate on those properties which our approach tells us we must measure. This method will continue to develop to incorporate more global aspects and no doubt will design more ingenious experiments. But one thing is certain: to understand the whole, you must look at the whole.

REFERENCES

Burns, J.A., Cornish-Bowden, A., Groen, A.K., Heinrich, R., Kacser, H., Porteous, J.W., Rapoport, S.M., Rapoport, T.A., Stucki, J.W., Tager, J.M., Wanders, R.J.A., and Westerhoff, H.V., 1985, Control analysis of metabolic systems, Trends Biochem. Sci., 10:16.

Dean, A.M., Dykhuizen, D.E. and Hartl, D.L., 1986, Fitness as a function of β-galactosidase activity in Escherichia coli, Genet. Res. Camb., in press.

Fell, D.A. and Sauro, H.M., 1985, Metabolic control and its analysis. Additional relationships between elasticities and control coefficients, Eur. J. Biochem., 148:555.

Flint, H.J., Tateson, R.W., Barthelmess, I.B., Porteous, D.J., Donachie, W.D., and Kacser, H., 1981, Control of the flux in the arginine pathway of Neurospora crassa: Modulations of enzyme activity and concentration, Biochem. J., 200:231.

Groen, A.K., Wanders, R.J.A., Westerhoff, H.V., van der Meer, R. and Tager, J.M., 1982, Quantification of the contribution of various steps to the control of mitochondrial respiration, J. Biol. Chem, 257:2754.

Heinrich, R. and Rapoport, T.A., 1974, A linear steady state treatment of enzymatic chains, Eur. J. Biochem., 42:97.

Heinrich, R. and Rapoport, T.A., 1975, Mathematical analysis of multi-enzyme systems. IL. Steady state and transient control, Biosystems, 7:130.

Hofmeyer, J.S., Kacser, H., and van der Meer, K.J., 1986, Metabolic control analysis of moiety conserved cycles, Eur. J. Biochem., in press.

Kacser, H., 1983, The control of enzyme systems in vivo: Elasticity analysis of the steady state, Biochem. Soc. Trans., a11:35.

Kacser, H. and Burns, J.A., 1973, The control of flux, in: "Rate Control of Biological Processes", D.D. Davies, ed., Cambridge University Press, Cambridge.

Kacser, H. and Burns, J.A., 1979, Molecular democracy: Who shares the controls?, Biochem. Soc. Trans., 7:1149.

Kacser, H. and Burns, J.A., 1981, The molecular basis of dominance, Genetics, 97:639.

Middleton, R.J. and Kacser, H., 1983, Enzyme variation, metabolic flux and fitness: alcohol dehydrogenase in Drosophila melanogaster, Genetics, 105:633.

Salter, M., Knowles, R.G., and Pogson, C.I., 1986, Quantitation of the importance of individual steps in the control of aromatic amino acid metabolism, Biochem. J., in press.

Westerhoff, H.V. and Chen, Y., 1984, How do enzyme activities control metabolite concentrations? An additional theorem in the theory of metabolic control, Eur. J. Biochem., 142:425.

Westerhoff, H.V., Groen, A.K., and Wanders, R.J.A., 1984, Modern theories of metabolic control and their application, Bioscience Reports, 4:1.

MAXWELL'S DEMONS IN CHANNELLED METABOLISM:

PARADOXES AND THEIR RESOLUTION

Hans V. Westerhoff and Frits Kamp

National Institute of Digestive
Metabolic and Kidney Diseases
National Institutes of Health
Bethesda, MD 20892, USA

INTRODUCTION

The second law of thermodynamics demands, among other things, that in an isolated system heat does not flow from a compartment at low to a compartment at high temperature. Maxwell (1871) pointed out that a "being" that could deal with individual molecules, when positioned at a trap door between two compartments at different temperatures, might open the door whenever a molecule with much higher than average kinetic energy would approach from the cold compartment, and whenever a kinetically "cold" molecule would approach from the "hot" compartment (see also Brush, 1976). Consequently, such a "Maxwell's demon" would cause heat to flow from the cold to the hot compartment and violate the second law. For quite some time, the validity of the second law of thermodynamics seemed to depend on the absence of Maxwellian demons from the systems under consideration. Also, it was deemed plausible that the special property long sought for living systems, was nothing but the presence of Maxwellian demons within them. Whilst Maxwell already argued that beings, other than supernatural, would lack the knowledge about the positions and velocities of the molecules, Szilard (1929), Demers (1945) and Brillouin (1956) have since resolved this paradox: the demon would require a continuous influx of information, which should be counted as a kind of work-input and would violate the required isolation of the system. In fact, the collection of this information would always cost more than could be obtained by having the displaced heat drive an engine. Thus, the second law of thermodynamics does apply to all physical chemical systems, living and inanimate alike. Whenever an apparent (or, even, potentially real) violation of the second law of thermodynamics arises, one may indicate this by invoking a Maxwellian demon.

Biological systems do not operate in isolation. Rather, they are open to heat and volume exchange with their surroundings; they operate at constant temperature and pressure. For such systems the second law of thermodynamics requires that any process absorbs more Gibbs free energy from its environment than it returns to it. For a chemical transformation from substance S to substance P, this law becomes the familiar rule that the reaction will only run from S to P if the

chemical potential of S exceeds that of P, i.e., if the free-energy
difference across the reaction is negative. Of course, this is only
required if the reaction occurs independently from other processes; if
the transformation from S to P were coupled to the hydrolysis of ATP,
then the reaction may still proceed if P is higher in chemical potential
than S, since under the usual conditions, the chemical potential of the
ATP greatly exceeds that of ADP and inorganic phosphate.

It is widely accepted (Boyer et al., 1977), that membrane-mediated
oxidative phosphorylation, the major ATP-production line in most cells,
consists of two independent reactions connected through an
electrochemical potential difference for protons ($\Delta\tilde{\mu}_H$) across the inner
mitochondrial membrane (in bacteria, the plasma membrane). This is the
so-called "chemiosmotic coupling" model (Mitchell, 1961). The first
reaction of the two (see Fig. 1A, the "dashed" enzyme) is a proton pump
driven by a redox reaction. It generates the $\Delta\tilde{\mu}_H$. The second ("H$^+$-
ATPase") reaction couples the reverse movement of protons to the
phosphorylation of ADP to yield ATP. From the reaction equation of the
latter:

$$ADP + P_i + n\cdot H^+_{out} \rightleftharpoons ATP + n\cdot H^+_{in} \tag{1}$$

we glean that the second law of thermodynamic requires the condition
(Wombacher, 1983):

$$\mu_{ADP} + \mu_{P_i} + n\cdot\tilde{\mu}_{Hout} > \mu_{ATP} + n\cdot\tilde{\mu}_{Hin} \tag{2}$$

or

$$n\cdot\Delta\tilde{\mu}_H > \Delta G_p \tag{3}$$

Here, the so-called "phosphate potential" is defined by:

$$\Delta G_p = \Delta G_p^{0\prime} + 17.0 + 5.7\cdot\log_{10}([ATP]/[ADP][P_i]) \tag{4}$$

and:

$$\Delta\tilde{\mu}_H = -5.7\ \Delta pH + 0.096\cdot\Delta\psi \tag{5}$$

(ΔG_p, $\Delta\tilde{\mu}_H$ in kJ/mol, ΔpH in units, $\Delta\psi$ in mV, total concentrations in mM,
ΔG_p can be read from [Rosing and Slater, 1972]. $\Delta G_p^{0\prime} + 17$ is a constant
at approximately 47 kJ/mol).

The experiments comparing $\Delta\tilde{\mu}_H$ to ΔG_p have been reviewed by a number
of authors (Ferguson, 1985; Slater et al., 1985; Westerhoff et al.,
1984a). It turns out that, in quite a few cases, the observed $\Delta\tilde{\mu}_H$ is
much lower than compatible with the observed ΔG_p for realistic values of
n, and that, where the H$^+$-ATPase supposedly attains equilibrium, the
ratio of these two quantities varies with $\Delta\tilde{\mu}_H$. The potential errors in
such experiments should not be underestimated, as has been shown by
Woelders and colleagues (Woelders et al., 1985). Yet, some cases are so

340

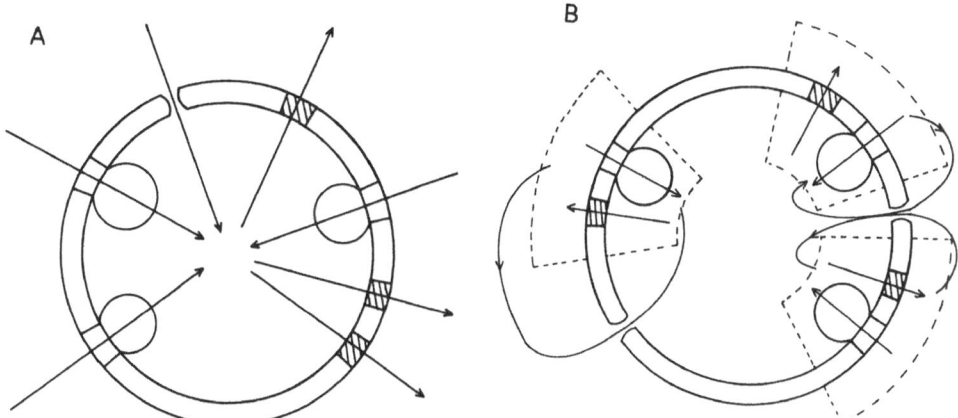

Fig. 1.   Models for membrane linked free-energy transduction.   A.   The
          delocalized chemiosmotic coupling model from Mitchell (1961).   A
          membrane separates two bulk aqueous phases and contains two
          types of proton pump, one (shaded) linked to a redox reaction,
          the other (the "mushrooms") to the synthesis of ATP through the
          so-called H^+-ATPase.   The former pump generates a pH gradient
          and a membrane potential, which drive protons back through the
          latter enzyme and force it to make ATP.   Note that protons
          pumped by the various redox-linked proton pumps are thermo-
          dynamically and kinetically equivalent with each other and with
          the protons in the adjacent bulk aqueous phase.   B.   The
          coupling unit version of the chemiosmotic coupling hypothesis
          (Westerhoff et al., 1984b).   This hypothesis differs from the
          delocalized chemiosomotic coupling hypothesis, in that the
          protons pumped by the different redox-linked proton pumps are no
          longer thermodynamically and kinetically indistinguishable.
          Because there is some (though not insurmountable) barrier (the
          dashed line) for proton movement from the region ("domain"),
          into which the proton is pumped by a redox-linked proton pump,
          to the aqueous bulk phase and to the domains where other protons
          are pumped into, the proton has an increased thermodynamic
          tendency to react with the H^+-ATPase that sits next to the
          redox-linked proton pump that pumped the proton.   Consequently,
          there is a multitude of extra proton compartments, one for every
          combination of redox proton pumps and H^+-ATPases.   The proton
          domains are very small, such that introduction of just a few
          protons into them already brings the local (apparent) proton
          potential to a magnitude sufficient to give rise to ATP
          synthesis by the bordering H^+-ATPase.   It should be noted that
          the coupling unit hypothesis invokes more than just that the
          proton, once pumped, would most frequently react with the H^+-
          ATPase that is nearest to it; it requires a difference in
          electrochemical activity for protons between the proton domain
          close to the redox-linked proton pump and the adjacent aqueous
          phase (this difference can persist because of the spatial
          inhomogeneity and the proton leak also present in the membrane),
          and even between the various proton domains.

flagrant that the insufficiency of $\Delta\tilde{\mu}_H$ in accounting for the observed ATP synthesis seems beyond doubt, e.g., in alcalophilic bacteria the external pH is a number of units above the internal one, and the transmembrane electric potential difference, $\Delta\psi$, is not very much higher than in other bacteria, such that $\Delta\tilde{\mu}_H$ amounts to only a few kJ/mol (Guffanti et al., 1983). It would seem that a Maxwellian demon must be invoked here in order to account for this ATP synthesis, in violation of the second law of thermodynamics.

In studies of the Na^+-K^+-ATPase in erythrocytes another such Maxwellian demon has turned up. Tsong and colleagues (Serpersu and Tsong, 1984; Tsong and Astumian, 1985) observed that, when an oscillating electric field is applied to erythrocytes, transport of Rb^+ could be induced against its electrochemical gradient, whilst any obvious source of free energy (such as the hydrolysis of ATP) was either absent or inoperative. Since the chemical potential difference of Rb^+ rose to a significant magnitude, whereas the time average of the applied electrical potential amounted to zero, this seemed to be yet another violation of the second law of thermodynamics.

Explanations for both demons have been offered. For the latter, Tsong and Astumian (1985) pointed out that the Na^+-K^+-ATPases may have a large dipole moment (by virtue of its α-helices), which may differ between two conformational states of the protein. An oscillating electric field would cause the protein to oscillate between the two conformations. If the conformational transitions of the protein were coupled to the catalytic cycle for transporting Rb^+, then free energy present in the electric field might be transduced into a Rb^+ electrochemical potential difference.

For oxidative phosphorylation, it has been proposed by several authors (for review, see Ferguson, 1985; Slater et al., 1985; Westerhoff et al., 1984a) that the protons relevant for biological free-energy transduction would not delocalize into the bulk aqueous phases in which the experimentor attempts to measure their electrochemical potential. Rather, they would remain in (or on) the membrane (in special channels, or domains), allowing for a locally higher $\Delta\tilde{\mu}_H$.

In recent versions of this model (Westerhoff et al., 1984b), it has been stressed that rather than one extra, membraneous proton compartment, there probably exists a multitude of tiny proton compartments, approximately one for each combination of redox proton pump and H^+-ATPase (see Fig. 1B) (cf. Slater et al., 1985). It turns out (Westerhoff and Chen, 1985), that in such a protonic-coupling-unit model the rate of ATP synthesis would vary with the (apparent) electrochemical potential difference for protons in a way different from that predicted by the usual (deterministic) rate equations. Moreover, this variation is not uniquely defined; it depends on whether the electrochemical potential difference for protons is altered through change in the activity of the redox proton pumps or by alteration of the proton permeability of the membranes. Finally, when calculating the (apparent) electrochemical potential difference for protons, by inserting the concentration of the protons averaged over all the proton domains into the usual formula relating chemical potential to concentration:

$$\mu_x = \mu_x^0 + RT\cdot\ln(\bar{N}/V) = \mu_x^0 + 5.7\cdot\log_{10}(\bar{N}/V) \tag{6}$$

suggests that ATP would be synthesized at values of the proton electro-chemical potential difference far below $\Delta G_p/n$. Thus, it would seem that in a system with a multitude of tiny coupling units, a Maxwell's demon would exist. The way out of the paradox is to realize that the above method of calculating the chemical potential of the protons is not really in line with its thermodynamical definition; Eq. 6 is only valid if the protons are distributed in an unconstrained equilibrium fashion, which implies that their distribution over the proton domains should follow the Poisson distribution (Hill and Chen, 1985). Indeed, the apparent violation of the second law in the model calculations depends on the proton distribution over the domains being non-Poissonian (Westerhoff and Chen, 1985). Yet, empirical use of the chemical potential is nearly always in the sense of Eq. 6. The conventional wisdom would have it, that we use Eq. 6 as the operational definition of the chemical potential of X, also when the distribution of X over the channels is not an equilibrium distribution. If this is done, then Maxwell's demon would persist in this coupling unit system.

The schematic drawing (Fig. 2A) of the protonic-coupling-unit hypothesis demonstrates that this model entails "channelling" (Davis, 1967; Friedrich, 1985; Welch, 1977; Wombacher, 1983; and other chapters in this book) for membrane-linked free-energy transduction: The proton pumped by any of the individual redox enzymes is not allowed to drive any arbitrary H^+-ATPase in the synthesis of ATP, but can only react at the H^+-ATPase that connects to the same proton domain as the redox enzyme that pumped the proton. A modification of the protonic-coupling-unit hypothesis corresponds to the dynamic type of channelling discussed by Friedrich (1985). It postulates that protons can only be transferred from redox proton pumps to H^+-ATPases, when these enzymes collide during their two-dimensional diffusion in the free-energy transducing membrane (Slater et al., 1985).

Noting this close correspondence between the protonic-coupling-unit model and models for metabolic channelling, we wondered whether the rather remarkable thermodynamic and kinetic properties of the former model would find parallel in that of metabolic channelling. Results obtained by Smeach and Gold (1975a,b) already hinted at this; for a channelling system of two enzymes in series, these authors calculated that the dependence of the pathway flux on the concentration of the pathway substrate would be different from that expected from the usual (deterministic) rate equations. Here we carry out some calculations for simple models of metabolic channelling to investigate further this question. We arrive at the conclusion that, when metabolic channelling occurs, the kinetics of the system in terms of the concentration of the channelled metabolite become anomalous (i.e., different from that predicted by deterministic rate equations). Moreover, if leakage out of the channels may occur, then also here Maxwell's demon may turn up, in the sense that a product of a channelled metabolic pathway may be synthesized even if the chemical potentials of the metabolic intermediates (when calculated from their concentrations averaged over the channels) would seem thermodynamically insufficient.

THEORETICAL ANALYSIS

We consider the model metabolic pathway of Fig. 2B, with each channel being elaborated in some more detail in Fig. 2C. In each channel there is an enzyme of the type i (for input) that produces metabolite X from substrate S at a stoichiometry of 3. There also is an output enzyme (e_o)

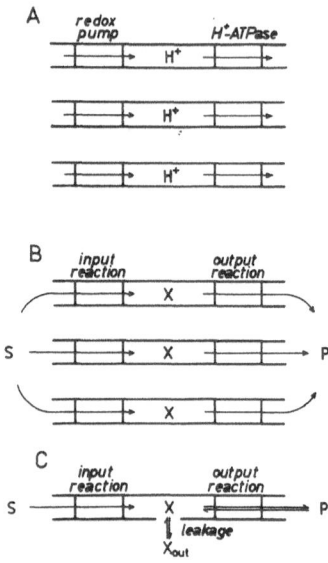

Fig. 2. Models of channelled metabolism. (A) Schematic version of the
protonic-coupling unit hypothesis. Protons pumped by a given
redox-linked proton pump are almost confined to reacting with
one particular H^+-ATPase. (B) Simple model of a channelled,
two-enzyme metabolic pathway leading from the substrate S to the
product P. The pathway is organized such that molecules of X
produced by an enzyme molecule of the type i (for "input") can
only react with one of the enzyme molecules o (for "output").
Thus, the X's depicted are chemically identical molecules, but
present in different channels and therefore potentially at different concentrations. (C) Elaboration of one of the channels
of (B): in addition to the input enzyme that produces X at a
stoichiometry of 3, and the output enzyme that consumes X at a
stoichiometry of two, there is a leak which allows molecules of
X to leave the channel one at a time. The channel volume, V,
and the concentrations [S], $[X]_{out}$, and [P], as well as the
kinetic constants are identical for all individual channels; N,
the number of molecules of X in the channel, may differ statistically between the different individual channels, giving rise
to a non-zero variance, σ^2, in that number of molecules.

which converts X to P at a stoichiometry of 2 per turnover. Then, in
some cases we shall assume that X may leak out of the channel. The
probability for this leakage is assumed to be proportional to the numbers
of molecules of X in the channel. In our calculations we shall assume
chemical, rather than enzyme, kinetics, because that makes our
explanations more translucent. For the analogous calculations using the
more realistic enzyme kinetics, we refer to Westerhoff and Chen (1985)
and Hill and Chen (1985). The stoichiometries 3 and 2 may seem rather
unreal. Again, they are chosen for clarity. For the results obtained
below, it will be sufficient if the molecularities of the two reactions
be different and the molecularity of the output reaction be not exactly
equal to 1 (it suffices for this reaction to follow enzyme kinetics; this
will bestow it with an effective kinetic order below 1). We shall assume
the input reaction to be irreversible. The concentrations of the (non-

channelled) S and P, as well as that of X in the compartment to which it
leaks reversibly, will be assumed to be constant.

For a bimolecular reaction, the usual (deterministic) chemical
kinetic rate equation is:

$$v_{det} = k \cdot (\overline{N})^2/V^2 \qquad\qquad (7)$$

where $\overline{N}$ is the average number of molecules of X per channel, V the volume
of the channel (all channels have equal volumes), and the deterministic
rate, v_{det}, is the rate per channel. The probability of finding a
molecule of X in a channel containing N molecules of X is proportional to
N. Once the first molecule of X has been set aside for the reaction, the
remaining number of molecules X is N-1. As a consequence, the
probability of finding two identical particles X is proportional to
N(N-1). Therefore, the actual ("stochastic") rate equation is as
follows (see McQuarrie, 1967):

$$v_{stoch} = k \cdot N(N-1)/V^2 \qquad\qquad (8)$$

The observed rate per channel is v_{stoch} averaged over all channels:

$$\overline{v}_{stoch} = \sum_{n=o}^{\infty} P(N=n) \cdot k \cdot (n^2-n)/V^2 = k \cdot [\overline{(N^2)-\overline{N})}]/V^2 =$$

$$= v_{det} + k \cdot (\sigma^2-\overline{N})/V^2 \qquad\qquad (9)$$

Here the summation is from n=0 to infinity, and P(N=n) is the probability
that the channel contains n molecules of X. It is crucial that, in
general the average of a square, such as (N^2), does not equal the square
of the average, $(\overline{N})^2$. σ^2 represents the statistical variance in the
number of molecules of X:

$$\sigma^2 = \overline{[N - (\overline{N})]^2} = \overline{(N^2)} - (\overline{N})^2 \qquad\qquad (10)$$

The relative deviation of the deterministic from the actual average rate
per channel is given by:

$$(\overline{v}_{stoch}-v_{det})/\overline{v}_{stoch} = 1/[\ 1 + (\overline{N})^2/(\sigma^2 - \overline{N})] \qquad\qquad (11)$$

Since (Hill, 1960; Westerhoff and Van Dam, 1986) the variance in number
of molecules, σ^2, tends to be of the order of the average number of
molecules, $\overline{N}$, this relative deviation will be small whenever the
average number of molecules of X per channel exceeds 10. This explains
why in unchannelled systems, where the number of molecules X in the one
collective "channel" usually exceeds 1 million, the deterministic rate
equations are sufficiently accurate to predict the actual reaction rates.

In channelled systems, the deviation of the actual average reaction rate from that predicted by the deterministic rate equation may become significant, provided that the variance in the distribution of the particle number of X deviates significantly from the average number of particles X. It can be shown (Hill, 1960) that, when either the reaction catalyzed by enzyme "i" or that catalyzed by enzyme "o" approaches equilibrium, σ^2 approaches $\overline{N}$ (the distribution of the molecule numbers over the channels becomes Poissonian). If, in the other extreme, both the input and the output reactions are irreversible, then for the stoichiometries we use here for the input and the output reaction, $\sigma^2/\overline{N}$ would become equal to 1.25 (Van Kampen, 1976; Nicolis and Prigogine, 1977), in the absence of leakage (if the leak rather than the output reaction were dominant, this ratio would even become equal to 2). In this case, the actual reaction rate would deviate significantly from that predicted by the deterministic rate equation, whenever the average number of molecules of X per channel would become low (the relative deviation between the stochastic and the deterministic rates will amount to $1/[1 + 4\overline{N}]$).

Our main question, however, is whether channelling may give rise to the appearance of a Maxwellian demon, i.e., whether in the scheme of Figs. 2 B and C, P could be produced whilst its chemical potential exceeded two times the (apparent) chemical potential of X. To examine this question further, we introduce the reverse output reaction (i.e., that producing X from P). For this simple case the rate of reaction, v_{-o}, just depends on the concentration of P:

$$v_{-o} = k_{-o}[P] \tag{12}$$

As a consequence the deterministic rate equation for the net reaction o becomes (from Eqs. 7 and 12):

$$v_{o,det} = k_{+o}(\overline{N})^2/V^2 - k_{-o}[P] \tag{13}$$

The stochastic net rate of reaction o becomes (froms Eqs. 8 and 12):

$$v_{o,stoch} = k_{+o}(N^2-N)/V^2 - k_{-o}[P] \tag{14}$$

We shall now consider the situation where the deterministic rate equation would predict the output reaction to be in equilibrium and ask the question whether the average of the stochastic rate equation (Eq. 14) may still predict synthesis of P. To this end we set $v_{o,det}$ to zero in Eq. 13, and eliminate $k_{-o}[P]$ from the average of Eq. 14:

$$\overline{v}_{o,stoch} = k_{+o}[\overline{(N^2)} - \overline{N}]V^2 - k_{+o}(\overline{N})^2/V^2 \tag{15}$$

Rearrangement of this equation and application of the definition of the variance (Eq. 10) leads to:

$$v_{o,stoch} = k_{+o}(\sigma^2 - \overline{N})/V^2 \tag{16}$$

From this equation one may expect synthesis of P if the deterministic rate equation predicts equilibrium, provided that the variance in the number of molecules of X exceeds the average [i.e., the distribution should be "broader" than the (Poisson) equilibrium distribution].

Above we have mentioned that a deviation from the Poissonian distribution, when the leak reaction is absent and the output reaction approaches equilibrium, is not expected. When leakage occurs, however, a Maxwellian demon may arise: in the extreme case where both the input and the leak reaction proceed at much faster rates than both unidirectional rates of the output reaction, the distribution of X will be mainly determined by the former two reactions. In that case (see above) the variance will approach two times the mean, and the forward actual output reaction rate may approach two times its reverse reaction under conditions where the deterministic rate equation would predict equilibrium and where the naive definition of the chemical potential of X (Eq. 6) would do the same. Actual equilibrium for the output reaction would be attained at an average concentration of X approximately half that expected from the deterministic rate equations, i.e., at a chemical potential of X approximately 1.7 kJ/mol lower than half the chemical potential of P.

SAMPLE CALCULATIONS

To illustrate and substantiate further the above conclusions, we carry out some sample calculations. We employ the Monte Carlo method, which is probably the closest to the actual process. It pictures one channel and follows step-by-step what may happen. A time average of this single channel will be representative of a space average of the multitude of uncorrelated channels. In our simple example, the state of the channel is completely determined by the number of molecules of X within it; all the other properties, such as the concentrations of S and P and the magnitudes of the kinetic constants, are invariant. If there are N molecules of X in the channel, the respective probabilities for a turnover of the input enzyme, a forward turnover of the output enzyme, a reverse turnover of the output enzyme, a forward turnover of the leak enzyme, and a reverse turnover of the leak enzyme are:

$$P_{+i} = k_{+i}[S]/\Sigma \tag{17}$$

$$P_{+o} = k_{+o}N(N-1)/V^2/\Sigma \tag{18}$$

$$P_{-o} = k_{-o}[P]/\Sigma \tag{19}$$

$$P_{+1} = k_1 N/V/\Sigma \tag{20}$$

$$P_{-1} = k_1[X]_{out}/\Sigma \tag{21}$$

In order to have the sum of these probabilities equal 1, Σ is defined by:

$$\Sigma = k_{+i}[S] + k_{+o}N(N-1)/V^2 + k_{-o}[P] + k_1 N/V + k_1[X]_{out} \tag{22}$$

The same k_1 occurs in both Eq. 20 and Eq. 21, because leakage is a passive process and microscopic reversibility must apply (Westerhoff and Van Dam, 1986). $[X]_{out}$ is the concentration of X outside the channels (which may, in an extreme case of channelled metabolism, correspond to the concentration of X outside the cell) and will be assumed to be constant and small.

Starting with our channel in the state where the number of molecules of X it contains is N, the above equations give us the relative probabilities of each of the five events that may occur next; but they do not predict with certainty what will happen next. One approach would be to assume just that the event with the greatest probability will take place next. However, this would obviously be erroneous. From our own experience with Murphy's laws, we know that the most probable thing often does not happen, especially if we would like it to. The proper ("Monte Carlo") solution therefore is to draw a random number between 0 and 1 and let this random number decide which event will take place next. If the number falls between 0 and P_i, then the next event will be a turnover of the input enzyme with a concomitant increase in the number of molecules of X in the channel by 3; if the number falls beteen P_{+i} and $P_{+i} + P_{+o}$, then the next event will be a forward turnover of the output enzyme with concomitant lowering of the number of molecules of X by 2; etc. This then brings the channel to its next state. For this state the number of molecules of X in the channel is known, and we can compute the transition probabilities for the next event from Eqs. 17-22. Again a random number between 0 and 1 will decide what happens next. This procedure is repeated over and over again, and after some time we will find the channel to fluctuate around a longer-time average. This will simulate the steady-state situation we are usually after. In the simulation of the actual process, we keep track of the total numbers of turnover of the three reactions and of the average and variance of the number of molecules of X in the channel. The latter is done by multiplying, for every state the system attains, its values of N and N^2 with the average lifetime of the state, τ, given by:

$$1/\tau = P_{+i} + P_{+o} + P_{-o} + P_{+1} + P_{-1} \qquad (23)$$

From the averages of N^2 and N, the variance is evaluated through application of Eq. 10. The (apparent) chemical potential of X is evaluated from Eq. 6.

The first line in Table 1 shows, respectively, the forward rate of the input reaction, the average number of molecules of X in the channel, the variance in that number taken relative to the mean, the average output reaction rate per channel, the same but as predicted by the deterministic rate equation, and the ratio of the leak reaction to three times the input reaction. It is seen that the variance in the number of molecules of X does deviate from the mean, by some 4%. As expected, this is accompanied by a difference between the actual rate at which P is produced and the rate predicted by the deterministic rate equation. Going down the rows of Table 1, one sees the effect of reducing the input reaction rate (e.g., through lowering the concentration of S): Initially the deviation of the actual rate from that predicted by the deterministic rate equation becomes more and more significant, until the input reaction is almost completely inhibited. That in Table 1 the variance deviates from the mean is analogous to, though not quite deducible from, considerations above (see Van Kampen, 1976). One might have expected the variance to exceed the mean, because the stoichiometry of the input reaction exceeds that of both output reactions.

Table 1. Relationship Between the Number of Molecules of X in the Channel of Fig. 2 B and C and the Output Reaction Rate at Different Activities of the Input Enzyme. (Shown are: the input reaction rate; the average number of molecules of X per channel, $\overline{N}$; the statistical variance in that number, taken relative to that average, $\sigma^2/\overline{N}$; the rate of the output reaction averaged over all channels, v_o; and the ratio of the rate at which X leaks from the channel to the rate at which X is injected into the channel by the input reaction, $v_1/(3v_i)$. Averages were over 19,000 iterations, after discarding the initial 1,000 iterations. Start was at N=0. $k_{+o} = 250$, $k_{-o}[P] = 2 \times 10^{-4}$, $k_1 = 6$, $[X]_{out} = 10^{-4}$, V = 1.)

$\overline{v_i}$	$\overline{N}$	$\sigma^2/\overline{N}$	$\overline{v_o}$	$v_{o,det}$	$\overline{v_1}/3\overline{v_i}$
100	0.79	0.96	148	154	0.016
75	0.71	0.91	111	127	0.018
50	0.63	0.84	73	99	0.025
20	0.50	0.73	28	62	0.050
10	0.42	0.72	14	43	0.081
5	0.33	0.75	6.5	27	0.132
2	0.20	0.84	2.4	10	0.207
0.1	0.016	1.01	0.10	0.06	0.322

For the calculations of Table 1, the actual reaction rate is lower than the one predicted from the deterministic rate equation. This is in line with Eq. 11, because the variance is lower than the mean. As a consequence, no Maxwell's demon is to be expected for this type of condition.

The implications of our observation (Westerhoff and Chen, 1985) and of the earlier observations of Smeach and Gold (1975a), that the reaction rates in channelled systems do not necessarily obey the canonical (deterministic) rate equations, might well be academic only, were it not for the subsequent observation we made (see also Westerhoff and Chen, 1985; Hill and Chen, 1985). We repeated the calculations of the relationship between the output rate and the average number of molecules of X in the channels, but this time we varied the average number by increasing the leakage out of the channel, rather than decreasing the rate at which the input reaction replenishes the channel. The result is shown by the full line in Fig. 3: not only do the actual output rate and the deterministically predicted rate ("dashed" line) differ, but the relationship between the rate and the average number of molecules of X in the channels is different from the relationship in the case where the input reaction was progressively inhibited (which is repeated as the "dotted" line). Apparently, with the output enzyme itself unaffected, the dependence of the output rate on the concentration of X is still not unique but depends on how [X] is varied.

Close inspection of Fig. 3 suggests that, also at low magnitudes of the average number of molecules of X in the channels, the output rate for the case of increased leakage significantly exceeds the rate predicted by the deterministic rate equation. This suggests that, perhaps in this case, Maxwellian demons may arise, i.e., synthesis of P may occur whilst the deterministic rate equation would predict reversal of the flux from X to P. Fig. 4 was drawn from the results of a more detailed calculation

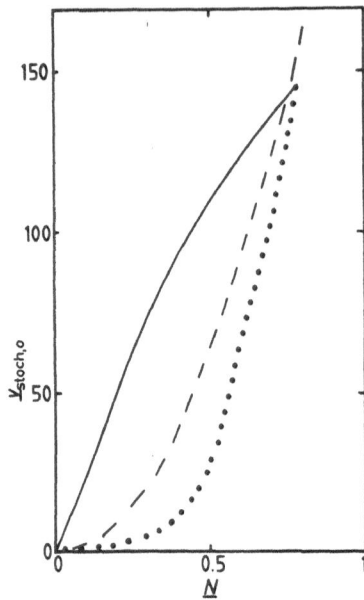

Fig. 3.   Relationships between output flux and intermediate concentration
          for the channelled metabolic system of Fig. 2 B and C. Dashed
          line: deterministic relationship, full line: relationship at
          varying leakiness of the channels, dotted line: relationship at
          varying activity of the input reaction. Parameter values were
          the same as in Table 1, except for the full line where $k_1[S]$
          was kept constant at 100 and k_1 was increased from 6 to 100,000.

for a similar case. It gives the rate of the output reaction versus the
difference between two times the apparent chemical potential of X and
the chemical potential of P (both calculated by applying Eq. 6 to the
average concentration). As demonstrated by the "dashed" line, the
deterministic rate equation predicts that flux can only flow from X to P
when two times the chemical potential of the former exceeds that of the
latter. The "full" line, however, shows that in a channelled system
flux may flow from X to P, even though the apparent free energy of
reaction, calculated from (cf. Eq. 13 and [27]):

$$\Delta G_{out} = 2 \cdot \mu_X - \mu_P = RT \cdot \ln\{k_{+o}(\overline{N})^2/V^2/(k_{-o}[P])\} \qquad (24)$$

is insufficient to account for it. This is an apparition of Maxwell's
demon.

     Eq. 16 indicates in which case one might find synthesis of P in
(apparent) violation of the second law of thermodynamics: if the variance
of the distribution of X over the channels exceeds the mean number of X
per channel, i.e., if the distribution is broader than the (Poissonian)
equilibrium distribution. for one of the cases where we observed an
apparent violation of the second law, we did not only keep track of the
averages of N and N^2, but also of the entire distribution of N. The
"full" line in Fig. 5A gives the relative frequency by which the channel
was found to contain N molecules. For comparison, the "dashed" line

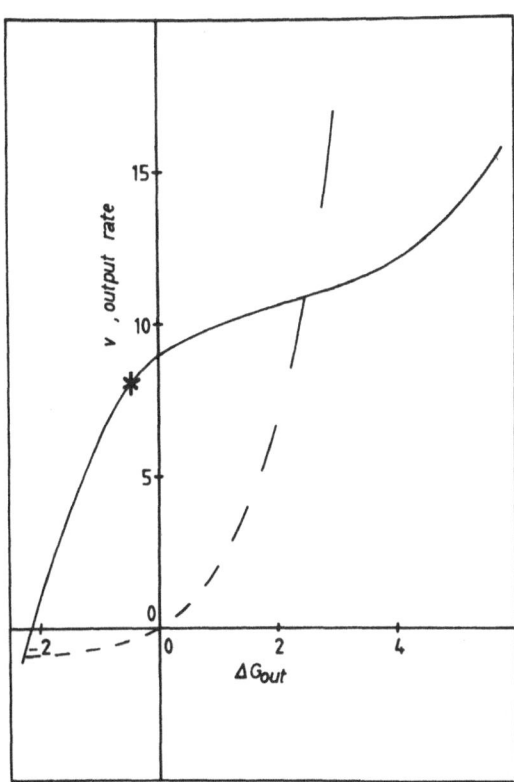

Fig. 4. Apparent violation of the second law of thermodynamics for the
output reaction in a channelled system. $\Delta G_{out} = 2 \cdot \mu_X - \mu_P$ was
calculated from the average number of molecules of X and Eq. 24.
Parameter values were: $k_{+i}[S] = 10$, $k_{-i} = 10-20$, $k_{+o} = 10^5$,
$k_{-o}[P] = 1$, $[X]_{out} = 10^{-3}$, $V = 10$, leak was varied; for the
point indicated by the *, $k_1 = 10^4$. Averages over 56,000 iterations, after discarding 1,999, start at N=0. The dashed line is
the result predicted by the deterministic rate equations (Eqs.
13 and 24).

gives the Poisson distribution with the same average. Because for these
small averages the Poisson distribution is a monotonically decreasing
function of N, the fact that the variance of the actual distribution is
greater than that of the Poisson distribution is most clearly observed in
the fact that the actual distribution has a longer tail; the frequencies
by which the channel contains 2, etc., molecules of X is higher than in
the Poisson distribution. The "full" line in Fig. 5B gives the output
rate as a function of N (obtained by multiplying the frequencies plotted
in Fig. 5A, "full" line, to the stochastic rates given by Eq. 14).
Because of the nonlinearity of the rate function (Eq. 14), the width of
the distribution on the right-hand side has a greater positive effect on
the rate of net synthesis of P, than the width on the left-hand side has
a negative effect. This explains why the broader distribution gives rise
to higher net synthesis of P. The dashed line in Fig. 5B shows that, if
the distribution would have been Poissonian, the outcome would be vastly
different, even such that negative output reaction would have dominated.
This accounts for the slightly negative deterministic output rate of

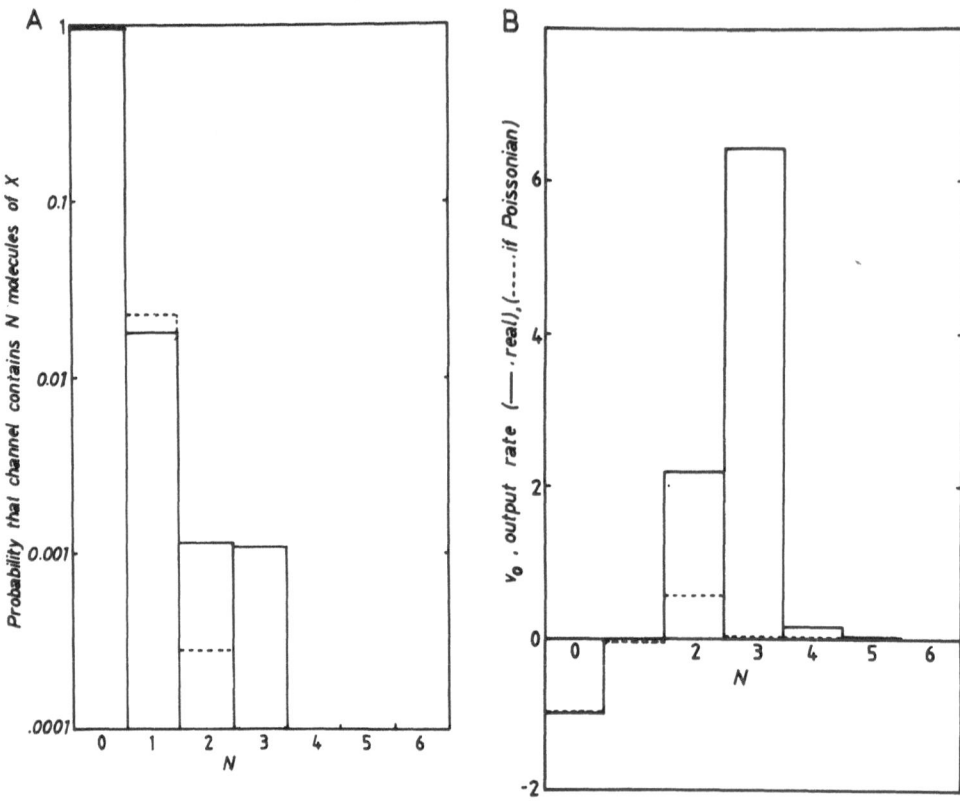

Fig. 5. Distribution of X over the different channels (A), and (B) the
implication for the flux through the output reaction in a
channelled system in a condition that gives rise to an apparent
violation of the second law of thermodynamics (the point "*" in
Fig. 4). A (note the log scale): The full line gives the actual
frequency at which the channels contain N molecules of X, the
dashed line gives the frequency that would occur if the
distribution of X over the channels would be Poissonian, with
the same average number of molecules (i.e., 0.024) per channel.
B: The full line indicates the output rate obtained by
multiplying the frequency at which a channel contains N
molecules of X with the corresponding stochastic rate (from Eq.
14). The dashed line gives the result obtained by multiplying
for each N the frequency that would have been obtained if the
distribution would have been Poissonian with the same average
(the dashed line in A) with the stochastic rate (from Eq. 14).

-0.43 under these conditions (see the point labelled "*" in Fig. 4).

A SECOND MAXWELLIAN DEMON: WORK FROM AN OSCILLATING ELECTRIC FIELD

The above findings for channelled systems may be summarized, by
stating that Maxwellian demons may arise because of spatial heterogeneity
in the concentration of X. Indeed, because the difference in [X] between
individual channels may exceed the difference allowed by an equilibrium
distribution of X over the channels, one may find that X can do more work

than thermodynamically predicted from its average concentration. We shall now briefly report on another way in which more work may be done, than would be predicted by the average of a thermodynamic potential. Here, however, the average will be a time average.

Inspired by the work of Tsong and colleagues (Serpersu and Tsong, 1984; Tsong and Astumian, 1985), observing that an oscillating electric field can cause uphill transport of Rb^+, we considered a naive model for an ion pump (Fig. 6A) in which a protein "arm" that carries an electric charge would move across the membrane, coupled to the catalytic cycle of the enzyme. For simplicity we consider the ion, that is normally pumped, to be absent. As a consequence, it would seem that the enzyme could only carry out its chemical reaction in the thermodynamically downhill direction, coupled to a futile cyclic translocation of the charged protein arm across the membrane.

In the presence of an oscillating electric field, however, this rather wasteful picture might change if, somehow, the protein arm could be made to move in phase with the oscillation of the field: when the field was directed outward, the protein arm (if positively charged) would cross the membrane in the outward direction, and as the electric field would reverse its sign, the protein arm would be pulled back across the membrane. In view of the coupling between the movement of the protein arm and the catalytic cycle of the enzyme, this might then cause a

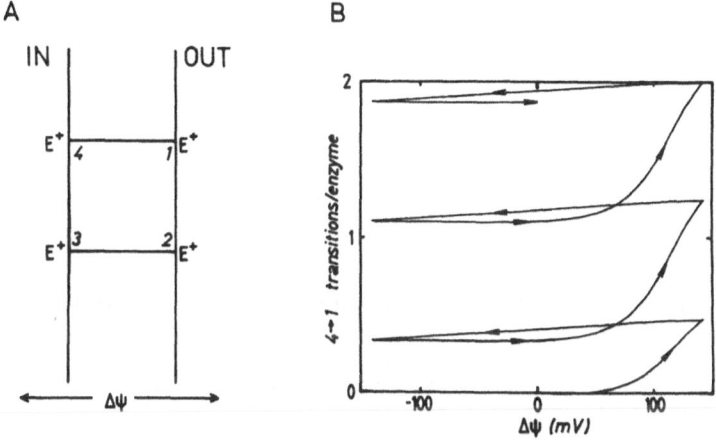

Fig. 6. An oscillating electric field may drive an enzyme's catalytic cycle. (A) The catalytic cycle of a membraneous enzyme. In going from states 1 to 4, or 3 to 2, a positively charged arm of the enzyme moves across the membrane. The clockwise cycle of the enzyme may be coupled to the chemical conversion of S to P. An oscillating electric field may then drive that thermodynamically-uphill reaction: (B) The net number of transitions through the branch 4 - 1 and the transmembrane electric potential difference as they vary with time (the independent variable). Calculations were described in (Westerhoff et al., 1986). Starting in equilibrium at zero field, the field sweeps sinusoidally, first to +142 mV. Time runs as indicated by the arrow heads. Three complete field cycles are followed.

reversal of that catalytic cycle, such that the oscillating electric field would drive an endergonic (i.e., free-energy requiring) chemical reaction.

To demonstrate the feasibility of this idea, we carried out some numerical calculations for the enzyme diagram of Fig. 6A. Fig. 6B shows some results: the number of transitions per number of enzymes along the top line in the diagram of Fig. 6A, plotted versus the transmembrane electric potential difference in mV. The calculation started at zero potential. As the electric field became directed to the outside, the population of enzyme molecules gradually started to move from state 4 to state 1, until at the maximum potential of 142 mV, some 35% of the enzymes had undergone this transition. As the field was reversed, some enzymes went back through the 4-1 transition. Most importantly, however, the total number of enzyme retracing their own steps was smaller than the number of enzymes that underwent the transition in the first place. Thus, after a complete cycle of the field, some 30% of the enzymes still had undergone the net 4 to 1 transition. In the subsequent field cycles, the net surplus of transitions from 4 to 1 continued to occur. Thus, the oscillating electric field (with a time average of zero) induced cyclic turnover of the enzymes. When coupled to an endergonic (i.e., free-energy requiring) reaction, this would lead to work being performed (this was actually shown for the calculations of Fig. 6 [Westerhoff et al., 1986]).

For the present paper the most important aspect of this calculation is that rather ordinary enzymes may carry out work driven by an electric field that has a time-average of zero. Again, this is a Maxwellian demon. In this case the solution of the paradox is that the oscillating component of the electric field should be taken into account when working out the free-energy balance sheet.

DISCUSSION

In our contribution to this book on metabolic organization, we have attempted to note the following peculiarities of channelled metabolic systems:

(i) To the observation of Smeach and Gold (1975a,b), that the kinetics of channelled systems in terms of the concentration of the substrate of the channelled pathway is anomalous, we added that the kinetics in terms of the average concentration of the metabolic intermediates is also anomalous.

(ii) Moreover, the dependence of the flux through the pathway on the concentration of that intermediate depends on how that concentration is altered, in ways not expected for unchannelled pathways. This goes as far as causing an apparent interaction between output enzyme and input enzyme: at the same concentration of pathway intermediate, the flux through the output enzyme depends on the activity of the input enzyme, as if the output enzyme feels what happens to the input enzymes through a direct conformational interaction.

(iii) Metabolic fluxes may occur, even though they are in apparent conflict with the second law of thermodynamics. Because the distribution of the intermediate over the different channels may deviate significantly from the distribution in equilibrium, its actual ability to do work may exceed its apparent thermodynamic potential calculated from its average concentration.

(iv) Even if the time average of a thermodynamic potential is low, it may be able to do a significant amount of work, because of regularly oscillating components. This observation may again be relevant for channelled systems, because of the spatial proximity of channelling enzymes. Conformational changes in enzymes with dipole moments may readily cause changes in the electric field surrounding them (Tsong and Astumian, 1985) and hence affect the catalytic cycles of the surrounding enzymes. Welch and Berry (1985) have proposed that metabolism may be organized in a quasi-electronic fashion. The electric coupling in their model had electrons or protons flowing through the, otherwise traditional, metabolic pathway. It would seem that coupling between enzymes through the mechanism described here is somewhat more realistic.

In recent papers, we have indicated that the conclusions from our calculations may be pertinent to membrane-linked free-energy transduction (Westerhoff et al., 1984a; Westerhoff and Chen, 1985). Literature data on other channelled systems (Welch, 1977; Srivastava and Bernhard, 1986; and other chapters in this book) suggest that these contentions may also be valid there.

Beyond interest in the kinetic aspects of channelled systems, there may be three take-home messages. The first is that thermodynamic and kinetic anomalies may be expected for channelled systems; and since they are rather rigidly absent from unchannelled systems, experimentally observed anomalies of this kind may be used as evidence for channelling (Kell and Westerhoff, 1985; Westerhoff and Kell, 1986). The second is that, in the modeling of channelled systems, one should be cautious not to limit one's analysis to the usual kinetic rate equations, but to invoke stochastic approaches. Finally, a question beyond the present paper, but the subject of intensive work by a number of colleagues [Kacser, Keleti, Kell, and Welch—personal communications], is that the special kinetic properties of channelling lead to a very peculiar control structure of channelled pathways (Westerhoff et al., 1983). It is probably a good guess that this will have important physiological consequences (see also Welch and Berry, 1985).

ACKNOWLEDGEMENTS

We wish to thank the organizers of the conference (Drs. Welch, Clegg, Kell, and Srere) for doing such a splendid job, Rick Welch for directing us to the references by Smeach and Gold (1975a,b) and Drs. Chen, Astumian, Tsong, and Chock for very instructive discussions concerning the stochastic protons and the potential effects of oscillating electric fields.

REFERENCES

Boyer, P.D., Chance, B., Ernster, L., Mitchell, P., Racker, E., and Slater, E.C., 1977, Annu. Rev. Biochem., 46:955.

Brillouin, L., 1956, "Science and Information Theory", Academic Press, New York.

Brush, S.G., 1976, "The Kind of Motion We Call Heat", North-Holland, Amsterdam.

Davis, R.H., 1967, in: "Organizational Biosynthesis", H.J. Vogel, J.O. Lampen, and V. Bryson, eds., Academic Press, New York.

Demers, P., 1945, Can. J. Research., 23:47.

Ferguson, S.J., 1985, Biochim. Biophys. Acta, 866:47.

Friedrich, P., 1985, in: "Organized Multienzyme Systems", G.R. Welch, ed., Academic Press, New York,

Guffanti, A.A., Fuchs, R.T., and Krulwich, T.A., 1983, J. Biol. Chem., 258:35.

Hill, T.L, 1960, "Statistical Thermodynamics", Addison-Wesley, Reading, MA.

Hill, T.L. and Chen, Y., 1985, Proc. Natl. Acad. Sci. USA, 82:3654.

Kell, D.B. and Westerhoff, H.V., 1985, In: "Organized Multienzyme Systems", G.R. Welch, ed., Academic Press, New York.

Maxwell, J.C., 1871, "Theory of Heat", Longmans Green, London.

McQuarrie, D.A., 1967, Suppl. Rev. Series Appl. Probability, 8:1.

Mitchell, P., 1961, Nature, 191:144.

Nicolis, G. and Prigogine, I, 1977, "Self-Organization in Nonequilibrium Systems", Wiley, New York.

Rosing, J. and Slater, E.C., 1972, Biochim. Biophys. Acta, 267:275.

Serpersu, E.H. and Tsong, T.Y., 1984, J. Biol. Chem., 259:7155.

Slater, E.C., Berden, J.A., and Herweijer, M.A., 1985, Biochim. Biophys. Acta, 811:217.

Smeach, S.C. and Gold, H.J., 1975a, J. Theor. Biol., 51:79.

Smeach, S.C. and Gold, H.J., 1975b, J. Theor. Biol., 51:59.

Srivastava, D.K. and Bernhard, S.A., 1986, Curr. Top. Cell. Regul., in the press.

Szilard, L., 1929, Z. Physik, 53:840.

Tsong, T.Y. and Astumian, R.D., 1985, in: "Proc. 8th Intl. Symp. Bioelectrochem. Bioenerg.", in the press.

Van Kampen, N.G., 1976, Adv. Chem. Phys., 34:245.

Welch, G.R., 1977, Prog. Biophys. Molec. Biol., 32:103.

Welch, G.R. and Berry, M.N., 1985, in: "Organized Multienzyme Systems", G.R. Welch, ed., Academic Press, New York.

Westerhoff, H.V. and Chen, Y., 1985, Proc. Natl. Acad. Sci. USA, 82:3267.

Westerhoff, H.V. and Kell, D.B., 1986, Comm. Molec. Cellul. Biophys., in the press.

Westerhoff, H.V. and Van Dam, K., 1986, "Mosaic Non-Equilibrium Thermodynamics and the Control of Biological Free-Energy Transduction", Elsevier, Amsterdam.

Westerhoff, H.V., Colen, A.-M. and Van Dam, K., 1983, Biochem. Soc. Trans., 11:81.

Westerhoff, H.V., Melandri, B.A., Venturoli, G., Azzone, G.F., and Kell, D.B., 1984a, Biochim. Biophys. Acta, 768:257.

Westerhoff, H.V., Melandri, B.A., Venturoli, G., Azzone, G.F. and Kell, D.B., 1984b, FEBS Lett., 165:1.

Westerhoff, H.V., Tsong, T.Y., Chock, P.B., Chen, Y., and Astumian, R.D., 1986, Proc. Natl. Acad. Sci., in the press.

Woelders, H., Van der Zande, W.J., Colen, A.-M.A.F., Wanders, R.J.A., and Van Dam, K., 1985, FEBS Lett., 179:278.

Wombacher, H., 1983, Molec. Cellul. Biochem., 56:155.

MOLECULAR MACHINES AND ENERGY CHANNELLING

Frits Kamp and Hans V. Westerhoff

National Institute of Digestive
Metabolic, and Kidney Diseases
National Institutes of Health
Bethesda, MD 20892, USA

INTRODUCTION

This Workshop deals with the phenomenon of channelling, being defined as the transfer of metabolites between enzymes before the former equilibrate with a pool. Several examples of channelling have been outlined. The question may be raised, whether also _energy_ can be channelled from locations in the cell producing energy towards locations consuming energy. This question is of particular significance in the field of bioenergetics, studying mechanisms of free-energy transduction.

What is "free-energy transduction"? The Gibbs free energy, or the isothermal-isobaric "work potential" (e.g., Van Dam and Westerhoff, 1984) of a system, is defined as

$$G = H - T \cdot S = U + P \cdot V - T \cdot S \tag{1}$$

The first term (U) is the only true energy term. ("True" energy is defined as the energy that is subject to the conservation requirement of the First Law of Thermodynamics). It represents the total potential plus kinetic energy stored in the system at all different levels of ordering: in chemical bonds, in macromolecules, in membranes, in vibrational, rotational and translational degrees of freedom, etc. The second term (PV) derives from the volume work. Most biochemical processes are carried out under isobaric conditions, so that changes in thia term reduce to $P \cdot \Delta V$. At the usual pressure of 1 atmosphere and usual volume changes, this term is negligible (van Dam and Westerhoff, 1984). The last term (TS) refers to the entropy. Thus, according to Boltzmann's H-theorem, S relates to the disorder of the system as follows: $S = k_B \cdot \ln \Omega$, Ω being the number of "complexions" of the system at present volume and energy content.

In free-energy transduction, free energy is converted from one way of ordering into another. For instance, free energy carried by NADH plus oxygen in excess over that carried by NAD^+ plus water may be converted to the free energy ATP has in excess over ADP and P_i (inorganic phosphate); or, alternatively, an electrochemical gradient of one ion across a membrane may be converted into the electrochemical gradient of another ion, or into ATP, and so on. Owing to the Second Law of Thermodynamics, the efficiency of such conversions must be less than 100% in order to make the process run (e.g., van Dam and Westerhoff, 1984). We are interested in how cells accomplish these conversions at sufficient efficiency.

At the beginning of the 1970's, McClare published several highly controversial papers in which he suggested that free energy can be transduced rather mechanically (and, therefore, extremely efficiently) by so-called "molecular machines". This was in contrast to the thermo-dynamical free-energy transducing mechanisms that were, he wrote, usually regarded in bioenergetics (McClare, 1970, 1974). Moreover, he claimed that these "molecular machines" can be the only effective free-energy transducers. He rejected a priori the contemporary models of muscle contraction, since they would entail a violation of the Second Law of Thermodynamics. McClare also claimed that the Second Law is not statistical but applies already to single molecules, and that living cells are so small that they cannot be described by way of thermodynamics but need a (quantum-)mechanical approach.

Much in the trail-blazing writings by McClare seems to be at fault (Kemeny, 1974; Hill, 1975). His papers are quite abstruse and his ideas hard to conceive. On the other hand, some of his ideas are intriguing and, if elaborated with appropriate statistical-mechanical tools, may nevertheless be quite relevant for biological free-energy transduction. In McClare's absence others (Gray and Gonda, 1977; Blumenfeld, 1983; Welch and Kell, 1986) have been further solidifying his concepts. This paper focuses on the definition and physical features of the notion of "molecular machine" and its potential relevance for channelled metabolism.

ENERGETICS OF COUPLING

What then is a "molecular machine", or rather, in McClare's wording, a "molecular energy machine"? To come to grips with this notion, we will discuss what happens with energy in a catalytic system that transduces free energy. Let us outline the following example of two coupled reactions, catalyzed by one enzyme, in which (chemical) free energy is transduced from one form (redox-free energy) into another (ATP):

$$
\begin{array}{ccc}
A_{ox} + B_{red} & \diagdown & ADP + P_i \\
& \diagup\diagdown & \\
A_{red} + B_{ox} & \diagup & ATP + H_2O
\end{array}
\qquad (2)
$$

What sort of mechanism accounts for this transduction? This can be nicely illustrated by way of a King-Altman (1956), or Hill (1977), diagram:

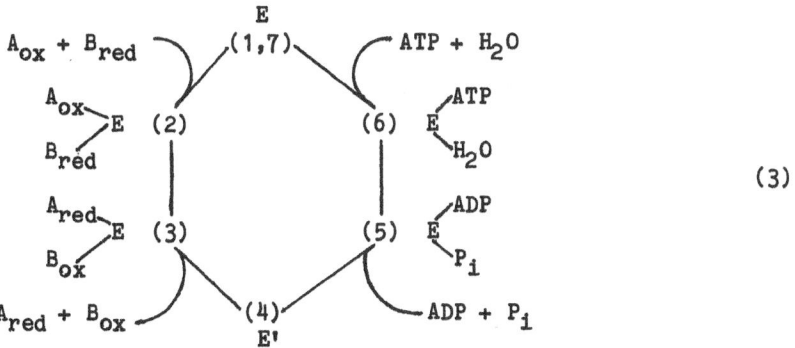

$$(3)$$

Every state (1 through 7; 7 is conformationally identical to 1, except that one molecule each of A_{ox}, B_{red}, ADP, and P_i has been transformed to one molecule each of A_{red}, B_{ox}, ATP, and H_2O) represents a configuration of the catalytic system, and every transition represents a change in at least one specific degree of freedom. Thus the 1→2 transition accounts for the binding of the redox-substrates, 2→3 for the redox reaction, 3→4 for the release of the redox products, 4→5 for the binding of ADP and inorganic phosphate, 5→6 for the hydrolysis of ADP + P_i into ATP + H_2O, and finally 6→7(=1) for the release of ATP from the enzyme. During every transition, the "basic" (i.e., state concentration-independent) free energy of the system may change (with "system" we mean the enzyme plus its substrates and products) (Hill and Eisenberg, 1981). Also, during every transition the system passes through a metastable transition state (or activated complex) with a higher basic free-energy content. This is illustrated by Figure 1. The horizontal axis repre-sents the different states of the system, and the vertical axis gives the concomitant basic free-energy changes of the system during the transi-tions. In addition, the relative contribution of U to G is depicted. If the change in U is downhill, heat is produced, which means that energy of the system, in as far as present in a specific degree of freedom (e.g.,

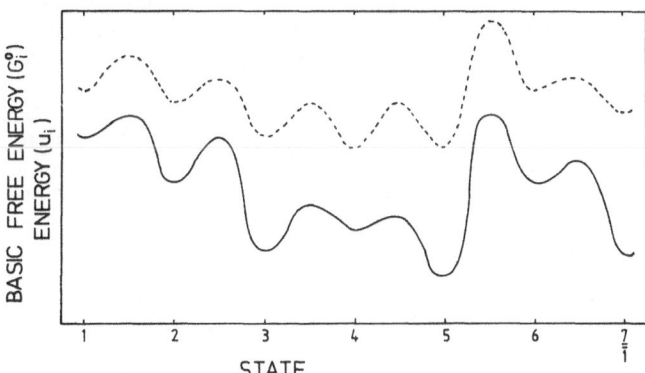

Fig. 1.   Basic free-energy and energy changes in traversing the Hill cycle of Eq. 3. X-axis: Different states of the cycle (1 through 7), Y-axis: (full line) Basic free-energy changes and (dashed line) energy changes.

the bond between the two electrons and NADH) is transferred to the environment and distributed over many thermal degrees of freedom. If ΔU is uphill, heat is absorbed from the environment and stored in a specific degree of freedom (e.g., the bond between ADP and P_i) of the system, increasing the energy of the system. We see that the environment functions as a reservoir in which thermal energy can be dumped at one transition, and from which thermal energy can be picked up at a following transition. Thus energy, represented by $A_{ox} + B_{red}$, is converted into a pool of thermal energy, and later on this pool provides energy for the formation of a chemical bond (ATP). Welch et al. (Welch et al., 1982; Somogyi et al., 1984) have discussed how enzymes may be devised to couple to the environment in this manner. This does not [pace McClare (1970)] mean that the Second Law is violated. For, the Second Law only states that in an open system the free energy must decrease. Therefore, exchange of thermal energy is allowed (Kemeny, 1974).

The basic free-energy change during a transition from a state i to a state j amounts to:

$$\Delta G^O_{i \to j} = G^O_i - G^O_j = - R \cdot T \ln(K) = R \cdot T \ln [(C_i/C_j)_{eq}] \qquad (4)$$

(C_i is the concentration of the enzyme in state i). This basic free-energy change contains all the components of the free-energy change of the system, except for the terms that depend on the concentrations of the states of the enzyme: it contains their difference in energy (or enthalpy, $H = U + PV$), their difference in (standard) entropy and their difference in free energy due to the binding and transformation of the ligands. For the 1 to 4 transition of the above Hill-diagram (Eq. 3):

$$\Delta G^O_{1 \to 4} = h_1 - h_4 - T \cdot (s_1 - s_4) + \mu_{Aox} + \mu_{Bred} - \mu_{Ared} - \mu_{Box} \qquad (5)$$

Here h_i is the partial molar enthalpy (= $u_i + Pv_i$) of state i, s_i is the partial molar entropy of that state (which has to do with the number of configurations the state may have). The μ's indicate the (concentration-dependent!) chemical potentials of the substrates and products of the reaction. Note that with "state 1" we really refer to the enzyme in state 1 plus A_{ox} plus B_{red} plus ADP plus P_i (Hill, 1977; Hill and Eisenberg, 1981). Likewise, with "state 7" we mean the enzyme in the conformation of state 1 plus A_{red} plus B_{ox} plus ATP plus water.

For a transition between any state i and any state j, the total free-energy change is:

$$\Delta G_{i \to j} = G^O_i - G^O_j + R \cdot T \ln (C_i/C_j) = R \cdot T \ln \left[\frac{C_i}{C_j} \middle/ \left(\frac{C_i}{C_j} \right)_{eq} \right] \qquad (6)$$

Hence, though the basic free-energy change $G^O_i - G^O_j$ might be uphill, the total ΔG for that step can still be downhill if $RT \cdot \ln(C_i/C_j)$ be sufficiently negative, that is if there be sufficient accumulation of state i. The only requirement for the basic free-energy change of the overall cycle is that it be negative. Only then will a net cycle flux, and thus free-energy transduction, take place (Hill, 1977; Hill and

Eisenberg, 1981). Free-energy transduction depends further on the absence of the direct E'→E transition. In order to traverse a complete cycle, the redox-reaction can only occur provided that it is followed by ATP synthesis.

Figure 2 depicts what may happen with the free energy of the system during the 1→4 transition (redox-reaction, X-direction, from A to B) and the 4→7 transition (ATP-synthesis, Y-direction, from B to C). For simplicity, the transitions 1→2, 2→3, and 3→4 have been grouped together into the A→B transition. The same has been done for the three subsequent transitions. The Z-axis gives the basic free energy of the system. We start from point A, which corresponds to state E in the Hill diagram. The curve in the Y-direction (from front to back) shows the basic free-energy change if ATP-synthesis would occur. We see that this reaction would be uphill. The coupling properties of the enzyme now are such that it will first let the redox-reaction take place, so that it will arrive at point B (where G^0 is at its minimum). To get to B, the system, after it has overcome the activation barrier in between points A and B, will have to lower its basic free energy to that characteristic of point B (otherwise the system would remain in an excited substate of point B). In fact, it will probably first end up in a quasi-state E^* which has approximately the same basic free energy as E (point A) and then decay to state E'. The excess energy of E^* over E' must be transferred, as heat, to the environment.

Subsequently, uphill ATP-synthesis occurs and we arrive at point C, which corresponds to state 7 of the Hill diagram (Eq. 3). We see that the overall free-energy change of the cycle is still downhill. However, if the basic free-energy change of the second reaction would be very uphill, the cycle flux rate might become very low. This can be explained as follows: according to Absolute Rate Theory (Glasstone et al., 1941), the forward flux rate for any transition i→j equals:

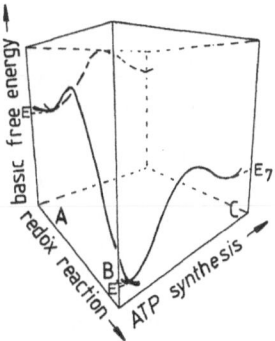

Fig. 2.   Three dimensional picture of the free-energy diagram of Eq. 3. X-axis (from left to right): Progress of the redox-reaction, Y-axis (from front to back): ATP-synthesis, Z-axis: Basic free-energy changes. The system starts at point A (zero progress in both redox-reaction and ATP-synthesis), where it is in state E (corresponding to state 1 in Eq. 3). It then develops to point B (completed redox-reaction, no progress in ATP-synthesis), where it is in state E' (state 4 in Eq. 3). Finally ATP-synthesis takes the system to point C (both reactions completed), where it is in state E_7 (state 7 in Eq. 3).

$$v_{i \to j} = k_0 \cdot \exp \, (\Delta S_a / R) \cdot \exp \, (-\Delta U_a / RT) \cdot C_i \qquad (7)$$

$k_0 \cdot \exp(\Delta S_a / R)$ is proportional to the frequency factor, or the frequency with which the substrate molecules (such as A_{ox} and B_{red}) collide. This factor is expected to be much higher for the bound substrates than for the substrate molecules in free solution. ΔS_a is the activation entropy, which accounts for the entropy difference between the activated complex and the substrate. ΔU_a is the activation energy which has to be overcome to arrive at the transition state.

If the basic free-energy change of the transition would become extremely uphill, ΔU_a could become very uphill as well, which means that large amounts of thermal energy need to be exchanged with the environment, making the reaction highly improbable. The enzyme might simply solve this problem by having a very high catalytic rate constant k_0, as well as a high activation entropy for the transition. If the enzyme could not achieve this, the high activation barrier could make this transition rate-limiting. Since the total enzyme concentration is conserved, this might seriously reduce the overall rate. Basically, this is the reason why Hill and Eisenberg (1981), as well as others (Albery and Knowles, 1976), require of enzymes that they make the basic free-energy changes small of every single step throughout the cycle.

THE ALTERNATIVE: THE "MOLECULAR MACHINE"

We already noted that, to get from state E to state E', the system would have to get rid of the excess energy E as compared with E'. A way to overcome the problem mentioned above, would be to prevent this energy disposal and concomitant evolution of the system to the very low basic free-energy level of state E'. In the situation described so far, the basic free energy of enzyme E' equals the basic free energy of E. However if $G_E^0 > G_{E'}^0$, the situation would become different. Here, the "molecular energy machine" comes in. McClare (1970) defined that, in such a device, the "energy stored in single molecules is released into a specific molecular form and then converted into another specific form (e.g., a chemical bond) so quickly that it never has time to become heat". Though this definition might not be very translucent, it can be illustrated by Figure 3. Again we start at point A. At this point, the ATP-synthesis would still be uphill. First, we let the redox-reaction occur. In contrast to the previous situation, this reaction is less downhill: After having climbed the activation barrier, the system now does not relax from state E* to E'; it does not release the excess energy to the environment. Hence, after the redox-reaction has taken place, ATP-synthesis now becomes completely downhill! In this case, change in one degree of freedom affects free-energy changes of another degree of freedom. In other words, free energy, represented by one specific degree of freedom, becomes directly available for another degree of freedom; or, energy is "channelled" from one degree of freedom to another.

In short, part of the free energy represented by NADH must be converted into that of a metastable enzyme form E*, that relaxes more slowly to its equilibrium state E' than that it reacts with ADP and P_i to form ATP (Blumenfeld, 1983). It may be noted that, although there may be an equilibrium between the substates of E* and an equilibrium between substates of E', there is, in this model, no equilibrium between E* and E'. One consequence is that the ratio between the concentrations of E*

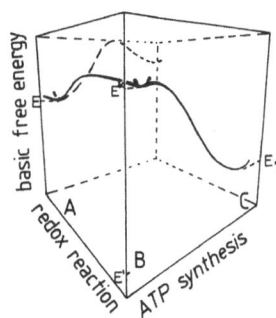

Fig. 3.   Free-energy diagram for a molecular machine, three dimensional
          picture. X-axis (from left to right): Redox-reaction, Y-axis
          (from front to back): ATP-synthesis, Z-axis: Basic free-
          energy changes. Note that in contrast to Figure 2, point B now
          corresponds to state E^*, which has a higher basic free energy
          than E'.

and E' will deviate from the Boltzmann factor $e^{(E'-E^*)/RT}$. Thus, a state
B that would comprise both E^* and E' would not be describable with
(equilibrium) thermodynamics. Perhaps this is what McClare (1970) and
Blumenfeld (1983) meant, when they claimed that "molecular machines"
cannot be described by conventional statistical thermodynamics. We would
here suggest that, provided that one properly disentangles B into two
substates E^* and E', one can again resort to conventional statistical
thermodynamics.

    We see that in this "molecular machine" business, we are in fact
dealing with energy channelling as opposed to metabolite channelling in
the other chapters of this book. Note that we use the term "channelling"
here to indicate, that energy is transferred without the possibility of
equilibrating with (escaping to) the environment. How this channelling
might occur, and the nature of the above-mentioned "specific molecular
form", are still controversial. For instance, this channelling could
just reside in conformational changes, e.g., the redox-reaction drives a
conformational change which enhances the binding of ADP and P_i and
reduces that of ATP. Thus, the concept of "molecular machine" does not
seem very special, but on the contrary, rather part-and-parcel of
existing considerations of biological free-energy transduction (e.g.,
Blumenfeld, 1983; Hill and Eisenberg, 1981; Boyer, 1965; Malstrom, 1985).
Welch and Kell (1986) reviewed suggestions, as to the involvement of
quasi-particles (phonons and solitons) in energy-transducing processes.
Somewhat surprisingly, Gray and Gonda (1977) emphasized that a catalytic
system functioning like a "molecular machine" needs to operate far from
equilibrium, so that it can attain self-organization and develop towards
a dissipative structure (Nicolis and Prigogine, 1977).

METABOLIC CHANNELLING

    Up to this point, we have been discussing reactions catalyzed by
single enzymes. The focal point of this Workshop has been channelling in
metabolic pathways, typically consisting of a number of enzymes. For
this channelling to occur, the enzymes of a pathway would have to be in
close, almost physical, contact. It would, therefore, be only a small

step in logic to consider also the possibility that these enzymes may transmit energy to each other through conformational interactions (Welch and Kell, 1986; Welch and Berry, 1983). Also, here the principles of the molecular machine may be relevant. Mechanisms for energy transfer between enzymes in a channelled pathway, other than conformational interaction mechanisms, have been proposed. Fröhlich (1982) has pointed out that proteins can have large numbers of coherently-excited giant dipoles, the features of which can be described as a phonon-condensation in analogy with the phenomenon of superconductivity (see also Del Guidice et al., 1982).

Westerhoff et al. (1985) and Westerhoff and Chen (1985) view this "specific molecular form", in the context of localized protonic coupling in oxidative phosphorylation, as a higher concentration of protons in small domains—which are not in thermodynamic equilibrium with the adjacent bulk-phases. Similarly, in the general case of metabolic channelling (see Westerhoff and Kamp in this volume), the concentration of metabolites cannot be described thermodynamically (like $\Delta\tilde{\mu}_{H^+}$ in delocalized chemiosmosis). The metabolites are in a metastable state, in the sense that their effective chemical potential can be much higher than expected on the basis of their average concentration. We see that there is a close analogy between "energy channelling" and "metabolite channelling".

We conclude, that the ideas of "molecular machines" are quite compatible with currently studied free-energy transducing mechanisms and can provide new ideas about possible free-energy converting devices in biology.

ACKNOWLEDGEMENT

We thank Douglas Kell, Dean Astumian, H.A. Smith, and Terrell Hill for illuminating discussions.

REFERENCES

Albery, W.J. and Knowles, J.R., 1976, Biochemistry, 15: 5627.
Blumenfeld, L.A., 1983, "Physics of Bioenergetic Processes", Springer-Verlag, Heidelberg.
Boyer, P.D., 1965, in: "Oxidases and Related Redox Systems", Vol. 2, T.E. King, H.S. Mason, and M. Morrison, eds., Wiley, New York.
Del Guidice, E., Doglia, S. and Milani, M., 1982, Phys. Letters, 90A(1,2): 104.
Fröhlich, H., 1982, Adv. Electron. Elect. Phys., 53: 85.
Glasstone, S., Laidler, K.J., and Eyring, H., 1941, "The Theory of Rate Processes", McGraw-Hill, New York.
Gray, B.F. and Gonda, I., 1977, J. Theor. Biol., 69: 167.
Hill, T.L., 1975, Prog. Biophys. Mol. Biol., 29: 149.
Hill, T.L., 1977, "Free Energy Transduction in Biology", Academic Press, New York.
Hill, T.L. and Eisenberg, E., 1981, Quart. Rev. Biophys., 14: 463.
Kemeny, G., 1974, Proc. Natl. Acad. Sci. USA, 71: 2655.
King, E.L. and Altman, C., 1956, J. Phys. Chem., 60: 1375.
Malmstrom, B., 1985, Biochim. Biophys. Acta, 811: 1.
McClare, C.W.F., 1970, J. Theor. Biol., 30: 1.
McClare, C.W.F., 1974, Ann. N. Y. Acad. Sci., 227: 74.
Nicolis, G. and Prigogine, I, 1977, "Self-Organization in Non-

Equilibrium Systems", Wiley, New York.

Somogyi, B., Welch, G.R., and Damjanovich, S., 1984, *Biochim. Biophys. Acta*, 768: 81.

van Dam, K. and Westerhoff, H.V., 1984, in: "Bioenergetics", L. Ernster, ed., Elsevier, Amsterdam.

Welch, G.R. and Berry, M.N., 1983, in: "Coherent Excitations in Biological Systems", H, Frohlich and F. Kremer, eds., Springer-Verlag, Heidelberg.

Welch, G.R. and Kell, D.B., 1986, in: "The Fluctuating Enzyme", G.R. Welch, ed., Wiley, New York.

Welch, G.R., Somogyi, B., and Damjanovich, S., 1982, *Prog. Biophys. Mol. Biol.*, 39: 109.

Westerhoff, H.V. and Chen, Y-D., 1985, *Proc. Natl. Acad. Sci. USA*, 82: 3222.

Westerhoff, H.V., Melandri, B.A., Venturoli, G., Azzone, G.F. and Kell, D.B., 1985, *Biochim. Biophys. Acta*, 768: 257.

CYTOSOCIOLOGICAL ASPECTS OF ENZYME ACTION

G. Rickey Welch

Department of Biological Sciences
University of New Orleans
New Orleans, LA 70148, U.S.A.

ENZYMOLOGY AND CYTOSOCIOLOGY

The scope of this Workshop has shown us something of the breadth of existing knowledge, regarding the metabolic infrastructure of the living cell. Over the years, many of us in attendance have been admonished and rebuked for advancing the kinds of concepts and principles at issue here. For, the opponents have argued, the marriage of cell biology and enzymology attained a consummate finality long ago — during the early days of differential centrifugation. The fruit of this early marriage has been a reductionistic period largely dominated by the "grind-and-find" study of isolated enzyme activities. According to this "classical" view, the organization of cell metabolism exhibits a simple bifurcation: whereas a certain (reproducible) fraction of the cellular enzyme constituency appears to be rather permanently associated in (or on) specific membranous elements (e.g., as "marker enzymes"), the majority of the enzymes of intermediary metabolism are homogeneously dissolved in the cytosol (i.e., the 100,000xg supernatant) or in the "plasm" of organelles (e.g., mitochondrion). Sadly to say, this picture continues to be promulgated in present-day biochemistry textbooks—which treat enzyme organization only as a passing fancy.

Now, it is safe to say, the "classical" view has become outmoded. This conclusion is manifestly evident in the proceedings of this Workshop, as well as in the research efforts of many workers not represented here. Indeed, we can assert resolutely that organization is the rule, rather than the exception, in reflecting the nature of intermediary metabolic processes. As can be gleaned from the presentations of Drs. Clegg, Fulton, Osborn, Penman, and Porter at this Workshop, cell biology presents us with a rather simple biphasic view of cellular infrastructure: a solid phase, encompassing extensive membrane surfaces and a fibrous lattice-work; and a soluble, aqueous phase (albeit containing a considerable amount of "structured" water). Our attention is being drawn more and more to the solid phase, as the primary site of intermediary metabolic processes. It is becoming ever-apparent that cell metabolism is the antithesis of the molecular chaos, under which we customarily study enzymes isolated in a homogeneous, isotropic aqueous medium in vitro. As stated metaphorically by Clegg (1984), "it appears unlikely that a messy alphabet soup would be used to spell out the elaborate prose of intermediary metabolism".

Any attempt to understand the nature of enzyme action in the living state must begin with the simple premise, that the enzyme is a _biological entity_ first and an object of physicochemical analysis second. What this implies is that there are certain biological principles which apply to the characterization of enzyme activity, above and beyond the basic physical and chemical properties thereof. Of course, this is not to say that biochemical processes contravene the laws of physics and chemistry. Rather, as formulated presciently some 100 years ago by the physiologist Claude Bernard (1878):

> _"Chemical phenomena in living organisms can never be fully equated with phenomena that take place outside them. This means to say, in other words, that the chemical phenomena of living beings, although they take place according to the general laws of chemistry, always have their own special apparatus and processes"._

Protoplasm, the living substance, appears superficially as a rather amorphous state of matter. Bernard (1878) termed protoplasm "non-determinate life", life in the "naked state". Accordingly, "here are to be found all the essential properties of which the manifestations of the higher beings are only diversified and definite expresssion, or higher modalities". Protoplasm is a microcosm of life we see in the macroscopic world around us.

It is a trite statement today to assert that a cell is not simply a "bag" containing an aqueous solution uniformly dispersed with enzymes and freely-diffusing metabolites. In fact, it is rather tautological to say that living systems possess a high degree of spatial order. But we are only beginning to grasp the implications of this tautology at the level of the single cell. If we accept Bernard's view of the nature of protoplasm, we should be able to look at the biological world around us - the world we can see, feel, and readily appreciate, to take relationships which we observe between interacting components in this macroscopic world and project those relationships into the function of protoplasm. Our license to act in this manner may be found in a principle enunciated by the late mathematical biologist, Nicholas Rashevsky: the _principle of relational invariance_, which expresses the fact that in spite of all the quantitative differences, all organisms are invariant with respect to some qualitative relations within them (Rashevsky, 1973). The qualitative relations between such phenomena as locomotion, attainment of nutrient, ingestion of nutrient, defense of nutrient supply, etc., remain the same throughout the spectrum of life. Hence, a living organism becomes completely analogous to a _society_ (Welch, 1977a).

Such a conceptualization is of significant heuristic value. The "cytosociological" approach to enzyme action leads us to perceive of the enzyme molecule at a higher hierarchical level than that of the isolated protein. The required transition in our perception of enzyme action can best be illustrated by an analogy from a sociobiological level much simpler than that of human society. Let us look at the termite, one of the most socially organized of all insects (Wilson, 1975). Begging the reader's indulgence, we would like to recount an actual naturalistic "experiment" performed by the author with the termite _Reticulitermes flavipes_ (which is endemic to the New Orleans area). A piece of dead wood infested with this termite was located in a lawn near the author's home. The presence of the termites was readily ascertained by observation of the tiny earthen tubes (channels) connecting the wood substratum to the underground nest. Gentle disruption of the walls of such a "channel" revealed inside a column of worker termites proceeding in an orderly fashion to and from the nest. Now, by carefully removing one of the workers to an isolated location far removed from this area, we observed a marked change in the insect's behavior. It was still "alive" by the usual criteria thereof, viz, it moved, it breathed, it responded to stimuli, it respired (i.e., it oxidized fuel molecules to CO_2 with the production of ATP), etc. However, its way of life was now disharmonious; the canalized orderliness of its motions in the insect society had dissipated into a peripatetic ambulation.

A closer inspection of the termite column, as the individuals proceed to and from the nest, shows that the "real life" of the termite does not lie in the biological processes of the individuals per se. Rather, it is the society which defines and dictates the vitality of each unit therein. In such insect societies as that of the termite, the nature of the societal contacts is readily perceivable (Wilson, 1975). The behavior of the workers in the channels connecting the nest to the wood substratum, for example, is governed by chemotaxis. They follow a course of action in response to the taste/smell of chemicals, either deposited in their path or exchanged between neighboring workers. [Note: The reader may perform a similar experiment with common ants.]

We submit, that there is a direct and immediate analogy to the comparison between enzyme action in vivo versus in vitro. It might be argued by those of a traditional reductionistic bent, that there is no empirical justification for invoking the "termite analogy" to the case of enzyme action. After all, enzymes isolated in vitro are in many instances quite stable; and we find them capable, generally, of catalyzing reactions at fantastic rates (when compared to the corresponding uncatalyzed processes, of course). Referring back to our termite study, if we had no a priori knowledge of the insect in its natural sociological setting, we might say that there is nothing out of the ordinary about the behavior of the "test termite" which we are observing in isolation; in fact, we would probably surmise that our termite is just a "bug" going about the usual motions of scouring for food. The empirical base of the conjoined sciences of enzymology and cell biology has now reached the point, whereby our knowledge of the "natural sociological setting" of many enzymes allows us to begin saying that the behavior of enzymes in vitro is actually the abnormal one.

In a recent essay (Ottaway, 1984), adherents to the holistic theme of enzyme organization were deemed as "vitalists". Actually, we welcome this appellation, in the true etymological sense of the word rather than in the pejorative anachronism of yesteryear. With many enzymes, the organized state corresponds more to "life" (vita) than the state of isolation in vitro.

As discussed by Srere (1984), an appreciation of the interactive nature of enzyme action in vivo answers in part the question of "why enzymes are so big". It was Albertsson (1978) who first used the term "social sites" to specify regions of the macromolecular surface which integrate the enzyme into the cellular infrastructure. In the same vein, McConkey (1982) has formally designated this level of organization as "quinary protein structure". And, Munkres and Woodward (1966) considered the cytogenetic aspects of such supramolecular enzyme organization, proposing the concept of "genetics of locational specificity" to describe the multiplicity of genetic loci which influence enzyme superstructure. The present author, with collaboration and constructive input from many colleagues, has been exploring the thermodynamic-kinetic features of the organized state. Here, we shall review briefly some particular aspects relevant to this Workshop.

VISCOSITY AND THE DYNAMICS OF ENZYME ACTION

Macroviscosity vs. Microviscosity

Viscosity is an important physical property of a medium. Basically, it relates to transfer of momentum (and therefore to particle mobility) in a given medium. The role of viscosity in biochemical processes has been largely ignored, owing primarily to the custom of extrapolating from dilute aqueous solution in vitro to the (supposedly similar) condition in vivo.

In principle, viscosity plays an intimate role in all aspects of enzyme action, in ligand association/dissociation processes as well as in energization of the protein for catalytic turnover. The basic issue is, whether or not medium viscosity is actually rate-determining in any of these processes. Under in vitro conditions its influence is usually quite insignificant, compared to the chemical events of ligand binding/release and catalysis. A discordant note to this status rerum comes with the increasing realization, that many enzymes of intermediary metabolism operate in vivo in heterogeneous states far different from that of isolated enzymes in vitro.

What with the ambience of abundant particulate elements (both membranous and fibrous) and "structured" water, as well as rather high protein concentration, one would intuit that the cytoplasmic viscosity must be large. Thus far, however, measurements (e.g., as reported by Drs. Clegg and Mastro at this Workshop) indicate values only 3-5X that of normal water (for small diffusing molecules the size of intermediary metabolites). Most of the experimental probes, used to determine cytoplasmic viscosity, sample widely different regions of the cellular space during the course of the measurement. Thus, the actual viscosity value so obtained is a statistical quantity, averaged over the bulk phase, i.e., a macroviscosity.

From the nature of the measurements (e.g., ESR), it is difficult to determine how much of the time the probe-molecule spends sticking to (or diffusing along the surface of) structural elements in the cell, and how much of the time it is freely moving in the bulk water. The relatively low experimental value obtained for the cytoplasmic macroviscosity would suggest to the present author, that the probe is spending considerable time free within the small aqueous cavities which are interspersed among the membranous and fibrous structures in the cell. If a significant portion of the cytoplasmic macromolecule (e.g., enzyme) constituency is associated with these structures, then diffusion in the surrounding "dilute" aqueous cavities could be quite rapid.

What is the actual magnitude of the "microviscosity" in the localized microenvironments at the interfaces near these membranous and filamentous structures? Unfortunately, we cannot assess this accurately with present-day methods. A microviscosity 10X that of normal water would seem to the present author to be a conservative estimate. The protein density on biomembranes (Sitte, 1980) and in organelle matrices (Srere, 1985) is very great. The solvent water in these dense microenvironments is certain to be ordered (Hagler and Moult, 1978). Although there is considerable disagreement on this issue (e.g., see contributions by Drs. Clegg, Hazlewood, Horowitz, and Mastro in this volume), this solvent "ordering" may extend several molecular diameters into the bulk phase. This "ordered water" increases the local (micro)viscosity, by virtue of the fact that the very concept of viscosity, extended down to the "local" level (Chandler, 1978), relates $1/\eta$ to the mobility of the individual particles constituting the medium (where η is medium viscosity). Moreover, the polyelectrolytic character of the membrane surface causes the establishment of a diffusive electrical double layer which, in itself, leads to an increase in local viscosity via the so-called "electro-viscous effect" (Booth, 1950; Jones, 1975).

Effect of Viscosity on Enzyme-Ligand Association/Dissociation Processes

The fundamental role of viscosity in the formulation of the rate constants for binding/release of substrate (or product) is well-known in enzymology (Eigen, 1974). Under diffusion-controlled steady-state conditions, the unitary rate constants reduce in the limit to the Smoluchowski rate coefficient, k_D, given by

$$k_D = \Omega \cdot f_e \cdot D \cdot r_c \qquad (1)$$

where Ω is a factor relating to geometric restrictions on the mode of approach of ligand to the enzyme active site ($0 < \Omega \leq 4\pi$), f_e a factor relating to electrostatic effects, D the diffusion coefficient for the ligand, and r_c a critical reaction radius. Using the Stokes-Einstein inverse relation between diffusion coefficient (D) and viscosity (η), we see that $k_D \propto (1/\eta)$.

The Smoluchowski theory shows how viscosity, in principle, influences the basic binding/release of ligand molecules to the protein. But, realistically, what does it tell us about viscosity and enzyme action? First, its potential is brought to bear only under conditions when the chemical events of ligand binding/release are faster than diffusive processes. Even in this case, it may be difficult to dissect the relative influence of diffusivity from the other parameters in Eq. (1). Obviously, a clear understanding of the importance of viscosity in the approach of ligands to (and escape from) the enzyme depends on our grasp of the complete mechanistic picture of the given enzyme, as well as a knowledge of the actual transport properties of the ambient medium. The Smoluchowski rate coefficient has come to play a more general role in enzyme kinetics, serving to delimit the overall dynamics of enzyme activity. Consider the reaction of free substrate with free enzyme to yield product. The apparent second-order rate constant for this reaction is k_{cat}/K_m, where k_{cat} is the catalytic turnover number. Rigorous reasoning (Fersht, 1985) shows that the upper (evolutionary) limit on k_{cat}/K_m is just k_D. That is to say, no matter how effective is the enzyme as a catalyst (from the standpoint of the speed and efficiency of the chemistry of substrate-binding and catalytic-turnover), the overall rate of catalysis can proceed no faster than the rate of diffusion of substrate to the enzyme active site. Recently, Keleti and Welch (1984) extended these arguments, to suggest that a more complete evolutionary picture is obtained by conjugating k_{cat}/K_m to the respective enzyme concentration (density) in situ. Accordingly, the previous authors defined the "kinetic power", $k_\Gamma = V_{max}/K_m$, as the real measure of the kinetic potential of an enzyme reaction in vivo. More detailed

discussion of the effects of viscosity on enzyme-ligand association/
dissociation processes is given elsewhere (Welch, 1977a; Welch et al.,
1983; Somogyi et al., 1987).

Viscosity and the Transient Time of Metabolic Processes

Since enzyme action involves a composite of unitary rate processes,
it may be difficult to perceive of the general effect of solvent
viscosity on a given enzyme by breaking it down into effects on the
individual rate constants. A holistic quantity which is directly related
to viscosity is the transient time. This is an important parameter in
the overall dynamics of a multienzyme system. Physically, it is defined
by the time required for diffusion and accumulation of intermediate
substrate/product species to levels sufficient to sustain a steady state.
Since it depends on mass transfer from one enzyme to another in the
sequence, one intuits that the transient time must be a function of
medium viscosity.

Theory (Welch, 1977b) shows that the transient time is directly
proportional to medium viscosity and inversely proportional to enzyme
concentration. Numerical calculations (Welch, 1977b), using typical
values of whole-cell enzyme concentration, yield unrealistically high
transient-time estimates—even with a viscosity as low as 3-5X that of
water. Such numerical estimates do not correspond to observed metabolic
transients. This situation has been used as theoretical evidence that
many enzymes of intermediary metabolism cannot be regarded as
homogeneously dissolved in the whole cell (Welch, 1977a,b).

Viscosity and the Protein-Dynamical Basis of Catalytic Turnover

The old notion, that the overall structure of an enzyme serves the
singular role of maintaining statically a requisite three-dimensional
arrangement of active-site residues, is now outmoded. Accumulating
evidence has revealed a wide variety of internal motions in proteins,
spanning the gamut of time domains and encompassing various structural
levels in the protein matrix. Realization of the dynamic nature of the
protein molecule has led to the development of a number of theoretical
models of enzyme action (reviewed in Somogyi et al., 1984; Welch, 1986;
Welch et al., 1982). These models propose that particular classes of
fluctuations in the protein provide for high free-energy events at the
active site. One finds an emerging view, that the enzyme molecule is an
intricate free-energy transducer, serving to link the molecular chaos of
the environment and the localized chemical-reaction coordinate.

Although the mechanistic picture is far from clear, it is apparent
that "something" (e.g., collisional processes, binding/relaxation of
solvent or ionic species) at the protein-solvent interface, acting
through the protein "mediator", is involved in generating the proper
phase-space correlation of active-site variables. Viscosity is an
appropriate parameter relating to the modality of thermal interaction (at
a given temperature) between the protein and solvent medium, since it
relates to particle mobility. The thermal "(de)energization" of an
enzyme-substrate complex must be determined by that "mobility"—whether
it specifies direct diffusional-collisional interaction of the solvent
molecules with the protein or binding/relaxation of the solvent/solute
particles with the protein.

The rate constant, k_{cat}, characterizing the catalytic conversion ES-->EP is the concern here. The overall free-energy change associated with the transition state for this process is due in part to the protein matrix interacting dynamically with the bound chemical subsystem. It is found, theoretically and empirically, that k_{cat} is inversely proportional to medium viscosity (Welch et al., 1982). We are just beginning to unravel the intricacies of the protein-dynamical basis of enzyme action; the theoretical and experimental foundation in this area is yet to be developed fully. However, it is apparent that a high viscosity can depress the rate of catalytic turnover for an ES complex.

Is Viscosity a Factor in the Evolutionary Design of Enzyme Micro-compartments?

From the empirical and theoretical information accumulated to date, what can we say, in conclusion, about viscosity and biochemical dynamics in vivo? First, there is the question of diffusional transit, especially as it applies to larger eukaryotic cells. Experimental measurements thus far (such as those reported by Drs. Clegg, Horowitz, and Mastro here) indicate that the translational motion of small molecules (e.g., metabolites) is retarded by less than an order of magnitude (compared to water) within the aqueous interstices of the cytomatrix elements. Questions of the physical properties of cell water, of the diffusional impedance of cytological structures, etc., hopefully will be answered more clearly in the near future. Generally, one cannot ascertain the specific influence of the rate of material (e.g., substrate) diffusion on an individual enzyme in vivo unless one knows the mechanistic details of the given enzyme reaction, as well as the nature of the microenvironment in situ.

It is known that the organizational state of many enzyme systems involves association with subcellular surface elements. But, merely binding the enzymes of a metabolic sequence, say to a membrane, does not, in itself, solve the diffusional-transit problem. As indicated above, the local viscosity at the cytosol-surface interface may be higher than that in the bulk. This situation would imply, theoretically, that the localized enzymes must be placed in rather close proximity on the surface. Interestingly, studies (Sitte, 1980) have shown protein densities on subcellular membranes to be very high (in some cases, crystal-like).

A problem of perhaps equal importance for these membrane-adsorbed enzymes concerns the effect of the locally-high viscosity on the protein-dynamical basis of catalytic turnover. Extrapolating the available data (viz., k_{cat} vs. η plots), from in vitro conditions to the kinds of viscosity values potentially extant in localized microenvironments in vivo, would suggest that many enzymes catalyze reactions at somewhat depressed rates--if activation of the catalytic complex is by purely thermal means (Somogyi et al., 1984). This raises the bizarre possibility that some membrane-associated enzymes may actually be activated by nonthermal means, i.e., are designed to "plug into" local nonequilibrium energy sources. One obvious source would be "active protons", which are generated by virtually all energy-transducing biomembranes. This possibility is rendered more palatable, with the realization of the central role of "mobile protons" in the structure, dynamics, function, and evolution of enzymes (Welch and Berry, 1983). While highly speculative at present, this idea seems to be physically plausible (Welch and Kell, 1986).

Despite the relative paucity of empirical and theoretical information presently accessible to us on viscosity and enzymology, it is imperative that this subject be pursued with exactitude. A knowledge of viscosity effects will provide us a real understanding of the thermal interaction of the enzyme molecule with its surroundings. The road to such an understanding is a convoluted one. Not only must we know more about the enzyme itself, but we must turn increasingly to cell biology to relate to us the physicochemical nature of the local microenvironments. And, we must rely on the rigor of physical chemistry and statistical mechanics, to give us a more exact grasp of the concept of "viscosity" at the microscopic level. This pursuit is sure to yield a fuller appreciation of the unity of the enzyme and its biological habitat in the living cell.

STRUCTURED MICROENVIRONMENTS AND GLOBAL METABOLIC OSCILLATORY DESIGNS

As seen from the presentations of Drs. Goldbeter and Hess at this Workshop, it is apparent that certain enzymatic processes can generate bulk-phase, oscillatory spatiotemporal behavior for metabolites and regulatory substances (e.g., adenine nucleotides). Frequently, this periodic regime is characterized as a limit-cycle oscillation. What can be said, regarding this global cytoplasmic phenomenon and the existence of enzyme microcompartments? Ostensibly, the two would seem to be discordant. After all, the basic theory of biochemical oscillations assumes that the enzyme catalysts and the metabolites are (initially) homogeneously dissolved in the bulk. Yet, there is an intimate regulatory connection potentially extant between localized (heterogeneous) catalysts and global oscillatory designs. Appreciation of this situation has come more recently with the finding that some regulatory enzymes can reversibly partition between the bulk and the localized states, as a function of metabolic conditions (e.g., see the presentation by Dr. Wilson here).

A basis for realizing the relationship between localized enzymes and metabolic oscillations may be found in a theoretical framework developed by Ross and coworkers (Ortoleva and Ross, 1975) some 10 years ago. Those workers analyzed theoretically various features of chemical reactions localized at particular sites, in media undergoing net reaction and transport. For example, they found that localized undulatory spatial patterns of chemical species can be induced in the region surrounding a given site. Notably, those authors emphasized the potential regulatory role of well-localized "pacemaker regions", which might generate chemical phase waves into an otherwise homogeneous reaction-diffusion medium (e.g., in a stable limit cycle). For such oscillatory designs the range of dynamic control of a heterogeneous catalyst on a bulk reacting system may far exceed that found for simple diffusion and chemical relaxation in the nonoscillatory state.

A qualitative connection between the Ross theory and the well-known glycolytic oscillatory system was suggested sometime ago by the present author (Welch, 1977a). Support of this model has come with the recent empirical finding that the key regulatory enzyme, phosphofructokinase, partitions between particulate structures and the bulk cytosol in eukaryotic cells. Importantly, only the "soluble" form of the enzyme is found to exhibit the kind of sigmoidal kinetics necessary for an oscillatory mode. Further details of this "heterogeneous glycolysis" model are given in the presentation by Dr. Kohen in this volume.

ENZYME ORGANIZATION AND THE THERMODYNAMIC-KINETIC CHARACTER OF

BIOCHEMICAL REACTION-RATE LAWS

Kinetic Description

What does "concentration" mean in vivo? Customarily, the laws of chemical kinetics tell us that the rate (velocity) of a reaction in bulk solution is given as the product of the concentrations and a unitary rate constant. Of course, this product is just the probability of finding a reactant molecule (or an encounter pair, in the case of a bimolecular reaction) in solution, multiplied by the probability that such a molecule (or encounter pair) actually reacts. We usually think of the term "concentration" very simplistically, as a scalar quantity reflecting the random statistical distribution of solutes in the bulk phase. It can be related to the configurational entropy of a system, which is defined according to the number of microscopic states compatible with the macroscopic configuration (e.g., positional degeneracy).

The concept of "concentration" in the usual macroscopic sense breaks down for heterogeneous enzyme systems. Notionally, we must define "concentration" according to the physical context in which a given reaction process occurs. In organized multienzyme systems the kinetically competent value of "concentration" for the intermediary metabolites takes on a local anisotropic character which is part-and-parcel of the organization itself. Elsewhere, the present author (Welch, 1977a, 1984) has discussed some mathematical considerations in the definition of "concentration" in heterogeneous states. For organized systems whose microenvironmental construction allows for the existence of sequestered metabolite "pools", the "concentration" of intermediates must be specified formally as the integral of a continuity equation over the relevant reaction-diffusion space. (The kernel [Green's function] of this integral characterizes the physical properties of the diffusion space.) For highly vectorialized systems, with very small "pool" volumes, an ordered-matrix formalism may be more apropos, where the matrices are mathematical operators specifying the action of directed, sequential reaction and diffusion steps. For multienzyme complexes which tightly "channel" the intermediates on a one-by-one basis, the notion of a "concentration" may not even apply, kinetically speaking, except for the initial substrate and the final product.

Many workers over the years have drawn attention to the problem of defining "concentration" in vivo and its relationship to biochemical kinetics (reviewed in Welch, 1977a; Welch and Kell, 1986). However, little in the way of substantive resolution has been suggested, owing to the following factors: 1) our lack of knowledge of the physicochemical properties of metabolic microenvironments in vivo; 2) the mathematical abstruseness of the issue; and 3) a lack of conviction among biochemists, in general, as to the seriousness of the problem. We measure K_m and k_{cat} values for isolated enzymes; we obtain whole-cell enzyme-concentration (or V_{max}) estimates; we do freeze-clamp analyses of tissues to determine in vivo metabolite concentrations; we follow the flow of labeled precursors, added to metabolizing cells; and we construct elaborate computer programs to simulate the system and sort the data. If the results do not fit our preconceived ideas as to how cell metabolism should behave (based on in vitro reasoning), we all too often twist and push the data otherwise — or, as a last resort, simply wave our hands. It is the opinion of the present author (and of a growing number of other concerned biochemists), that much of the numerology of contemporary biochemical kinetics has fallen ill to human contrivance, spawned by a myopia which prevents us from seeing the enzyme reaction beyond its test-tube existence.

"Structural" rate constants. The second major factor determining the reaction rate, in the kinetic description, is the rate constant itself. Advances in experimental methodologies and techniques have enabled us to get a good quantitative handle on the unitary rate constants for many enzymes. Knowledge of these constants has provided us a window into the physical chemistry of binding and catalytic events at the active center. Of course, this knowledge is vital to the sciences of enzymology and biochemistry. Yet, when we imbed the enzyme within its "cytosociological" setting, we find that these unitary rate constants (and their respective processes) are not absolute, unalterable enzymic parameters. Perhaps the first clue to this effect came with the realization that the protein molecule is, itself, a dynamic entity, i.e., the macromolecular matrix is not a static scaffolding serving simply to maintain a rigid three-dimensional arrangement of active-center residues. Among the first enzymological fruits of this finding were the "induced-fit" model of enzyme action by Koshland and the "allosteric" models of enzyme modulation. Most readers fail to note, that Monod, Wyman, and Changeux (1965) in their classic paper on allosterism stressed the relevance of their findings to higher levels of enzyme organization in vivo. Indeed, the concept of allosterism illustrates beautifully how the state of organization engenders modification of binding/catalytic properties of the individual enzymes.

Canvassing the wealth of present-day information on enzyme organization, Welch and Keleti (1981) were led to propose a generalized, composite representation of enzyme rate constants which reflects the kinds of "cytosociological" (or "structural") factors potentially brought to bear on enzyme action in organized regimes. (These "structural" rate constants can be formalized according to the transition-state theory or to the collision theory of chemical kinetics [Welch, 1977a; Welch and Keleti, 1981].) This representation shows the manner by which the kinetic barriers in a sequence of enzyme reactions can be smoothened and regulated as a unit. And, this approach leads to a refinement in current thinking on the kinetic basis of enzyme evolution (Keleti and Welch, 1984).

Thermodynamic Description

The idea of "rate" aside, we must also consider the thermodynamic driving force on chemical processes in organized enzyme systems. The latter concept is embodied in the chemical affinity, A. For a reaction such as S—>P, the affinity is given as

$$A = -(\partial G/\partial \xi)_{T,P} = \mu_S - \mu_P \qquad (2)$$

where the partial derivative (at constant temperature and pressure) is the free-energy (G) change per extent of reaction (ξ). The μ's are the respective chemical potentials of S and P. (The affinity is sometimes erroneously designated as " ΔG" [Welch, 1985].)

As is well known in physical chemistry, the chemical potential, μ_i, of a molecular species i is composed of two terms, one dependent upon, and one independent of, its concentration. The concentration-dependent part (which has no relation to the particular chemical nature of the molecule) is obtained from the configurational entropy of the system. For molecules dissolved in aqueous solution, this part is approximated crudely as $RT \cdot \ln C_i$, where C_i is the molar concentration of i. For bound ligand, this contribution is usually calculated statistically, according to the number of possible ways of distributing (randomly) a given amount of ligand over a number of binding sites. Multienzyme aggregates which "channel" metabolites manifest a much reduced configurational entropy, as a result of the aforementioned notion of, for want of a better term, "localized concentration". This feature obviously leads to a higher effective chemical potential for reaction processes in organized states.

The concentration-independent part, designated μ_i^0, is a function of the actual molecular species i under consideration, including the free energy contributed by interaction with its solvent environment. In "channelled" systems the intermediary metabolites do not equilibrate with the bulk solution. For these compartmentalized substrates (products), the "environment" is a microcavity whose properties differ significantly from the bulk phase. Moreover, the microscopic thermodynamic "phase" (and, hence, μ_i^0) of the substrate/product molecule varies from point to point along the reaction/diffusion coordinate. In some types of multienzyme complexes, the chemical potential of the "channeled" metabolite may even blend with that of the macromolecular superstructure (Welch and Kell, 1986).

Evidently, the issue of enzyme organization casts an ominous shadow of doubt on our current quantitative understanding of flow-force relationships in biochemical reactions — including such matters as the rate, nonequilibrium character, and efficiency of such processes (Welch, 1985). This unfortunate situation is due primarily to our penchant for 1) attempting to establish bulk solution mass-action ratios for "ΔG" calculations (while neglecting the reality of metabolic microenvironments) and 2) estimating standard-state free-energy changes under highly artificial, solution conditions in vitro. (Some of these points are treated also in the presentations of Drs. Kamp and Westerhoff.)

ENZYME ORGANIZATION AND THE THEORY OF METABOLIC CONTROL

An understanding of the control of cellular metabolism as a whole is perhaps the most central issue in the science of cell physiology, as it draws on an edifice of knowledge from such fields as biochemistry, enzymology, cytology, and cytogenetics. Despite the wealth of empirical information which has accumulated in these various subdisciplines, our grasp of the functioning "whole" is of a meager and incomplete nature. Perception of the regulation of individual metabolic processes has come from such notable discoveries as that of protein-conformational flexibility (entailing such designs as allosterism, hysteresis, etc.) and genetic control (e.g., induction, repression). Yet, it would be a futile effort, indeed, to attempt a construction of metabolic control as a whole by simple additive extrapolation from the findings on individual processes (usually assessed under artificial in vitro conditions).

A "metabolic control theory" developed over the past ten years or so by two independent groups (Heinrich et al., 1977; Kacser and Burns, 1979) has become a major focal point in discussions of regulatory networks in vivo (see the presentation by Dr. Kacser herein). Importantly, this theory (which we shall call the "KBHR theory") seeks to imbed the enzyme molecule in its natural setting, as it participates in the dynamic metabolic system. The mathematical framework is geared toward the parametrization of enzyme action according to the complete spectrum of metabolic factors relevant to a given flow process.

In this control theory, Kacser and Burns (1979) have employed the idea of a "molecular democracy" to characterize the nature of the interacting population of enzyme molecules and metabolites (and effectors). Whereby, each enzyme in a metabolic process is considered as an autonomous entity, and control is a sort of linear superposition of the effects of the individual enzymes. The milieu of this "molecular society" is regarded to be a bulk aqueous phase, with enzymes and metabolites homogeneously dispersed therein. The "links" in such a metabolic network are simply the intermediate substrate/product pools.

The developing KBHR theory has been hailed for its potential as a unifying principle in biochemistry. We share the enthusiasm and optimism for this sort of theoretical approach; for, it is refreshingly holistic and synthetic. Yet, the theory, in its present form, is fraught with some limitations which must be removed before it can be regarded as a general paradigm of metabolic control. While the homogeneous, bulk-phase, "pool" view of metabolic control (as an "enzymo-democracy") may be valid for some enzymatic process in vivo, it is not valid for many others. This conclusion is rather obvious when we recount the kinds of structural, kinetic, and thermodynamic properties of organized enzyme systems detailed at this Workshop. Recently, Welch, Keleti, and Vértessy (submitted) suggested a way by which the mathematicl formalism of the KBHR theory might be extended to heterogeneous states.

The present considerations allow us to return full circle to the sociological discourse at the beginning of this chapter. To wit, is cell metabolism, by its nature, a "molecular democracy" (should we say, more appropriately, a "free enterprise") or a "supramolecular socialism"? Based on a marshalling of information on the milieu intérieur of the cell, we would submit that metabolism (and its control) involves both ideologies. The form might vary from pathway to pathway in the cell and, indeed, might even vary from time to time (depending on changing regulatory inputs) within the same pathway.

In a sense, the situation is quite analogous to the multifariousness in the human world. We see that the form of socio-politico-economic government varies from country to contry. (Generally, most such ideologies fall into either some category of socialism or some category of democracy [or free enterprise].) And if we follow the past history of any one country, we would probably see its governmental form vary from time to time. It is apparent, though, that underlying all governance (even in the "democratic" nations) must be at least some degree of socialism. Total and complete freedom to the individuals provides no fabric to the society. Particularly at times of national emergency, all individuals must be conscripted and subserve for the common good of the society.

Rather striking example of this very sociological phenomenon of "national emergency", at the cellular level, is seen in glycolysis. As we alluded above, glycolytic enzymes in larger eukaryotic cells can reversibly partition between particulate structures and the bulk cytosol, depending on metabolic conditions (e.g., see presentations by Drs. Kohen and Wilson). For example, a fraction of the hexokinase activity associates with the mitochondrial membrane, the degree of organization (should we say "socialization"?) being influenced by key metabolites and effector substances. Binding to the mitochondrion is enhanced under stringent conditions, as seen for example with ischemia and with hormonal stimulation. Similarly, the binding of various glycolytic enzymes to filamentous structures in muscle is enhanced by such stresses as electrical stimulation. The association of the regulatory enzyme phoshofructokinase with erythrocyte membrane is dependent on the cellular energy status (see Welch, 1977a, for examples from other pathways).

The human endeavor of viewing the enzyme in isolation in vitro (in conjunction with the seemingly amorphous appearance of the cytoplasm) and asserting prima facie that cell metabolism is, say, a "molecular democracy" perhaps bespeaks the observer's predilection toward a particular politico-economic trapping in our own world. As indicated above, one can counterargue quite strongly that a "supramolecular socialism" is a more realistic picture of metabolism. This is not to say that one ideology is better than the other in human society itself; such contention is beyond the realm of biochemistry and cell biology. Notwithstanding, we will chance to say that, long after we humans have ceased to argue the merits of political ideologies, protoplasm, the real "stuff of life", will persist — as it has long before our level of life came on the scene. Such is so, because it is protoplasm — not metabolites, nor DNA, nor enzymes — which evolves and endures. Perhaps this Workshop is a witness to the effect, that we are learning to put the enzyme in its place.

REFERENCES

Albertsson, P.A., 1978, Trends Biochem. Sci., 3: N37.
Bernard, C., 1878, "Lecons sur les Phénomènes de la Vie communs aux Animaux et aux Végétaux", translated from the French by H.E. Hoff, R. Guillemin, and L. Guillemin, (Charles C. Thomas Publisher, Springfield, Illinois, 1974).
Booth, F., 1950, Proc. Roy. Soc. Lond. A, 203: 533.
Chandler, D., 1978, Ann. Rev. Phys. Chem., 29: 441.
Clegg, J.S., 1984, Amer. J. Physiol., 246: R133.
Eigen, M., 1974, in: "Quantum Statistical Mechanics in the Natural Sciences", B. Kursunoglu, S.L. Mintz, and S.M. Widmayer, eds., Plenum, New York.
Fersht, A., 1985, "Enzyme Structure and Mechanism" (2nd ed.), Freeman, San Francisco.
Hagler, A.T. and Moult, J., 1978, Nature (Lond.) 272: 222.
Heinrich, R., Rapoport, S.M., and Rapoport, T.A., 1977, Prog. Biophys. Mol. Biol., 32: 1.
Jones, M.N., 1975, "Biological Interfaces", Elsevier, New York.
Kacser, H. and Burns, J.H., 1979, Biochem. Soc. Trans., 7: 1149.
Keleti, T. and Welch, G.R., 1984, Biochem. J., 223: 299.
McConkey, E.H., 1982, Proc. Nat. Acad. Sci. USA, 79: 3236.
Monod, J., Wyman, J., and Changeux, J.P., 1965, J. Mol. Biol., 12: 88.
Munkres, K.D. and Woodward, D.O., 1966, Proc. Nat. Acad. Sci. USA, 55: 1217.
Ortoleva, P. and Ross, J., 1975, Adv. Chem. Phys., 29: 49.
Ottaway, J.H., 1984, BioEssays, 1: 283.
Rashevsky, N., 1973, in: "Foundations of Mathematical Biology", Vol. 3, R. Rosen, ed., Academic Press, New York.

Sitte, P., 1980, in: "Cell Compartmentation and Metabolic Channelling",
    L. Nover, F. Lynen, and K. Mothes, eds., Elsevier, New York.
Somogyi, B., Welch, G.R., and Damjanovich, S., 1984, Biochim. Biophys.
    Acta, 768: 81.
Somogyi, B., Rosenberg, A., Welch, G.R., and Damjanovich, S., 1987, in:
    "Towards A Cellular Enzymology", A. Klyosov, S. Varfolomeev, and
    G.R. Welch, eds., Plenum, New York.
Srere, P.A., 1984, Trends Biochem. Sci., 9: 387.
Srere, P.A., 1985, in: "Organized Multienzyme Systems", G.R. Welch, ed.,
    Academic, New York.
Welch, G.R., 1977a, Prog. Biophys. Mol. Biol., 32: 103.
Welch, G.R., 1977b, J. Theor. Biol., 68: 267.
Welch, G.R., 1984, in: "Dynamics of Biochemical Systems", J. Ricard and
    A. Cornish-Bowden, eds., Plenum, New York.
Welch, G.R., 1985, J. Theor. Biol., 114: 433.
Welch, G.R., ed., 1986, "The Fluctuating Enzyme", Wiley, New York.
Welch, G.R. and Berry, M.N., 1983, in: "Coherent Excitations in
    Biological Systems", H. Fröhlich and F. Kremer, eds., Springer-
    Verlag, New York.
Welch, G.R. and Keleti, T., 1981, J. Theor. Biol., 93: 701.
Welch, G.R. and Kell, D.B., 1986, in: "The Fluctuating Enzyme", G.R.
    Welch, ed., Wiley, New York.
Welch, G.R., Somogyi, B., and Damjanovich, S., 1982, Prog. Biophys. Mol.
    Biol., 39: 109.
Welch, G.R., Somogyi, B., Matkó, J., and Papp, S., 1983, J. Theor. Biol.,
    100: 211.
Wilson, E.O., 1975, "Sociobiology: The New Synthesis", Harvard University
    Press, Cambridge, Massachusetts.

PARTICIPANTS

Dr. T. Bak
Kemisk Laboratorium III
H.C. Orsted Institutet
2100 Copenhagen, Denmark

Ms. S. Barnes
Biochemistry Department, 4-West
University of Bath
Bath, England BA2 7AY, U.K.

Dr. S. Beeckmans
Laboratorium voor Chemie
    Proteinen
Vrije Universiteit Brussel
Paardenstraat, 65
B-1640 Sint-Genesius-Rode, Belgium

Dr. S. Bernhard
Institute of Molecular Biology
University of Oregon
Eugene, Oregon 97403, U.S.A.

Dr. M.N. Berry
Department of Clinical
    Biochemistry
School of Medicine
Flinders University of South
    Australia
Bedford Park, S.A. 5042,
    Australia

Dr. H. Bisswanger
Physiologisch-Chemisches Institut
Universität Tübingen
Hoppe-Seyler-Str., 1
D-7400 Tübingen, F.R.G.

Dr. J.S. Clegg
Department of Biology
University of Miami
Coral Gables, Florida 33124
    U.S.A.

Dr. J.A. DeMoss
Department of Biochemistry
University of Texas Medical School
P.O. Box 20708
Houston, Texas 77025, U.S.A.

Dr. A. Friboulet
Laboratoire de Technologie
    Enzymatique
Université de Technologie de
    Compiègne
60206 Compiègne, France

Dr. A.B. Fulton
Department of Biochemistry
University of Iowa College of
    Medicine
Iowa City, Iowa 52242, U.S.A.

Dr. A. Goldbeter
Service de Chimie Physique 2
Université Libre de Bruxelles
Campus Plaine, C.P. 231
1050 Brussels, Belgium

Dr. C. Hazlewood
Department of Physiology
Baylor College of Medicine
Houston, Texas 77030, U.S.A.

Dr. B. Hess
Max-Planck-Institut für
    Ernährungsphysiologie
Rheinlanddamm, 201
D-4600 Dortmund, F.R.G.

Dr. S.B. Horowitz
Department of Physiology and
    Biophysics
Michigan Cancer Foundation
Detroit, Michigan 48201, U.S.A.

Dr. H. Kacser
Department of Genetics
University of Edinburgh
Edinburgh, Scotland EH9 3JN, U.K.

Dr. F. Kamp
Department of Biochemistry
University of Amsterdam
B.C.P. Jansen Institute
1018 TV Amsterdam
The Netherlands

Dr. L. Kanarek
Laboratorium voor Chemie Proteinen
Vrije Universiteit Brussel
Paardenstraat, 65
B-1640 Sint-Genesius-Rode, Belgium

Dr. T. Keleti
Institute of Enzymology
Biological Research Center
Hungarian Academy of Sciences
Budapest H-1502, Hungary

Dr. D.B. Kell
Department of Botany and
    Microbiology
University College of Wales
Aberystwyth, Dyfed SY23 3DA, U.K.

Dr. E. Kohen
Department of Biology
University of Miami
Coral Gables, Florida 33124,
    U.S.A.

Dr. R. Marco
Instituto de Investigaciones
    Biomedicas del CSIC
Universidad Autonoma
Madrid 28029, Spain

Dr. A.M. Mastro
Department of Molecular and
    Cell Biology
Pennsylvania State University
University Park, Pennsylvania
    16802, U.S.A.

Dr. V. Moses
Department of Microbiology
Queen Mary College
University of London
London, England E1 4NS, U.K.

Dr. M. Osborn
Max-Planck-Institut für
    Biophysikalische Chemie
D-3400 Göttingen, F.R.G.

Dr. S. Penman
Department of Biology
Massachusetts Institute of
    Technology
Cambridge, Massachusetts 02139,
    U.S.A.

Dr. K.R. Porter
Department of Biological Sciences
University of Maryland —
    Baltimore County
Catonsville, Maryland 21228,
    U.S.A.

Dr. A. Reith
Norsk Hydro's Institute for
    Cancer Research
The Norwegian Radium Hospital
Montebello
0310 Oslo 3, Norway

Dr. G.L. Rossi
Institute of Molecular Biology
University of Parma
43100 Parma, Italy

Mr. Harry Smith
Department of Chemistry
Florida State University
Tallahassee, Florida 32306, U.S.A.

Dr. B. Somogyi
Department of Biophysics
Medical University of Debrecen
Debrecen H-4012, Hungary

Dr. P.A. Srere
Department of Pre-Clinical Sciences
Veterans Administration Medical
    Center
4500 S. Lancaster Road
Dallas, Texas 75216, U.S.A.

Dr. T. Ureta
Departamento de Biologia
Seccion de Bioquimica y Biologia
    Molecular
Universidad de Chile
P.O. Box 653
Santiago, Chile

Mr. R. Walter
Department of Botany and
    Microbiology
University College of Wales
Aberystwyth, Dyfed SY23 3DA, U.K.

Dr. P.D.J. Weitzman
Biochemistry Department, 4-West
University of Bath
Bath, England BA2 7AY, U.K.

Dr. G.R. Welch
Department of Biological Sciences
University of New Orleans
New Orleans, Louisiana 70148,
    U.S.A.

Dr. H.V. Westerhoff
National Institutes of Health
NIDDK
Bldg. 2, Rm. 319
Bethesda, Maryland 20892, U.S.A.

Dr. D.N. Wheatley
Department of Pathology
University of Aberdeen
Aberdeen, Scotland AB9 2ZD, U.K.

Dr. J.E. Wilson
Department of Biochemistry
Michigan State University
East Lansing, Michigan 48824,
    U.S.A.

Dr. H.G. Wittmann
Max-Planck-Institut für
    Molekuläre Genetik
Ihnestrasse, 63-73
D-1000 Berlin 33 (Dahlem),
    F.R.G.

INDEX